De Gruyter Studies in Mathematical Physics 18

Michael V. Sadovskii

Statistical Physics

De Gruyter

Physics and Astronomy Classification 2010: 05.20.-y; 05.20.Dd; 05.20.Gg; 05.30.-d; 05.30.Ch; 05.30.Fk; 05.30.Pr; 05.70.Ph; 68.18.Jk; 68.18.Ph

ISBN 978-3-11-027031-0
e-ISBN 978-3-11-027037-2

Library of Congress Cataloging-in-Publication Data

A CIP catalog record for this book has been applied for at the Library of Congress.

Bibliographic information published by the Deutsche Nationalbibliothek

The Deutsche Nationalbibliothek lists this publication in the Deutsche Nationalbibliografie; detailed bibliographic data are available in the internet at http://dnb.dnb.de.

Typesetting: Da-TeX Gerd Blumenstein, Leipzig, www.da-tex.de
Printing and binding: Hubert & Co. GmbH & Co. KG, Göttingen
♾Printed on acid-free paper

Printed in Germany

www.degruyter.com

MIX
Papier aus verantwortungsvollen Quellen
FSC www.fsc.org FSC® C016439

Preface

This book is essentially based on the lecture course on "Statistical Physics", which was taught by the author at the physical faculty of the Ural State University in Ekaterinburg since 1992. This course was intended for all physics students, not especially for those specializing in theoretical physics. In this sense the material presented here contains the necessary minimum of knowledge of statistical physics (also often called statistical mechanics), which is in author's opinion necessary for every person wishing to obtain a general education in the field of physics. This posed the rather difficult problem of the choice of material and compact enough presentation. At the same time it necessarily should contain all the basic principles of statistical physics, as well as its main applications to different physical problems, mainly from the field of the theory of condensed matter. Extended version of these lectures were published in Russian in 2003. For the present English edition, some of the material was rewritten and several new sections and paragraphs were added, bringing contents more up to date and adding more discussion on some more difficult cases. Of course, the author was much influenced by several classical books on statistical physics [1, 2, 3], and this influence is obvious in many parts of the text. However, the choice of material and the form of presentation is essentially his own. Still, most attention is devoted to rather traditional problems and models of statistical physics. One of the few exceptions is an attempt to present an elementary and short introduction to the modern quantum theoretical methods of statistical physics at the end of the book. Also, a little bit more attention than usual is given to the problems of nonequilibrium statistical mechanics. Some of the more special paragraphs, of more interest to future theorists, are denoted by asterisks or moved to Appendices. Of course, this book is too short to give a complete presentation of modern statistical physics. Those interested in further developments should address more fundamental monographs and modern physical literature.

Ekaterinburg, 2012 M. V. Sadovskii

Preface

[illegible]

Contents

Preface v

1 Basic principles of statistics 1

1.1 Introduction 2

1.2 Distribution functions 2

1.3 Statistical independence 7

1.4 Liouville theorem 9

1.5 Role of energy, microcanonical distribution 13

1.6 Partial distribution functions* 17

1.7 Density matrix 21

1.7.1 Pure ensemble 22

1.7.2 Mixed ensemble 24

1.8 Quantum Liouville equation 26

1.9 Microcanonical distribution in quantum statistics 28

1.10 Partial density matrices* 29

1.11 Entropy 32

1.11.1 Gibbs entropy. Entropy and probability 32

1.11.2 The law of entropy growth 35

2 Gibbs distribution 43

2.1 Canonical distribution 43

2.2 Maxwell distribution 48

2.3 Free energy from Gibbs distribution 50

2.4 Gibbs distribution for systems with varying number of particles 52

2.5 Thermodynamic relations from Gibbs distribution 55

3 Classical ideal gas 60

3.1 Boltzmann distribution 60

3.2 Boltzmann distribution and classical statistics 61

3.3 Nonequilibrium ideal gas 63

3.4 Free energy of Boltzmann gas 66

3.5 Equation of state of Boltzmann gas 67
3.6 Ideal gas with constant specific heat 69
3.7 Equipartition theorem 70
3.8 One-atom ideal gas 72

4 Quantum ideal gases 75
4.1 Fermi distribution 75
4.2 Bose distribution 76
4.3 Nonequilibrium Fermi and Bose gases 77
4.4 General properties of Fermi and Bose gases 79
4.5 Degenerate gas of electrons 82
4.6 Relativistic degenerate electron gas* 85
4.7 Specific heat of a degenerate electron gas 86
4.8 Magnetism of an electron gas in weak fields 88
4.9 Magnetism of an electron gas in high fields* 92
4.10 Degenerate Bose gas 94
4.11 Statistics of photons 97

5 Condensed matter 101
5.1 Solid state at low temperature 101
5.2 Solid state at high temperature 104
5.3 Debye theory 105
5.4 Quantum Bose liquid 109
5.5 Superfluidity 113
5.6 Phonons in a Bose liquid* 117
5.7 Degenerate interacting Bose gas 121
5.8 Fermi liquids 124
5.9 Electron liquid in metals* 130

6 Superconductivity 133
6.1 Cooper instability 133
6.2 Energy spectrum of superconductors 135
6.3 Thermodynamics of superconductors 144
6.4 Coulomb repulsion* 148
6.5 Ginzburg–Landau theory 151

7 Fluctuations 160
7.1 Gaussian distribution 160
7.2 Fluctuations in basic physical properties 164
7.3 Fluctuations in ideal gases 167
8 Phase transitions and critical phenomena 170
8.1 Mean-field theory of magnetism 170
8.2 Quasi-averages* 177
8.3 Fluctuations in the order parameter 180
8.4 Scaling 186
9 Linear response 195
9.1 Linear response to mechanical perturbation 195
9.2 Electrical conductivity and magnetic susceptibility 201
9.3 Dispersion relations 204
10 Kinetic equations 208
10.1 Boltzmann equation 208
10.2 H-theorem 214
10.3 Quantum kinetic equations* 216
10.3.1 Electron-phonon interaction 218
10.3.2 Electron–electron interaction 222
11 Basics of the modern theory of many-particle systems 225
11.1 Quasiparticles and Green's functions 225
11.2 Feynman diagrams for many-particle systems 233
11.3 Dyson equation 237
11.4 Effective interaction and dielectric screening 241
11.5 Green's functions at finite temperatures 244
A Motion in phase space, ergodicity and mixing 248
A.1 Ergodicity 248
A.2 Poincare recurrence theorem 254
A.3 Instability of trajectories and mixing 256
B Statistical mechanics and information theory 259
B.1 Relation between Gibbs distributions and the principle of maximal information entropy 259
B.2 Purging Maxwell's "demon" 263

C Nonequilibrium statistical operators 270
C.1 Quasi-equilibrium statistical operators 270
C.2 Nonequilibrium statistical operators and quasi-averages 273
Bibliography 277
Index 279

Chapter 1

Basic principles of statistics

We may imagine a great number of systems of the same nature, but differing in the configurations and velocities which they have at a given instant, and differing not merely infinitesimally, but it may be so as to embrace every conceivable combination of configuration and velocities. And here we may set the problem not to follow a particular system through its succession of configurations, but to determine how the whole number of systems will be distributed among the various conceivable configurations and velocities at any required time, when the distribution has been given at some specific time. The fundamental equation for this inquiry is that which gives the rate of change of the number of systems which fall within any infinitesimal limits of configuration and velocity. Such inquiries have been called by Maxwell statistical. *They belong to a branch of mechanics which owes its origin to the desire to explain the laws of thermodynamics on mechanical principles, and of which Clausius, Maxwell and Boltzmann are to be regarded as principal founders.*

J. Willard Gibbs, 1902 [4]

1.1 Introduction

Traditionally, statistical physics (statistical mechanics) deals with systems consisting of large numbers of particles, moving according to the laws of classical or quantum mechanics. Historically it evolved, by the end of XIX century, from attempts to provide mechanistic derivation of the laws of thermodynamics in the works by J. Maxwell and L. Boltzmann. The formalism of statistical mechanics was practically finalized in the fundamental treatise by J. W. Gibbs [4], which appeared at the beginning of the XX century. The remarkable advantage of Gibbs method, which was created long before the appearance of modern quantum theory, is its full applicability to the studies of quantum (many-particle) systems. Nowadays, statistical physics has outgrown the initial task of justification of thermodynamics, its methods and ideology actually penetrating all the basic parts of modern theoretical physics. Still being understood mainly as the theory of many (interacting) particle systems, it has deep connections with modern quantum field theory, which is at present the most fundamental theory of matter. At the same time, it is now also clear that even the description of mechanical motion of relatively few particles moving according to the laws of classical mechanics often requires the use of statistical methods, as this motion, in general (nontrivial) cases, is usually extremely complicated (unstable). The ideas and methods of statistical mechanics form the basis of our understanding of physical processes in solids, gases, liquids and plasma, while the modern theory of elementary particles (based on the quantum field theory) is, from the very beginning, actually the theory of systems with an infinite number of degrees of freedom, where statistical methods are at the heart of the problem. Unfortunately, due to the lack of space we will not be able to discuss in detail all of these deep interconnections and just limit ourselves to the studies of more or less traditional models of statistical mechanics [1, 2, 3], which provide the foundation for understanding of much more complicated problems.

1.2 Distribution functions

Consider a system of N (for simplicity) identical interacting particles, moving in a finite but macroscopically large volume V. For simplicity, we also assume that these particles do not possess internal degrees of freedom. If we describe the motion of particles by classical mechanics, the state of the motion of the k-th particle is completely characterized by the values of its coordinates $\mathbf{q}_k$ and momentum $\mathbf{p}_k$, and the state of the system as a whole is determined by the values of all particles' coordinates $\mathbf{q}_1, \mathbf{q}_2, \dots, \mathbf{q}_N$ and momenta $\mathbf{p}_1, \mathbf{p}_2, \dots, \mathbf{p}_N$. Thus, the state of the system may be described by the point in $6N$-dimensional *phase* space: $(\mathbf{q}_1, \mathbf{q}_2, \dots, \mathbf{q}_N, \mathbf{p}_1, \mathbf{p}_2, \dots, \mathbf{p}_N)$ – the so called *phase point*. Dynamical evolution (motion) of the system is determined by Hamilton's equations of motion:[1]

$$\frac{d\mathbf{q}_k}{dt} = \frac{\partial H}{\partial \mathbf{p}_k} \qquad \frac{d\mathbf{p}_k}{dt} = -\frac{\partial H}{\partial \mathbf{q}_k}, \tag{1.1}$$

where

$$H = H(\mathbf{q}_1, \mathbf{q}_2, \ldots, \mathbf{q}_N, \mathbf{p}_1, \mathbf{p}_2, \ldots, \mathbf{p}_N) \equiv H(p, q) \tag{1.2}$$

is the full Hamiltonian of the system.

Consider the simplest case of particles interacting with each other via the two-particle spherically symmetric potential $U(|\mathbf{q}_i - \mathbf{q}_k|)$, so that the Hamiltonian takes the form:

$$H = \sum_{k=1}^{N} \frac{\mathbf{p}_k^2}{2m} + \frac{1}{2}\sum_{i \neq k} U(|\mathbf{q}_i - \mathbf{q}_k|). \tag{1.3}$$

The equations of motion are written as:

$$\dot{\mathbf{q}}_k = \frac{\mathbf{p}_k}{m} \qquad \dot{\mathbf{p}}_k = -\sum_{i \neq k} \frac{\partial U(|\mathbf{q}_i - \mathbf{q}_k|)}{\partial \mathbf{q}_k} = \mathbf{F}_k, \tag{1.4}$$

where $\mathbf{F}_k$ is the force enacted upon the k-th particle by the rest. It is clear that for any significantly large value of N the complete solution of the system of equations (1.4) is not feasible even numerically. Also, such a solution (in the improbable case we find it) would be of no real use. The real trajectory of each particle will most probably be quite complicated (chaotic). More so, we have to solve Equations (1.4), with appropriate initial conditions, and this solution is, as a rule, quite sensitive to the choice of initial velocities and coordinates, which are actually not known precisely in any realistic situation. As the motion of particles is in most cases unstable, the trajectories corresponding even to quite close initial values become quite different in a rather short time (and this difference grows exponentially with time), so that they do not have anything in common anymore. Thus, from such solutions we have almost nothing to learn about *macroscopic* properties of the system with large number N of particles, which are of main interest to us. In fact, due to the instability of mechanical motion, these problems usually appear even for systems consisting of rather few particles. This inevitably leads us to use statistical analysis.

Thus, the equations of motion (1.4) determine the trajectory of the phase point in the phase space, defining the mechanical state of the system. This trajectory in phase space is called the *phase trajectory*. For conservative systems with fixed energy we can write:

$$H(q, p) = E. \tag{1.5}$$

[1] It is interesting to note that Gibbs' approach is completely based on the use of Hamilton form of mechanics and not on that of Lagrange.

This means that the phase trajectory belongs to the surface of constant energy in the phase space, defined by Equation (1.5) – the so called *ergodic surface*[2].

When a macroscopic system is in (thermodynamic) equilibrium, its macroscopic characteristics (temperature, volume, pressure etc.) remain constant in time, though its microscopic state continuously changes and we do not know it at all (i.e. where precisely is its phase point on the ergodic surface at the given moment in time). The statistical approach attempts to determine only the *probability* of the realization of some set of microstates, corresponding to the given macrostate of our system. In fact, following Gibbs, we shall consider not the fixed system, but an *ensemble* i.e. the set of the large number (in the limit of $N \to \infty$ the infinite!) of its copies, all remaining in macroscopically equivalent conditions (states). This is usually called the *Gibbs ensemble*, describing the macroscopic state of the system. Macroscopic equivalence of external conditions (states) means that all the systems within the ensemble are characterized by the same values of the appropriate macroscopic parameters (neglecting small fluctuations) and the same types of contacts with surrounding bodies (energy or particle reservoirs, pistons, walls etc.). This leads to certain limitations on coordinates and momenta of particles, which otherwise remain rather arbitrary.

A statistical ensemble is defined by a *distribution function* $\rho(p,q,t)$, which has the meaning of the probability density of systems in the phase space, so that:

$$dw = \rho(p,q,t)dpdq \tag{1.6}$$

gives the probability to find a system (from the Gibbs ensemble!) in the element of phase space $dpdq$ around the point $(p,q) \equiv (\mathbf{p}_1,\ldots,\mathbf{p}_N,\mathbf{q}_1,\ldots,\mathbf{q}_N)$ at time t. The distribution function must satisfy the obvious normalization condition:

$$\int dpdq\rho(p,q,t) = 1, \tag{1.7}$$

as the sum of the probabilities of all possible states must be unity. Such a normalization condition is used e.g. in the famous book by Landau and Lifshitz [1]. However, this is not the only possible form of the normalization condition. In fact, we understand from the very beginning, that classical statistics is the limiting case of quantum statistics (below, we shall see that transition from the quantum case to the classical one takes place at high enough temperatures, when quantum effects become negligible) From quantum mechanics we know [5] that a notions of coordinate and momenta of the particles can be introduced only within the limits of a quasiclassical approximation. The minimal size of the phase space cell for the one-dimensional motion of

[2] We must stress here the important role of the Cauchy theorem on the uniqueness of the solution of the system of usual differential equations. Under the rather weak requirements for the r.h.s. of Equations (1.4), there exists a unique (at any moment in time) solution, which automatically excludes the possibility of the crossing of two *different* phase trajectories in any regular point of the phase space (except some fixed points, corresponding to the zeroes of the r.h.s. of (1.4)).

the i-th particle in quasiclassical approximation is given by $h = 2\pi\hbar$:[3]

$$\Delta q_i^x \Delta p_i^x \geq h. \tag{1.8}$$

Thus the minimal size of the cell in the phase space of one particle (for three-dimensional motion) is equal to $h^3 = (2\pi\hbar)^3$, and $(2\pi\hbar)^{3N}$ in the phase space of N particles. The value of $(2\pi\hbar)^{3N}$ is the natural volume unit in the phase space. Accordingly, it is often convenient to introduce the distribution function normalized to unity after integration over the dimensionless phase space $\frac{dpdq}{(2\pi\hbar)^{3N}}$.

For the system consisting of N identical particles, we have to take into account the fact that taking different permutations of identical particles does not change the quantum state of the system. The number of permutations of N identical particles is equal to $N!$ and the volume of the phase space cell should be divided by $N!$ if we wish to take into account only physically distinguishable states.

Thus it is convenient to define the distribution function by the relation:

$$dw = \rho(p,q,t)\frac{dpdq}{N!(2\pi\hbar)^{3N}}, \tag{1.9}$$

and write the normalization condition as:

$$\int d\Gamma \rho(p,q,t) = 1, \tag{1.10}$$

where:

$$d\Gamma = \frac{dpdq}{N!(2\pi\hbar)^{3N}} \tag{1.11}$$

is the *dimensionless* phase space element. Integration in (1.10) corresponds to the summation over all distinguishable quantum states of the system[4].

Knowing the distribution function $\rho(p,q,t)$ we can, in principle, calculate the average values of arbitrary physical characteristics, which depend on the coordinates and momenta of particles forming our system. The average value of any such function of dynamic variables $f(p,q)$ is defined as:

$$\langle f \rangle = \int d\Gamma \rho(p,q,t) f(p,q) \tag{1.12}$$

---

[3] Quasiclassical quantization condition for Bohr and Sommerfeld in the one-dimensional case takes the form: $\oint pdq = (n + \frac{1}{2})h$. The integral here represents an area of the closed orbit in phase space. Dividing this area into cells of area $2\pi\hbar$ we obtain n cells. But here, n is the number of the quantum state, with energy below the given value, corresponding to this orbit. Thus, for any quantum state there is a corresponding cell in the phase space with an area $2\pi\hbar$. Introducing the wave vector of a particle as $k = p/\hbar$ we get $\frac{\Delta p \Delta q}{2\pi\hbar} = \frac{\Delta k \Delta q}{2\pi}$, which corresponds to the well known relation for the number of (eigen)modes of the wave field [6].

[4] Remarkably, the necessity to divide the phase space volume by $N!$ for the system of identical particles was stressed by Gibbs long before the discovery of quantum mechanics as a recipe to avoid the so called Gibbs paradox – the growth of entropy during the mixing of *identical* gases at the fixed temperature, volume and pressure [7].

and is sometimes called the *phase* average (ensemble average). Averaging with the distribution function (over the phase space) comes here instead of another possible procedure, when we follow the precise time evolution of $f(p,q)$ and calculate its average behavior in time. This last approach reduces to performing measurements at different moments in time, producing explicit time dependence $f = f(t)$, and calculating its average value as:

$$\widetilde{f} = \lim_{T\to\infty} \frac{1}{T} \int_0^T dt f(t) \tag{1.13}$$

i.e. as *time* average.

In the general case, the proof of the equivalence of phase and time averaging is the very difficult (and still not completely solved) problem of the so called *ergodic* theory, which is a developing branch of modern mathematics [8, 9]. In recent decades significant progress was achieved, but this material is definitely outside the scope of this book. Below we shall only give a brief and elementary discussion of this problem. The physical meaning of the Gibbs approach may be qualitatively illustrated as follows: let us consider a small, but still macroscopic, subsystem within our closed (isolated) system. This subsystem is also described by the laws of classical mechanics, but it is not isolated and influenced by all possible interactions with the other parts of the (big) system. Under these conditions, the state of our subsystem will change in time in a very complicated and chaotic way. Due to this chaotic motion, during a long enough time interval T the subsystem will "visit" all its possible states many times. Or in more rigorous terms, let us define $\Delta p \Delta q$ to be some small volume in the phase space of the subsystem. It can be assumed that during a large enough time interval T the complicated (chaotic) phase trajectory of the subsystem will pass this volume many times. Let Δt be that part of time T during which the subsystem is somewhere within this phase space volume $\Delta p \Delta q$. As T grows to infinity, the value of $\Delta t / T$ will tend to some limit:

$$\Delta w = \lim_{T\to\infty} \frac{\Delta t}{T}, \tag{1.14}$$

which can be considered as the probability of finding our subsystem within this volume of the phase space at a given moment in time. Now going to the limit of an infinitesimally small phase space volume we introduce the distribution function $\rho(p,q,t)$ and by definition of (1.14) statistical (phase) averaging (1.12) seems to be physically equivalent to time averaging (1.13). This simple justification is usually sufficient for physicists. In particular Landau claimed [1] that the importance of ergodic theory is overestimated by mathematicians. Though discussions of this problem are still continuing, from a pragmatic point of view the Gibbs approach is in no doubts correct, as all conclusions obtained within statistical mechanics are getting full *experimental* confirmation.

Finally, we shall mention one more qualitative point, which is very important for understanding the foundations of statistical mechanics. The distribution function of a given subsystem is, in general, independent of the initial state of any other part of the same system, as the influence of this initial state during a long enough time interval is completely smeared by the influence of many other parts of the system. It is also independent of the initial state of the subsystem under consideration, as it passes through all possible states during its long time evolution and actually each of these states can be considered as initial ("memory" loss).

1.3 Statistical independence

Let us consider some simple facts from mathematical statistics, which will be useful in the following. In many cases, the closed macroscopic system can be "divided" into a number of subsystems, which interact rather weakly with each other, and during long enough time intervals behave (approximately) as closed (isolated) systems. We shall call such subsystems quasi closed (or quasi isolated). Statistical independence of such subsystems means that the state of a given subsystem does not influence the probability distributions of other subsystems.

Consider two such subsystems with infinitesimal volume elements of phase spaces $dp^{(1)}dq^{(1)}$ and $dp^{(2)}dq^{(2)}$. If we consider the composite system consisting of both subsystems then, from a mathematical point of view, the statistical independence of subsystems means that the probability for the composite system to be found in the element of its phase space volume $dp^{(12)}dq^{(12)} = dp^{(1)}dq^{(1)}dp^{(2)}dq^{(2)}$ factorizes into the product of probabilities:

$$\rho_{12}dp^{(12)}dq^{(12)} = \rho_1 dp^{(1)}dq^{(1)}\rho_2 dp^{(2)}dq^{(2)}, \tag{1.15}$$

so that

$$\rho_{12} = \rho_1\rho_2, \tag{1.16}$$

where ρ_{12} is the distribution function of the composite system, while ρ_1 and ρ_2 are distribution functions of subsystems.

The inverse statement is also valid – the *factorization* of the distribution function means that the system can be decomposed into statistically independent subsystems. If f_1 of f_2 are two physical characteristics of two subsystems, from Equations (1.15) and (1.12) it follows immediately that the average value of the product $f_1 f_2$ is equal to the product of the averages:

$$\langle f_1 f_2\rangle = \langle f_1\rangle\langle f_2\rangle. \tag{1.17}$$

Consider some physical quantity f characterizing the macroscopic body or a part of it. As time evolves, it changes (fluctuates) around its average value $\langle f\rangle$. As a measure of these fluctuations we can not take just the difference $\Delta f = f - \langle f\rangle$, as due to the possibility of fluctuations in both signs it averages to zero: $\langle \Delta f\rangle = 0$. Thus, as a measure of fluctuation it is convenient to take its mean square: $\langle(\Delta f)^2\rangle$. We then always obtain $\langle(\Delta f)^2\rangle \geq 0$, and the average here tends to zero only as $f \to \langle f\rangle$, i.e. when the deviation of f from $\langle f\rangle$ appears with small probability. The value of

$$\sqrt{\langle(\Delta f)^2\rangle} = \sqrt{\langle(f - \langle f\rangle)^2\rangle} \tag{1.18}$$

is called *mean square* fluctuation in f. It is easily seen that:

$$\begin{aligned} \langle(\Delta f)^2\rangle &= \langle f^2 - 2f\langle f\rangle + \langle f\rangle^2\rangle \\ &= \langle f^2\rangle - 2\langle f\rangle\langle f\rangle + \langle f\rangle^2 = \langle f^2\rangle - \langle f\rangle^2, \end{aligned} \tag{1.19}$$

so that the mean square fluctuation is determined by the difference between the average square and the square of the average of the physical characteristic under study. The ratio $\sqrt{\langle(\Delta f)^2\rangle}/\langle f\rangle$ is called the *relative* fluctuation in f. It can be shown that the relative fluctuations in the typical physical characteristics of macroscopic systems drop fast with the growth of the size (the number of particles) of the body. In fact, most of the physical quantities are additive (due to the quasi isolated nature of different parts of the system): the value of such a quantity for the whole body (system) is the sum of its values for different parts (subsystems). Let us divide our system into a large number N of more or less similar (or equal) subsystems (often this may be just the number of particles in the system). Then for the additive characteristic we can write:

$$f = \sum_{i=1}^{N} f_i, \tag{1.20}$$

where f_i characterizes the i-th part (subsystem or particle). Obviously, for the average value we get:

$$\langle f\rangle = \sum_{i=1}^{N} \langle f_i\rangle. \tag{1.21}$$

With the growth of N the value of $\langle f\rangle$ grows approximately proportionally N: $\langle f\rangle \sim N$. Let us calculate the mean square fluctuation in f:

$$\langle(\Delta f)^2\rangle = \left\langle\left(\sum_i \Delta f_i\right)^2\right\rangle. \tag{1.22}$$

Due to the statistical independence of different parts (subsystems) we have:

$$\langle\Delta f_i \Delta f_k\rangle = \langle\Delta f_i\rangle\langle\Delta f_k\rangle = 0 \qquad (i \neq k) \tag{1.23}$$

as each $\langle\Delta f_i\rangle = 0$. Then:

$$\langle(\Delta f)^2\rangle = \sum_{i=1}^{N} \langle(\Delta f_i)^2\rangle. \tag{1.24}$$

Then it is clear that with the growth of N we also get $\langle(\Delta f)^2\rangle \sim N$. Then the relative fluctuation is estimated as:

$$\frac{\sqrt{\langle(\Delta f)^2\rangle}}{\langle f\rangle} \sim \frac{\sqrt{N}}{N} = \frac{1}{\sqrt{N}}. \tag{1.25}$$

Now we see that the relative fluctuation in any additive characteristic is inversely proportional to the square root of the number of independent parts of the macroscopic body (e.g. number of particles), so that for a large enough value of N (e.g. for $N \sim 10^{22}$ for a typical number of particles per cubic centimeter) the value of f may be considered practically constant and equal to its average value. If N is not big enough, e.g. $N \sim 10^6$, the relative fluctuations are not small and quite observable. Such systems sometimes are called *mesoscopic*.

1.4 Liouville theorem

Introduction of the distribution function for mechanical systems as probability density in the phase space is based on the Liouville theorem – a purely mechanical statement, which does not contain any statistical assumptions. According to this theorem, for systems with motion described by Hamilton equations:

$$\frac{dq_k}{dt} = \frac{\partial H}{\partial p_k} \qquad \frac{dp_k}{dt} = -\frac{\partial H}{\partial q_k} \tag{1.26}$$

the phase volume (of an ensemble) remains constant in time. If at the initial moment in time the phase points (p^0, q^0) of systems forming the Gibbs ensemble continuously fill some region G_0 in the phase space, while at the moment t they fill the region G_t, then the volumes of these regions in the phase space are the same:

$$\int_{G_0} dp^0 dq^0 = \int_{G_t} dp dq \tag{1.27}$$

or, for infinitesimal elements of the phase space:

$$dp^0 dq^0 = dp dq. \tag{1.28}$$

In other words, *the motion of phase points representing systems of the ensemble is like that of a non compressible liquid*, as is shown in Figure 1.1 – the "drop", formed by phase points, representing an ensemble, can deform in a rather complicated way in the process of motion, but its volume is conserved.

To prove the Liouville theorem we transform the integral on the r.h.s. of Equation (1.27) by changing integration variables from p, q to p^0, q^0. Then, according to the well known rules for multiple integrals we get:

$$\int_{G_t} dp dq = \int_{G_0} \frac{\partial(p,q)}{\partial(p^0,q^0)} dp^0 dq^0, \tag{1.29}$$

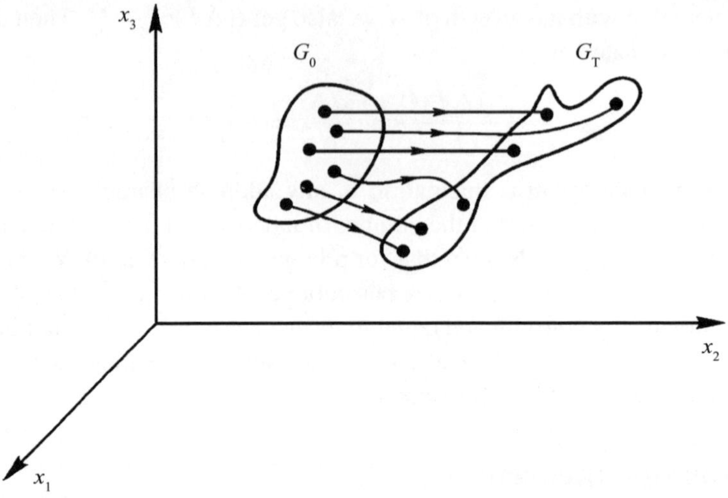

Figure 1.1. The change of initial volume G_0 in the phase space due to the motion of phase points representing an ensemble according to the Liouville theorem.

where $\frac{\partial(p,q)}{\partial(p^0,q^0)}$ is the appropriate Jacobian. We remind that the Jacobian is a determinant of the following form (for simplicity we write the explicit expression below for the two-dimensional case, generalization for multiple dimensions is direct):

$$\frac{\partial(u,v)}{\partial(x,y)} = \begin{vmatrix} \frac{\partial u}{\partial x} & \frac{\partial u}{\partial y} \\ \frac{\partial v}{\partial x} & \frac{\partial v}{\partial y} \end{vmatrix}. \tag{1.30}$$

The following general properties of the Jacobian are:

$$\frac{\partial(u,v)}{\partial(x,y)} = -\frac{\partial(v,u)}{\partial(x,y)}, \tag{1.31}$$

$$\frac{\partial(u,y)}{\partial(x,y)} = \frac{\partial u}{\partial x}. \tag{1.32}$$

Also it is easy to see that:

$$\frac{\partial(u,v)}{\partial(x,y)} = \frac{\partial(u,v)}{\partial(t,s)}\frac{\partial(t,s)}{\partial(x,y)}, \tag{1.33}$$

$$\frac{d}{dt}\frac{\partial(u,v)}{\partial(x,y)} = \frac{\partial(\frac{du}{dt},v)}{\partial(x,y)} + \frac{\partial(u,\frac{dv}{dt})}{\partial(x,y)}. \tag{1.34}$$

Let us now show that the Jacobian in Equation (1.29) is unity if systems move according to Hamilton's equations:

$$\frac{\partial(p,q)}{\partial(p^0,q^0)} = 1. \tag{1.35}$$

To prove this we show that the total time derivative of the Jacobian is equal to zero:

$$\frac{d}{dt}\frac{\partial(p,q)}{\partial(p^0,q^0)} = 0. \tag{1.36}$$

Then it follows that the Jacobian is a constant, more precisely just unity, because it was equal to unity at the initial moment in time.

For simplicity let us write down the proof for the case of a two-dimensional phase space, when there is only one coordinate q and one momentum p. According to Equation (1.34) we can write:

$$\frac{d}{dt}\frac{\partial(p,q)}{\partial(p_0,q_0)} = \frac{\partial(\dot{p},q)}{\partial(p_0,q_0)} + \frac{\partial(p,\dot{q})}{\partial(p_0,q_0)}. \tag{1.37}$$

Then, according to Equations (1.32) and (1.33) we have:

$$\frac{\partial(p,\dot{q})}{\partial(p_0,q_0)} = \frac{\partial(p,\dot{q})}{\partial(p,q)}\frac{\partial(p,q)}{\partial(p_0,q_0)} = \frac{\partial\dot{q}}{\partial q}\frac{\partial(p,q)}{\partial(p_0,q_0)}, \tag{1.38}$$

$$\frac{\partial(\dot{p},q)}{\partial(p_0,q_0)} = \frac{\partial(\dot{p},q)}{\partial(p,q)}\frac{\partial(p,q)}{\partial(p_0,q_0)} = \frac{\partial\dot{p}}{\partial p}\frac{\partial(p,q)}{\partial(p_0,q_0)}, \tag{1.39}$$

$$\frac{d}{dt}\frac{\partial(p,q)}{\partial(p_0,q_0)} = \left(\frac{\partial\dot{p}}{\partial p} + \frac{\partial\dot{q}}{\partial q}\right)\frac{\partial(p,q)}{\partial(p_0,q_0)}. \tag{1.40}$$

It is seen that the sum in the r.h.s. is equal to zero, due to the equations of motion:

$$\dot{q} = \frac{\partial H}{\partial p};\quad \dot{p} = -\frac{\partial H}{\partial q} \tag{1.41}$$

so that

$$\frac{\partial\dot{q}}{\partial q} = \frac{\partial^2 H}{\partial q\partial p} = -\frac{\partial\dot{p}}{\partial p} \tag{1.42}$$

and accordingly

$$\left(\frac{\partial\dot{p}}{\partial p} + \frac{\partial\dot{q}}{\partial q}\right) = 0, \tag{1.43}$$

which proves everything.

The Liouville theorem is a purely mechanical statement and up to now, we used the distribution function nowhere. However, with the help of the distribution function we may give another formulation of the Liouville theorem. As the "drop" representing the Gibbs ensemble moves through the phase space (Figure 1.1), the number of phase points in it (the number of systems in the ensemble) obviously does not change and all phase points belonging at time t to volume element $dpdq$ at time t' move to element $dp'dq'$. Then we can write[5]:

$$\rho(p,q,t)dpdq = \rho(p',q',t')dp'dq', \tag{1.44}$$

and from the Liouville theorem we have $dpdq = dp'dq'$, so that:

$$\rho(p,q,t) = \rho(p',q',t'). \tag{1.45}$$

Now we see that the distribution function ρ is *constant along phase trajectories* – this is an alternative formulation of the Liouville theorem, using the notion of the distribution function. But still it is simply a mechanical statement, not using any probability (statistical) considerations.

Using these results, we can now derive the Liouville *equation*, which is actually the equation of motion for the distribution function. Assuming the moment t to be infinitesimally close to $t' = t + dt$ we get from Equation (1.45):

$$\rho(p,q,t) = \rho(p + \dot{p}dt, q + \dot{q}dt, t + dt) \tag{1.46}$$

so that (if ρ is differentiable) we obtain a differential equation:

$$\frac{d\rho}{dt} = \frac{\partial\rho}{\partial t} + \sum_{k=1}^{3N}\left(\frac{\partial\rho}{\partial p_k}\dot{p}_k + \frac{\partial\rho}{\partial q_k}\dot{q}_k\right) = 0 \tag{1.47}$$

and taking into account the Hamilton equations:

$$\frac{\partial\rho}{\partial t} = \sum_k\left(\frac{\partial H}{\partial q_k}\frac{\partial\rho}{\partial p_k} - \frac{\partial H}{\partial p_k}\frac{\partial\rho}{\partial q_k}\right). \tag{1.48}$$

The sum in the r.h.s. of Equation (1.48) is the so called *Poisson bracket* [11] for H and ρ:

$$\{H,\rho\} = \sum_k\left(\frac{\partial H}{\partial q_k}\frac{\partial\rho}{\partial p_k} - \frac{\partial H}{\partial p_k}\frac{\partial\rho}{\partial q_k}\right) \tag{1.49}$$

so that the Liouville equation can be written as:

$$\frac{\partial\rho}{\partial t} = \{H,\rho\}. \tag{1.50}$$

[5] Distribution function ρ can obviously be treated just as the density of phase points in the ensemble!

This equation is the basic *equation of motion* for the distribution function, which remains valid for both equilibrium and nonequilibrium problems. In principle, it allows one to calculate ρ at any moment in time t if it is known in an initial moment $t = t_0$. It can also be used, as we shall show later, to find the response of statistical systems to an *external perturbation*.

It is easy to see that the Liouville equation can be written as a continuity equation for the phase points moving in the phase space. Consider the motion of phase points in $6N$-dimensional phase space as the motion of a "phase liquid" with density ρ. The velocity of this motion is represented by the vector $(\dot{\mathbf{p}}_1, \dot{\mathbf{p}}_2, \ldots, \dot{\mathbf{p}}_N; \dot{\mathbf{q}}_1, \dot{\mathbf{q}}_2, \ldots, \dot{\mathbf{q}}_N)$ in this space. Accordingly, the appropriate continuity equation takes the form:

$$\frac{\partial \rho}{\partial t} + \sum_k \left[\frac{\partial}{\partial p_k}(\rho \dot{p}_k) + \frac{\partial}{\partial q_k}(\rho \dot{q}_k) \right] = 0, \tag{1.51}$$

where an expression in parentheses is just the divergence of the appropriate current. Performing differentiations we can write this term as:

$$\sum_k \left[\dot{p}_k \frac{\partial \rho}{\partial p_k} + \dot{q}_k \frac{\partial \rho}{\partial q_k} \right] + \rho \sum_k \left[\frac{\partial \dot{p}_k}{\partial p_k} + \frac{\partial \dot{q}_k}{\partial q_k} \right]. \tag{1.52}$$

Because of the Hamilton equations, the second term in this expression is identically zero, so that Equation (1.51) reduces to:

$$\frac{\partial \rho}{\partial t} + \sum_k \left[\dot{p}_k \frac{\partial \rho}{\partial p_k} + \dot{q}_k \frac{\partial \rho}{\partial q_k} \right] = 0, \tag{1.53}$$

which coincides with Equation (1.47). From here it follows, in particular, that the motion of the "phase liquid" is *incompressible*. For the case of systems in statistical (thermodynamic) equilibrium both ρ and H do not depend *explicitly* on time[6], so that Equation (1.50) reduces to:

$$\{H, \rho\} = 0 \tag{1.54}$$

and the distribution function ρ becomes an *integral of motion*. As we shall see, this fact alone (based upon an *assumption* of the existence of thermodynamic equilibrium!) immediately leads to a radical simplification of the whole analysis of equilibrium statistical ensembles.

1.5 Role of energy, microcanonical distribution

Thus we convinced ourselves that for the system in thermodynamic equilibrium the distribution function should be an integral of motion, i.e. it should be expressed via

[6] In this case there also is no explicit time dependence of appropriate *averages* of any physical characteristics, considered as functions of coordinates and momenta of particles of our system, which is an obvious property of an equilibrium state.

such combinations of coordinates and momenta p and q that remain constant in time as the (closed) system performs its motion in phase space. The number of independent integrals of motion for the closed (conserved) mechanical system with s degrees of freedom is equal to $2s - 1$ [11]. For the system consisting of N particles moving in three-dimensional space we have $2s = 6N$ (i.e., the number of all components of particle coordinates and momenta), so that the number of integrals of motion is immensely large. However, we can drastically reduce the number of integrals of motion on which the distribution function can actually depend. To do this we shall use *statistical* (not mechanical!) arguments. We have seen above that the distribution function ρ_{12} of the composite system consisting of two independent (non interacting) subsystems is equal to the product of distribution functions ρ_1 and ρ_2 of these subsystems: $\rho_{12} = \rho_1\rho_2$. Thus:

$$\ln \rho_{12} = \ln \rho_1 + \ln \rho_2 \tag{1.55}$$

i.e. the logarithm of the distribution function is additive. Accordingly, the logarithm of the distribution function of the system in equilibrium should be not just be an integral of motion, but an *additive* integral of motion.

In mechanics it is shown [11], that from all of the integral of motion of a closed (isolated) system only a few are in fact additive. These are the integrals of motion connected with basic properties of space and time – homogeneity and isotropy: energy, momentum and angular momentum [7]. Let us denote these integrals of motion for the a-th subsystem as $E_a(p,q)$, $\mathbf{P}_a(p,q)$ and $\mathbf{M}_a(p,q)$. The only additive combination of these integrals is the linear combination of the following form:

$$\ln \rho_a = \alpha_a + \beta E_a(p,q) + \boldsymbol{\gamma}\mathbf{P}_a(p,q) + \boldsymbol{\delta}\mathbf{M}_a(p,q) \tag{1.56}$$

with constant coefficients α_a, β, $\boldsymbol{\gamma}$, δ, and where β, $\boldsymbol{\gamma}$, δ should be the same for all subsystems – only in this case additivity (1.55) is satisfied. The coefficient α_a is just a normalization constant and can be determined from the requirement that $\int d\Gamma_a \rho_a = 1$. The coefficients β, $\boldsymbol{\gamma}$ and δ can be similarly determined via the constant values of corresponding additive integrals of motion (calculating the appropriate averages with the distribution function (1.56)).

Thus we come to a most important conclusion: the values of additive integrals of motion – energy, momentum and angular momentum – *completely* determine the statistical properties of a closed (isolated) system and statistical distributions of its (independent) subsystems, as well as the average values of its arbitrary physical characteristics in the state of thermodynamic (statistical) equilibrium. These seven (taking

[7] Additivity of energy follows from its general expression via the Lagrange function: $E = \sum_k \dot{q}_k \frac{\partial L}{\partial \dot{q}_k} - L$, and from additivity of the Lagrange function itself, which follows from the fact that the equations of motion of each of the non interacting parts of the system can not contain any parameters from other parts. Additivity of momentum of the many particle system is obvious: $\mathbf{P} = \sum_k m_k \mathbf{v}_k$ and, unlike energy, momentum is simply the sum of the momenta of different particles, despite the possibility of their interaction. An analogous property is valid also for the angular momentum: $\mathbf{M} = \sum_k [\mathbf{r}_k \mathbf{p}_k]$.

into account the number of vector components) additive integrals of motion replace the immense number of variables on which the distribution function can depend in the general (non equilibrium) case and which are necessary for a "complete" mechanical description of the many particle system.

The number of relevant integrals of motion diminishes, if from the very beginning we consider systems at rest. Then, both full momentum and angular momentum are obviously zero and the distribution function of the equilibrium state depends only on *one* variable – the total energy of the system:

$$\rho = \rho(E). \tag{1.57}$$

Thus the introduction of the simplest statistical considerations for systems at equilibrium immediately leads to a radical reduction in the number of relevant variables on which the distribution function depends and opens the way for the formulation of equilibrium statistical mechanics. Let us stress that these radical conclusions are based on the introduction of statistics and are "non derivable" from classical mechanics. Of course, in the general case, the distribution function can depend on some "external" parameters, which define macroscopic conditions for an ensemble and which are considered the same for all copies of the system within the ensemble (e.g. on volume, number of particles etc.).

Let us now explicitly construct the distribution function for a *closed* (adiabatically isolated) system in equilibrium[8]. It was first proposed by Gibbs. Consider the statistical ensemble of closed *energetically isolated* systems with a constant volume V, i.e. the ensemble of systems with a constant number of particles N, which are surrounded by *adiabatic* (in the thermodynamic sense) boundaries and possessing the same energy E, fixed up to some small uncertainty $\Delta E \ll E$. Following Gibbs we *assume* that the distribution function $\rho(p,q)$ for such an ensemble is just a constant within some layer of the phase space between two iso-energetic surfaces, corresponding to energies E and $E + \Delta E$ and zero outside this layer:

$$\rho(p,q) = \begin{cases} [\mathcal{W}(E,N,V)]^{-1} & \text{for } E \le H(p,q) \le E + \Delta E \\ 0 & \text{outside this layer.} \end{cases} \tag{1.58}$$

Such a distribution (ensemble) is called *microcanonical*. The distribution function (1.58) expresses the principle of *equal probability* of all microscopic states of a closed system. In fact it is the simplest possible assumption – we suppose that there is no preferable microscopic state (all are equally probable) so that systems of an ensemble, during the motion in phase space, just randomly "visit" all the microscopic states within the energy layer of the width ΔE, to which belong all the phase trajectories. The distribution function (1.58) represents simply the statistics of a "gambling die"

[8] Equation (1.56) in fact already represents an explicit form of the distribution function of an arbitrary *subsystem* weakly interacting with an environment of a much larger closed system. We shall return to this case later.

with $\mathcal{W}$ sides. Naturally, this distribution cannot apparently be *derived* from purely mechanical considerations, it can be justified only by comparing the results obtained from experiments, with its help.

The macroscopic state of microcanonical ensemble is characterized by three extensive parameters E, N, V. The constant $\mathcal{W}(E, N, V)$ is called the *statistical weight* and is determined by the normalization condition:

$$\int \frac{dpdq}{N!(2\pi\hbar)^{3N}} \rho(p,q) = 1$$

$$\int\limits_{E \le H(p,q) \le E+\Delta E} \frac{dpdq}{N!(2\pi\hbar)^{3N}} \frac{1}{\mathcal{W}(E, N, V)} = 1 \tag{1.59}$$

and is in fact the dimensionless phase volume of our energy layer ΔE, i.e. the *number of quantum states* in it (which is just the number of sides of our "gambling die"):

$$\mathcal{W}(E, N, V) = \frac{1}{N!(2\pi\hbar)^{3N}} \int\limits_{E \le H(p,q) \le E+\Delta E} dpdq. \tag{1.60}$$

In case of classical statistics we can always take the limit of $\Delta E \to 0$ and write:

$$\rho(p,q) = \mathcal{W}^{-1}(E, N, V)\delta(H(p,q) - E), \tag{1.61}$$

where

$$\mathcal{W}(E, N, V) = \frac{1}{N!(2\pi\hbar)^{3N}} \int dpdq\delta(H(p,q) - E). \tag{1.62}$$

Now it is obvious that $\mathcal{W}$ can also be considered as the density of states on the surface of constant energy in phase space. In the quantum case all this is limited by the well known uncertainty relation for time and energy: $\Delta E \Delta t \sim \hbar$. In the following we always, even in the classical limit, use microcanonical distribution in the form (1.58), assuming the quasiclassical limit of quantum mechanics.

The *hypothesis* that the microcanonical ensemble describes the macroscopic state of a closed (adiabatically isolated) system, i.e. the averages calculated with the distribution function (1.58) give experimentally observable values of all physical characteristics of the system, is one of the major postulates of equilibrium statistical mechanics. We already mentioned above that the observable values of arbitrary physical quantity $f(p,q)$ can also be calculated as an average over some observation time, and the problem of justification of our replacements of time averages by phase averages over the ensemble is called the ergodic problem. From this point of view, the problem of justification of microcanonical distribution reduces to the proof that for the closed (isolated) systems in equilibrium we actually have:

$$\lim_{T\to\infty} \frac{1}{T} \int_0^T dt f(p(t), q(t)) = \frac{1}{N!(2\pi\hbar)^{3N}} \int dpdq\rho(p,q) f(p,q), \tag{1.63}$$

where $\rho(p,q)$ is defined by the microcanonical distribution (1.58). This problem is very difficult and, despite some important achievements by mathematicians, is still unsolved. Physically it is usually justified by the so called *ergodic hypothesis* that the phase trajectory of a closed system during a rather long time period necessarily passes infinitesimally close to any given point on the ergodic surface. In Appendix A we shall present some elementary considerations related to this problem. Rigorous mathematical analysis can be found in [8], while the modern situation is discussed in [9]. Here we only briefly note that in recent years the problem of the conceptual foundations of statistical mechanics obtained new developments related to the discovery of stochastic instability (chaotization) of mechanical motion in different, more or less simple dynamical systems with a pretty *small* number of degrees of freedom [12]. Now it is clear that a statistical description is actually necessary even for such systems, which naively appear to be quite “solvable” within classical mechanics. This is also briefly discussed on an elementary level in Appendix A. In this sense, from the modern point of view, the requirement of a large number of degrees of freedom to justify the statistical approach is unnecessary and we cannot ignore them even in rather “simple” systems, where typically we observe an extreme sensitivity of phase trajectories to initial conditions, which leads to chaotic instability of the motion in phase space. Thus, the notorious Laplace determinism is rather illusory even in classical mechanics of such systems.

1.6 Partial distribution functions*

Knowledge of the general distribution function (1.6), depending on dynamical variables (coordinates and momenta) of all N particles, allows us to determine different macroscopic characteristics of the system. For example, the density of particles at point $\mathbf{r}$, by definition, is given by:

$$\rho(t,\mathbf{r}) = \int \hat{\rho}(\mathbf{r})\rho(t,\mathbf{r}_1,\ldots,\mathbf{p}_N)d\mathbf{r}_1\cdots d\mathbf{p}_N, \tag{1.64}$$

where $\hat{\rho}(\mathbf{r})$ is a density operator (here it is convenient to introduce operators of physical quantities even in the classical case):

$$\hat{\rho}(\mathbf{r}) = \sum_{i=1}^{N} m_i\delta(\mathbf{r}-\mathbf{r}_i), \tag{1.65}$$

where m_i is the mass of the i-the particle. Analogously, the current density at point $\mathbf{r}$ is:

$$\mathbf{J}(\mathbf{r}) = \int \hat{\mathbf{J}}(\mathbf{r})\rho(t,\mathbf{r}_1,\ldots,\mathbf{p}_N)d\mathbf{r}_1\cdots d\mathbf{p}_N, \tag{1.66}$$

* For explanation see preface.

where $\hat{\mathbf{J}}(\mathbf{r})$ is the current density operator:

$$\hat{\mathbf{J}}(\mathbf{r}) = \sum_{i=1}^{N} \mathbf{p}_i \delta(\mathbf{r} - \mathbf{r}_i). \tag{1.67}$$

The density of kinetic energy at point $\mathbf{r}$ is equal to:

$$E(t, \mathbf{r}) = \int \hat{E}(\mathbf{r}) \rho(t, \mathbf{r}_1, \ldots, \mathbf{p}_N) d\mathbf{r}_1 \cdots d\mathbf{p}_N, \tag{1.68}$$

where $\hat{E}(\mathbf{r})$ is the kinetic energy operator:

$$\hat{E}(\mathbf{r}) = \sum_{i=1}^{N} \frac{p_i^2}{2m_i} \delta(\mathbf{r} - \mathbf{r}_i). \tag{1.69}$$

For charged particles we can introduce the electric current density as:

$$\mathbf{j}(t, \mathbf{r}) = \int \hat{\mathbf{j}}(\mathbf{r}) \rho(t, \mathbf{r}_1, \ldots, \mathbf{p}_N) d\mathbf{r}_1 \cdots d\mathbf{p}_N, \tag{1.70}$$

where $\hat{\mathbf{j}}(\mathbf{r})$ is the electric current density operator:

$$\hat{\mathbf{j}}(\mathbf{r}) = \sum_{i=1}^{N} \frac{e_i}{m_i} \mathbf{p}_i \delta(\mathbf{r} - \mathbf{r}_i), \tag{1.71}$$

where e_i is the charge of the i-th particle.

The distribution function $\rho(t, \mathbf{r}_1, \ldots, \mathbf{p}_N)$ is the function of a practically infinite number of variables. However, expressing macrovariables via microscopic characteristics using the general formula:

$$A(t, \mathbf{r}) = \int \hat{A}(\mathbf{r}) \rho(t, \mathbf{r}_1, \ldots, \mathbf{p}_N) d\mathbf{r}_1 \cdots d\mathbf{p}_N \tag{1.72}$$

we have to take into account that the majority of physical operators of interest to us can be written as:

$$\hat{A}(\mathbf{r}) = \sum_{j=1}^{N} \hat{A}(\mathbf{r}_j, \mathbf{p}_j) \delta(\mathbf{r} - \mathbf{r}_j), \tag{1.73}$$

expressed as the sum of operators acting on dynamical variables of one particle (single particle operators). Examples of such operators are $\hat{\rho}$, $\hat{\mathbf{J}}$, $\hat{E}$ and $\hat{\mathbf{j}}$ introduced above. Much more rarely we are dealing with two particle operators of the form:

$$\hat{A}(\mathbf{r}, \mathbf{r}') = \frac{1}{2} \sum_{i \neq j} \hat{A}(\mathbf{r}_i, \mathbf{r}_j, \mathbf{p}_i, \mathbf{p}_j) \delta(\mathbf{r} - \mathbf{r}_i) \delta(\mathbf{r}' - \mathbf{r}_j). \tag{1.74}$$

An example of such an operator is the operator of potential energy of the system of particles interacting via some central potential:

$$\hat{U}(\mathbf{r}',\mathbf{r}'') = \frac{1}{2}\sum_{i\neq j} U(|\mathbf{r}_i - \mathbf{r}_j|)\delta(\mathbf{r}' - \mathbf{r}_i)\delta(\mathbf{r}'' - \mathbf{r}_j). \tag{1.75}$$

Operators consisting of linear combinations of operators acting on dynamical variables of three, four and larger numbers of particles almost never appear in any practical tasks of interest.

Thus, for solving the majority of problems we actually do not need to know the full N-particle distribution function:

$$F_N(t,\mathbf{r}_1,\ldots,\mathbf{p}_N) \equiv \rho(t,\mathbf{r}_1,\ldots,\mathbf{p}_N), \tag{1.76}$$

depending on the dynamic variables of the enormous number of particles, it is sufficient to somehow determine only the one particle $F_1(t,\mathbf{r}_i,\mathbf{p}_j)$ and two particle $F_2(t,\mathbf{r}_i,\mathbf{r}_j,\mathbf{p}_i,\mathbf{p}_j)$ distribution functions, which are defined as (V is the volume of the system) [13, 14]:

$$\begin{aligned} &F_1(t,\mathbf{r}_i,\mathbf{p}_i) \\ &+ V\int F_N(t,\mathbf{r}_1,\ldots,\mathbf{p}_N)d\mathbf{r}_1\cdots d\mathbf{r}_{i-1}d\mathbf{r}_{i+1}\cdots d\mathbf{r}_N d\mathbf{p}_1\cdots d\mathbf{p}_{i-1}d\mathbf{p}_{i+1}\cdots d\mathbf{p}_N, \end{aligned} \tag{1.77}$$

$$\begin{aligned} &F_2(t,\mathbf{r}_i,\mathbf{r}_j,\mathbf{p}_i,\mathbf{p}_j) \\ &= V^2\int F_N(t,\mathbf{r}_1,\ldots,\mathbf{p}_N)d\mathbf{r}_1\cdots d\mathbf{r}_{i-1}d\mathbf{r}_{i+1}\cdots d\mathbf{r}_{j-1}d\mathbf{r}_{j+1}\cdots d\mathbf{r}_N d\mathbf{p}_1 \\ &\cdots d\mathbf{p}_{i-1}d\mathbf{p}_{i+1}\cdots d\mathbf{p}_{j-1}d\mathbf{p}_{j+1}\cdots d\mathbf{p}_N \end{aligned} \tag{1.78}$$

or, in the general case, the s-particle distribution function (with $s \ll N$):

$$\begin{aligned} &F_s(t,\mathbf{r}_1,\ldots,\mathbf{r}_s,\mathbf{p}_1,\ldots,\mathbf{p}_s) \\ &= V^s\int F_N(t,\mathbf{r}_1,\ldots,\mathbf{p}_N)d\mathbf{r}_{s+1}\cdots d\mathbf{r}_N d\mathbf{p}_{s+1}\cdots d\mathbf{p}_N. \end{aligned} \tag{1.79}$$

From an obvious normalization condition

$$\frac{1}{V^s}\int F_s(t,\mathbf{r}_1,\ldots,\mathbf{p}_s)d\mathbf{r}_1\cdots d\mathbf{p}_s = 1 \tag{1.80}$$

it follows that $\frac{1}{V^s}F_s(t,\mathbf{r}_1,\ldots,\mathbf{p}_s)$ gives the probability for s particles in the system of N particles to be present at the moment t in the elementary phase space volume $d\mathbf{r}_1\cdots d\mathbf{p}_s$ of $6s$-dimensional phase space near the point $(\mathbf{r}_1,\ldots,\mathbf{p}_s)$. There are the following relations between these partial distribution functions, which are directly derived from their definition:

$$F_s(t,\mathbf{r}_1,\ldots,\mathbf{p}_s) = \frac{1}{V}\int F_{s+1}(t,\mathbf{r}_1,\ldots,\mathbf{p}_{s+1})d\mathbf{r}_{s+1}d\mathbf{p}_{s+1}. \tag{1.81}$$

The use of these distribution functions allows us to calculate the average values of single particle, two particle etc. operators of different physical quantities. For example, for a macrovariable described by operator (1.73) we have:

$$A(t,\mathbf{r}) = \frac{1}{V}\sum_{j=1}^{N}\int \hat{A}_j(\mathbf{r},\mathbf{p}_j)F_1(t,\mathbf{r},\mathbf{p}_j)d\mathbf{p}_j. \tag{1.82}$$

If all $\hat{A}_j$ are the same, i.e. $\hat{A}_j = \hat{a}(j = 1, 2, \ldots, N)$, we have:

$$A(t,\mathbf{r}) = \frac{N}{V}\int \hat{a}(\mathbf{r},\mathbf{p})F_1(t,\mathbf{r},\mathbf{p})d\mathbf{p}. \tag{1.83}$$

For macrovariables described by two particle operators of the type of (1.74) we get:

$$A(t,\mathbf{r}',\mathbf{r}'') = \frac{1}{2}\sum_{i\neq j}\frac{1}{V^2}\int \hat{A}_{ij}(\mathbf{r}',\mathbf{p}_i,\mathbf{r}'',\mathbf{p}_j)F_2(t,\mathbf{r}',\mathbf{r}'',\mathbf{p}_i,\mathbf{p}_j)d\mathbf{p}_i d\mathbf{p}_j. \tag{1.84}$$

If all $\hat{A}_{ij}$ are the same, i.e. $\hat{A}_{ij} = \hat{a}$, we have:

$$A(t,\mathbf{r}',\mathbf{r}'') = \frac{N(N-1)}{2V^2}\int \hat{a}(\mathbf{r}',\mathbf{p}',\mathbf{r}'',\mathbf{p}'')F_2(t,\mathbf{r}',\mathbf{r}'',\mathbf{p}',\mathbf{p}''),d\mathbf{p}'d\mathbf{p}'', \tag{1.85}$$

where obviously we can take $(N-1) \approx N$.

Thus we obtain the following expressions for the main macroscopic characteristics of systems consisting of identical particles:

$$\rho(t,\mathbf{r}) = m\frac{N}{V}\int F_1(t,\mathbf{r},\mathbf{p})d\mathbf{p}, \tag{1.86}$$

$$\mathbf{J}(t,\mathbf{r}) = \frac{N}{V}\int \mathbf{p}F_1(t,\mathbf{r},\mathbf{p})d\mathbf{p}, \tag{1.87}$$

$$E(t,\mathbf{r}) = \frac{1}{2m}\frac{N}{V}\int \mathbf{p}^2F_1(t,\mathbf{r},\mathbf{p})d\mathbf{p}, \tag{1.88}$$

$$\mathbf{j}(t,\mathbf{r}) = \frac{e}{m}\frac{N}{V}\int \mathbf{p}F_1(t,\mathbf{r},\mathbf{p})d\mathbf{p}. \tag{1.89}$$

The problem now is to find an explicit form of the single particle distribution function.

The general approach to find partial distribution functions can be formulated as follows. An arbitrary N-particle distribution function (1.76) satisfies the Liouville equation (1.47),(1.48), (1.50):

$$\frac{\partial F_N}{\partial t} = \{H, F_N\}. \tag{1.90}$$

Integrating Equation (1.90) over phase spaces of $N - s$ particles and taking into account Equation (1.79) we get:

$$\frac{1}{V^s}\frac{\partial F_s(t,\mathbf{r}_1,\ldots,\mathbf{p}_s)}{\partial t} = \int \{H, F_N\}d\mathbf{r}_{s+1}\cdots d\mathbf{p}_N. \tag{1.91}$$

For the Hamiltonian of the system of *interacting* particles:

$$H = \frac{1}{2m}\sum_{i=1}^{N}\mathbf{p}_i^2 + \frac{1}{2}\sum_{i\neq j} U(|\mathbf{r}_i - \mathbf{r}_j|), \tag{1.92}$$

after some direct, but rather tedious calculations [14], we obtain from Equation (1.91):

$$\frac{\partial F_s}{\partial t} = \{H^{(s)}, F_s\} + \frac{N}{V}\sum_{i=1}^{s}\int \frac{\partial U(|\mathbf{r}_i - \mathbf{r}_{s+1}|)}{\partial \mathbf{r}_i}\frac{\partial F_{s+1}}{\partial \mathbf{p}_i} d\mathbf{r}_{s+1} d\mathbf{p}_{s+1} \tag{1.93}$$

where $H^{(s)}$ denotes the Hamiltonian of the subsystem consisting of s particles.

The most important property of Equation (1.93) is that the equation of motion for the s-particle distribution function contains the term, describing the interaction of the subsystem of s particles with the rest of the N-particle system, which depends on the $s+1$-particle distribution function F_{s+1}. Thus, during the construction of the equations of motion for partial distribution functions, we necessarily obtain a practically infinite system of integro – differential equations, which is usually called *Bogolyubov's chain.* Strictly speaking, now we have to solve this whole chain of equations, which is certainly not easier than solving the general Liouville equation for the N-particle distribution function. However, in many cases, using some physical assumptions and models, we can "decouple" this chain, reducing the problem to a finite number of equations, e.g. expressing F_{s+1} via F_s, F_{s-1} etc. Then we remain with the closed system of s equations for $F_1, F_2, \ldots, F_s$. Most interesting is, in particular, the possibility to obtain the closed equation for a single particle distribution function:

$$\frac{\partial F_1}{\partial t} = L(F_1), \tag{1.94}$$

where L is some operator. Constructing and solving this so-called *kinetic* equation is the central problem of kinetic theory or *physical kinetics* [15]. We shall briefly discuss it in Chapter 10. In most cases, kinetic equations can be derived and solved only by some approximate methods. As a result, we can calculate the behavior of the average physical characteristics of our system, including their time dependence in the nonequilibrium case. The formalism of partial distribution functions can also serve as a ground for constructing the equilibrium statistical mechanics [16, 17], but in the following we shall use more traditional approaches.

1.7 Density matrix

Up to now we considered the classical statistical mechanics in which the state of a system was described by the point (p, q) in $6N$-dimensional phase space of coordinates and momenta of all particles and the time evolution was determined by Hamilton's equations. In quantum mechanics such a description becomes impossible as, due to uncertainty relations, we can not simultaneously measure both the spatial coordinates and momentum of a quantum particle. It is clear that we have to construct a special formalism of quantum statistical mechanics. However, the remarkable fact is that the main principles of the Gibbs approach remain valid also in quantum case.

1.7.1 Pure ensemble

In quantum mechanics, the state of a many particle system is described by the wave function $\psi(\mathbf{x}_1, \ldots, \mathbf{x}_N, t)$, which depends on time and on the coordinates of the particles $\mathbf{x}_1, \ldots, \mathbf{x}_N$ (or on another set of simultaneously measurable variables, e.g. momenta). Time evolution is determined by the Schroedinger equation:

$$i\hbar \frac{\partial \psi}{\partial t} = H\psi. \tag{1.95}$$

For example, for the system of N identical particles with mass m, without internal degrees of freedom and interacting with a two particle potential $U(|\mathbf{x}|)$, the Schroedinger equation can be written as:

$$i\hbar \frac{\partial \psi}{\partial t} = \left\{ -\frac{\hbar^2}{2m} \sum_{j=1}^{N} \nabla_j^2 + \frac{1}{2} \sum_{j \neq k} U(|\mathbf{x}_j - \mathbf{x}_k|) \right\} \psi. \tag{1.96}$$

The Schroedinger equation fully determines the wave function ψ at the moment t, if it was known at some initial moment $t = 0$. For example, for an isolated system with time independent H we can write down its formal solution as:

$$\psi(t) = e^{\frac{i}{\hbar} H t} \psi(0). \tag{1.97}$$

In quantum mechanics, the physical characteristics of a system are represented by linear Hermitian (self-adjoint) operators acting in a Hilbert space of the wave functions. Eigenvalues of such operators define the possible values of physical observables. The knowledge of the quantum state of the system ψ (a vector in Hilbert space), in the general case, does not lead to the precise knowledge of physical characteristics. It only allows us to calculate the average value of a dynamic variable represented by an operator A in the state ψ as:

$$\langle A \rangle = (\psi^\star, A\psi), \tag{1.98}$$

where, as usual, we assume wave functions to be normalized:

$$(\psi^\star, \psi) = 1, \tag{1.99}$$

where parenthesis denote the scalar product of vectors in Hilbert space:

$$(\psi^\star, \phi) = \int dx \psi^\star(x) \phi(x), \tag{1.100}$$

where for brevity we denote by x the whole set of coordinates $\mathbf{x}_1, \ldots, \mathbf{x}_N$. Only in a special case, when ψ is an eigenfunction of operator A, Equation (1.98) gives the precise value of the physical quantity A in the state ψ.

The state described by the wave function is usually called the *pure* state. The corresponding statistical ensemble, i.e. the large number of non interacting "copies" of the system, belonging to the same quantum state, is called the pure ensemble. Pure state (ensemble) gives the most *complete* description of the system within the quantum mechanics.

Expressions for the averages of physical quantities in the pure ensemble can be conveniently written, using the notion of the projection operator. Let us write down the linear operator A as a matrix in x-representation, defining it by matrix elements:

$$A\psi(x) = \int dx' A(x, x')\psi(x'). \tag{1.101}$$

Substituting (1.101) into (1.98) we get[9]:

$$\langle A \rangle = \int dx dx' A(x, x')\mathcal{P}(x', x) = \mathrm{Sp}\,(A\mathcal{P}), \tag{1.102}$$

where:

$$\mathcal{P}(x, x') = \psi(x)\psi^\star(x') \tag{1.103}$$

is the projection operator on the state ψ. It can be said that the pure ensemble is described by the projection operator (1.103), while the averages over this ensemble are calculated according to (1.102). Naturally this description is completely equivalent to the standard formalism of quantum mechanics, using the wave function.

The name "projection operator" is connected with the nature of the action of $\mathcal{P}$ on an arbitrary vector φ in Hilbert space – it projects it onto the "direction" of the vector ψ:

$$\mathcal{P}\varphi = \int dx' \mathcal{P}(x, x')\varphi(x') = (\psi^\star, \varphi)\psi(x). \tag{1.104}$$

The projection operator is Hermitian, as can be seen from its definition (1.103):

$$\mathcal{P}^\star(x, x') = \mathcal{P}(x', x). \tag{1.105}$$

It also has the following property:

$$\mathcal{P}^2 = \mathcal{P} \tag{1.106}$$

[9] In the following we use the notation of Sp for the sum of the diagonal elements of the matrix (trace), which is traditional in European and Russian literature. In English literature it is usually denoted as Tr.

which follows from (1.104) – after the first projection the following projections on the same "direction" change nothing. We also have:

$$\mathrm{Sp}\,\mathcal{P} = 1 \tag{1.107}$$

which follows from (1.102) after the replacement of A by the unity operator or from the definition (1.103) taking into account the normalization (1.99).

1.7.2 Mixed ensemble

Quantum mechanics is an inherently statistical theory of pure ensembles, which provides the complete description of quantum reality. Quantum statistical mechanics considers more general *mixed* ensembles, which are based on *incomplete* information about the quantum system. Let us consider the bug number of identical non interacting copies of the given system, which can be in different quantum states. In the mixed ensemble, we only know the probabilities $w_1, w_2, \dots$ to find a system in its exact quantum states $\psi_1, \psi_2, \dots$. We do not know precisely in which state the system is in reality. In this sense our knowledge is incomplete, since we only know these probabilities. However, in the mixed ensemble we can certainly calculate the average value of an arbitrary physical quantity, represented by an operator A as:

$$\langle A \rangle = \sum_k w_k (\psi_k^\star, A\psi_k), \tag{1.108}$$

where

$$\sum_k w_k = 1; \quad w_k \le 1. \tag{1.109}$$

These relations are, in fact, quite obvious, as $(\psi_k^\star, A\psi_k)$ represents the quantum mechanical average of an operator A in the state ψ_k. The pure ensemble is just the limiting case of the mixed ensemble, when all probabilities w_k are zero, except the only one equal to unity. Then (1.108) reduces to (1.98).

To study mixed ensembles it is convenient to use the statistical operator first introduced, independently, by Landau and von Neumann. Let us return to the linear operator A in x-matrix representation (1.101). Substituting (1.101) into (1.108), we get:

$$\langle A \rangle = \int dx dx' A(x, x') \rho(x', x) \tag{1.110}$$

or

$$\langle A \rangle = \mathrm{Sp}\,(\rho A), \tag{1.111}$$

where

$$\rho(x, x') = \sum_k w_k \psi_k(x) \psi_k^\star(x') \tag{1.112}$$

is the *statistical operator* in the x-matrix representation or the so called *density matrix*.

The density matrix depends on $2N$ variables $\mathbf{x}_1, \dots, \mathbf{x}_N, \mathbf{x}'_1, \dots, \mathbf{x}'_N$ and satisfies the normalization condition:

$$\mathrm{Sp}\, \rho = 1, \tag{1.113}$$

which is evident from its definition:

$$\mathrm{Sp}\, \rho = \int dx \rho(x, x) = \sum_k w_k (\psi_k^\star, \psi_k) = 1, \tag{1.114}$$

where the last equality follows from $(\psi_k^\star, \psi_k) = 1$ and $\sum_k w_k = 1$. Equation (1.113) is the direct analogue of the normalization condition for the distribution function in classical statistical mechanics.

The general relation (1.111) is most convenient, as the trace of the matrix is invariant with respect to unitary transformations. Thus, Equation (1.111) is independent of the representation used for the operators A and ρ; it is valid for an arbitrary representation, not only for x-representation used above. For example in some discrete n-representation we have:

$$\langle A \rangle = \sum_{mn} A_{mn} \rho_{nm}, \tag{1.115}$$

where A_{mn} are the matrix elements of operator A in n-representation, ρ_{nm} is the density matrix in n-representation.

The density matrix (statistical operator) is Hermitian:

$$\rho^\star(x, x') = \rho(x', x) \tag{1.116}$$

which follows directly from its definition (1.112). Using the projection operator (1.103) we can write the statistical operator (1.112) as:

$$\rho = \sum_k w_k \mathcal{P}_{\psi_k}; \quad \sum_k w_k = 1; \quad w_k \le 1, \tag{1.117}$$

where $\mathcal{P}_{\psi_k}$ is the projection operator on the state ψ_k. In case of all w_k zero except one, which is unity, the statistical operator (1.117) simply coincides with the projection operator (1.103).

To conclude this discussion, we show that statistical operator is positive definite, i.e. its eigenvalues are non negative. As ρ is Hermitian, positive definiteness of its eigenvalues is equivalent to:

$$\langle A^2 \rangle = \mathrm{Sp}\, (\rho A^2) \ge 0, \tag{1.118}$$

where A is an arbitrary Hermitian operator. It becomes obvious, if we diagonalize ρ (which is possible due to its Hermiticity) and write Equation (1.118) as:

$$\sum_{nk} \rho_{nn} A_{nk} A_{kn} = \sum_{nk} \rho_{nn} |A_{nk}|^2 \geq 0, \tag{1.119}$$

which leads to conclusion that $\rho_{nn} \geq 0$. For the density matrix (1.112) the property (1.118) is satisfied as:

$$\langle A^2 \rangle = \sum_k w_k (A^2)_{kk} = \sum_{km} w_k A_{km} A_{mk} = \sum_{km} w_k |A_{km}|^2 \geq 0 \tag{1.120}$$

so that the statistical operator is positively definite. It can also be shown that all matrix elements of the density matrix are limited by [3]:

$$\mathrm{Sp}\,\rho^2 = \sum_{mn} |\rho_{mn}|^2 \leq 1. \tag{1.121}$$

1.8 Quantum Liouville equation

Let us consider the time evolution of the density matrix (statistical operator) of an ensemble of systems described by the Hamiltonian H, which has no explicit time dependence. At the moment t the density matrix (1.112) has the form:

$$\rho(x, x') = \sum_k w_k \psi_k(x, t) \psi_k^{\star}(x', t), \tag{1.122}$$

where all the time dependence is contained in wave functions, while probabilities w_k do not depend on t, as they correspond to the distribution of systems in the ensemble at $t = 0$. Wave functions $\psi_k(x, t)$ are the solutions of the time dependent Schroedinger equation with initial conditions:

$$\psi_k(x, t)|_{t=0} = \psi_k(x), \tag{1.123}$$

where $\psi_k(x)$ is some system of wave functions defining the density matrix at $t = 0$:

$$\rho(x, x') = \sum_k w_k \psi_k(x) \psi_k^{\star}(x'). \tag{1.124}$$

If at the initial moment the relative number w_k of dynamical systems were in the state $\psi_k(x, 0)$, then at the moment t the same number of systems will be in the state $\psi_k(x, t)$.

Time dependence of $\psi_k(x, t)$ is determined by the Schroedinger equation:

$$i\hbar \frac{\partial \psi_k(x, t)}{\partial t} = H \psi(x, t) \tag{1.125}$$

or in x-matrix representation:

$$i\hbar\frac{\partial\psi_k(x,t)}{\partial t} = \int dx' H(x,x')\psi_k(x',t). \tag{1.126}$$

Accordingly, the density matrix satisfies the equation:

$$\begin{aligned} i\hbar\frac{\partial\rho(x,x',t)}{\partial t} &= \int dx'' \sum_k \left[H(x,x'')w_k\psi_k(x'',t)\psi_k^\star(x',t) - w_k\psi_k(x,t)\psi_k^\star(x'',t)H(x'',x')\right] \\ &= \int dx'' \left[H(x,x'')\rho(x'',x',t) - \rho(x,x'',t)H(x'',x')\right], \end{aligned} \tag{1.127}$$

where we have used the Hermiticity of the Hamiltonian $H^\star(x,x') = H(x',x)$. Thus we obtain the equation of motion for the density matrix – the so called quantum Liouville equation. In operator form it is written as:

$$i\hbar\frac{\partial\rho}{\partial t} = [H,\rho], \tag{1.128}$$

where

$$\frac{1}{i\hbar}[H,\rho] = \frac{1}{i\hbar}(H\rho - \rho H) \equiv \{H,\rho\} \tag{1.129}$$

are the quantum Poisson brackets.

For systems in statistical (thermodynamic) equilibrium ρ does not explicitly depend on time and the quantum Liouville equation takes the form:

$$[H,\rho] = 0 \tag{1.130}$$

so that ρ commutes with the Hamiltonian and is an integral of motion, similar to the case of classical statistical mechanics. Commutativity of operators ρ and H and their Hermiticity allows them to have a common system of eigenfunctions. Thus, the statistical operator of the equilibrium system can be written as:

$$\rho(x,x') = \sum_k w(E_k)\psi_k(x)\psi_k^\star(x'), \tag{1.131}$$

where the wave functions are eigenfunctions of the Hamiltonian (stationary Schroedinger equation):

$$H\psi_k = E_k\psi_k. \tag{1.132}$$

1.9 Microcanonical distribution in quantum statistics

The main ideas of the Gibbs approach based on the concept of statistical ensembles can be directly generalized from the classical to the quantum case. In equilibrium, the state density matrix can depend only on *additive* integrals of motion, for the same reasons as in the classical case (factorization of the density matrix for statistically independent systems and additivity of its logarithm). In quantum mechanics these integrals of motion are the same as in the classical case: total energy of the system (Hamiltonian H), total momentum $\mathbf{P}$ and total angular momentum $\mathbf{M}$ (the corresponding operators acting in the space of wave functions). Accordingly, the equilibrium density matrix can be a function of H, $\mathbf{P}$, $\mathbf{M}$ only:

$$\rho = \rho(H, \mathbf{P}, \mathbf{M}). \tag{1.133}$$

If the number of particles in systems of an ensemble N is not fixed, it has to be taken into account as an additional integral of motion:

$$[N, H] = 0, \tag{1.134}$$

where N is the particle number operator with positive integer eigenvalues $0, 1, 2, \dots$. Then:

$$\rho = \rho(H, \mathbf{P}, \mathbf{M}, N). \tag{1.135}$$

For the system at rest we have $\mathbf{P} = \mathbf{M} = 0$ and:

$$\rho = \rho(H) \quad \text{or} \quad \rho = \rho(H, N). \tag{1.136}$$

Besides that, the statistical operator can depend on external parameters fixed for all systems in an ensemble, e.g. on volume V.

The microcanonical distribution in quantum statistics can be introduced in the same way as in classical statistics. Consider an ensemble of closed, energetically (adiabatically) isolated systems with constant volume V and total number of particles N, which possess the same energy E up to some small uncertainty $\Delta E \ll E$. Let us suppose that all quantum states in an energy layer $E, E + \Delta E$ are *equally probable*, i.e. we can find a system from an ensemble in either of these states with equal probability. Then:

$$w(E_k) = \begin{cases} [\mathcal{W}(E, N, V)]^{-1} & \text{for} \quad E \le E_k \le E + \Delta E \\ 0 & \text{outside this layer} \end{cases} \tag{1.137}$$

and this is what we call the microcanonical distribution of quantum statistics. Here everything is similar to the classical case, though the statistical weight $\mathcal{W}(E, N, V)$ is not equal to the phase volume, but from the very beginning is just the number of quantum states in the energy layer $E, E + \Delta E$, for the system of N particles and volume V. This follows directly from the normalization condition $\sum_k w(E_k) = 1$. The microcanonical distribution corresponds to the density matrix of the form:

$$\rho(x, x') = \mathcal{W}^{-1}(E, N, V) \sum_{1 \le k \le \mathcal{W}} \psi_k(x)\psi_k^\star(x'), \tag{1.138}$$

which can also be written in operator form as:

$$\rho = \mathcal{W}^{-1}(E, N, V)\Delta(H - E), \tag{1.139}$$

where $\Delta(x)$ is the function, which is unity on the interval $0 \le x \le \Delta E$, and zero otherwise.

Let us stress once again that the assumption of equal probability of quantum states with the same energy for the closed (isolated) system is the simplest one, but not obvious. The justification of this hypothesis is the task of quantum ergodic theory.

1.10 Partial density matrices*

Similar to the classical case, in practice we do not need knowledge of the full density matrix of the N-particle system. In fact, the most important information about the system can be obtained from the study of statistical operators for (rather small) complexes of particles or so called partial density matrices [18]. Consider again an arbitrary system of N identical particles. Let us denote as $x_1, x_2, \ldots, x_N$ the variables of these particles (these may be coordinates, momenta etc.). Wave functions of the whole system are functions of these variables:

$$\psi_n(x, t) = \psi_n(x_1, \ldots, x_N, t), \tag{1.140}$$

where n denotes the "number" (the set of quantum numbers) of the given state of the system. Operators of physical quantities are represented by generalized matrices of the following form:

$$\mathcal{A} = \mathcal{A}(x_1, \ldots, x_N; x_1', \ldots, x_N'). \tag{1.141}$$

Consider the statistical operator for an N-particle system:

$$\rho(x_1, \ldots, x_N; x_1', \ldots, x_N'; t) = \sum_n w_n \psi_n(x_1, \ldots, x_N, t)\psi_n^\star(x_1', \ldots, x_N', t). \tag{1.142}$$

For a system of Bosons:

$$P\psi_n(x_1, \ldots, x_N, t) = \psi_n(x_1, \ldots, x_N, t), \tag{1.143}$$

where P is permutation operator of variables x_i ($i = 1, 2, \ldots, N$). For a system of Fermions:

$$P\psi_n(x_1, \ldots, x_N, t) = (-1)^P \psi_n(x_1, \ldots, x_N, t), \tag{1.144}$$

where $(-1)^P = 1$ for even permutations and $(-1)^P = -1$ for odd permutations. Thus, in both cases we have:

$$P\rho = \rho P \quad \text{or} \quad P\rho P^{-1} = \rho. \tag{1.145}$$

While calculating the average values of physical characteristics we usually deal with operators depending on variables of one, two,..., s-particles:

$$\mathcal{A}_1 = \sum_{1 \le \mathbf{r} \le N} \mathcal{A}(\mathbf{r}), \tag{1.146}$$

$$\mathcal{A}_2 = \sum_{1 \le \mathbf{r}_1 < \mathbf{r}_2 \le N} \mathcal{A}(\mathbf{r}_1, \mathbf{r}_2), \tag{1.147}$$

$$\vdots$$

$$\mathcal{A}_s = \sum_{1 \le \mathbf{r}_1 < \mathbf{r}_2 < \cdots < \mathbf{r}_s \le N} \mathcal{A}(\mathbf{r}_1, \mathbf{r}_2, \ldots, \mathbf{r}_s), \tag{1.148}$$

where $\mathbf{r}_i$ denotes the dependence on the coordinates of the i-th particle. The average values of such operators can be calculated with the help of density matrices, obtained from the general ρ, taking the Sp over (most of the) the independent variables. Taking into account the symmetry of ρ with respect to particles permutations (1.145), we get:

$$\langle \mathcal{A}_1 \rangle = N \mathrm{Sp}_1 \{\mathcal{A}(1)\rho_1(1)\}. \tag{1.149}$$

$$\langle \mathcal{A}_2 \rangle = \frac{N(N-1)}{2!} \mathrm{Sp}_{1,2} \{\mathcal{A}(1,2)\rho_2(1,2)\}, \tag{1.150}$$

$$\vdots$$

$$\langle \mathcal{A}_s \rangle = \frac{N(N-1)\ldots(N-s+1)}{s!} \mathrm{Sp}_{1,2,\ldots,s} \{\mathcal{A}(1,2,\ldots,s)\rho_s(1,2,\ldots,s)\}, \tag{1.151}$$

where we have introduced the notations:

$$\rho_1(1) = \mathrm{Sp}_{2,\ldots,N}\, \rho(1, 2, \ldots, N), \tag{1.152}$$

$$\rho_2(1,2) = \mathrm{Sp}_{3,\ldots,N}\, \rho(1, 2, \ldots, N), \tag{1.153}$$

$$\vdots$$

$$\rho_s(1, 2, \ldots, s) = \mathrm{Sp}_{s+1,\ldots,N}\, \rho(1, 2, \ldots, s, s+1, \ldots, N) \tag{1.154}$$

and for brevity we used the notations $\rho_2(1,2) \equiv \rho_2(x_1, x_2; x_1', x_2', t)$, $\mathrm{Sp}_2\rho_2(1,2) = \mathrm{Sp}_{x_2}\rho_2(x_1, x_2; x_1', x_2'; t)$, etc. The density matrices ρ_s are called statistical operators of complexes of s-particles or s-particle density matrices.

For operators ρ_s, due to (1.145), we have the following relations:

$$P_s \rho_s P_s^{-1} = \rho_s, \tag{1.155}$$

where P_s is permutation operator of s particles, and

$$\rho_s(1,2,\ldots,s) = \mathrm{Sp}_{s+1}\rho_{s+1}(1,\ldots,s,s+1) \tag{1.156}$$

which gives the expression for the s-particle density matrix via the $s+1$-particle density matrix.

Instead of ρ_s, let us use the operators F_s defined as:

$$F_s(1,\ldots,s) = N(N-1)\cdots(N-s+1)\rho_s(1,\ldots,s). \tag{1.157}$$

From (1.156) we obtain similar relations:

$$F_s(1,\ldots,s) = \frac{1}{N-s}\mathrm{Sp}_{s+1}F_{s+1}(1,\ldots,s,s+1). \tag{1.158}$$

In the limit of $N \to \infty$ and for fixed s, we can neglect s in the denominator, so that:

$$F_s(1,\ldots,s) = \frac{1}{N}\mathrm{Sp}_{s+1}F_{s+1}(1,\ldots,s,s+1). \tag{1.159}$$

Analogously to the classical case we shall call F_s the s-particle distribution functions. Under permutations we obviously have: $P_s F_s P_s^{-1} = F_s$. The averages of the physical quantities (1.151) are now written as:

$$\langle \mathcal{A}_s \rangle = \frac{1}{s!}\mathrm{Sp}_{1,\ldots,s}\{\mathcal{A}(1,\ldots,s)F_s(1,\ldots,s)\}. \tag{1.160}$$

Let us write an operator $\mathcal{A}_s$ in the standard second quantized form:

$$\mathcal{A}_s = \frac{1}{s!}\sum_{\{f,f'\}} A(f_1,\ldots,f_s; f_s',\ldots,f_1')a_{f_1}^{+}\cdots a_{f_s}^{+}a_{f_s'}\cdots a_{f_1'}, \tag{1.161}$$

where a_f^{+}, a_f are operators of creation and annihilation of particles in some single particle states, characterized by quantum numbers f, and $A(f_1,\ldots,f_s; f_s',\ldots,f_1')$ is the appropriate matrix element of an operator of the dynamical variable A. Then, calculating the ensemble averages we have:

$$\langle \mathcal{A}_s \rangle = \frac{1}{s!}\sum_{\{f,f'\}} A(f_1,\ldots,f_s; f_s',\ldots,f_1')\langle a_{f_1}^{+}\cdots a_{f_s}^{+}a_{f_s'}\cdots a_{f_1'}\rangle. \tag{1.162}$$

Comparing this expression with Equation (1.160) we obtain the following general expression for the s-particle distribution function in second quantized form:

$$F_s(1,\ldots,s) = \langle a^+_{f_1} \cdots a^+_{f_s} a_{f'_s} \cdots a_{f'_1} \rangle, \tag{1.163}$$

which is very convenient in practical calculations of quantum statistical mechanics and is widely used in the modern theory of many-particle systems[10]. In fact, calculation of these averages in different physical situations is the main task of this theory. One of the methods to perform such calculations is to construct the system of coupled equations for such distribution functions (Bogolyubov's chain), similar to the classical case and its approximate solution by some method. This approach is used e.g. during the derivation of quantum kinetic equations (see Chapter 10).

For a number of problems, as well as to discuss the correspondence with the classical case, it is often convenient to introduce the so called Wigner's distribution function in the "mixed" coordinate-momentum representation. Consider the single particle density matrix $\rho_1(1) = \rho(\mathbf{x}, \mathbf{x}')$, where $\mathbf{x}$ are coordinates of the particle, and define Wigner's distribution function as:

$$f(\mathbf{x},\mathbf{p}) = \frac{1}{(2\pi)^3} \int d\boldsymbol{\xi}\, e^{\frac{i}{\hbar}\mathbf{p}\boldsymbol{\xi}} \rho\left(\mathbf{x} + \frac{\boldsymbol{\xi}}{2}, \mathbf{x} - \frac{\boldsymbol{\xi}}{2}\right) \tag{1.164}$$

i.e. via Fourier transform over the difference of coordinates $\boldsymbol{\xi} = \mathbf{x} - \mathbf{x}'$. Integrating this function by $\mathbf{x}$ and $\mathbf{p}$ we obtain the diagonal elements of the density matrix in $\mathbf{x}$ and $\mathbf{p}$ representations:

$$\rho(\mathbf{x},\mathbf{x}) = \int d\mathbf{p} f(\mathbf{x},\mathbf{p}) \quad \rho(\mathbf{p},\mathbf{p}) = \int d\mathbf{x} f(\mathbf{x},\mathbf{p}), \tag{1.165}$$

which is easily obtained from the definition of Wigner's function, after proper change of variables. Of course, this distribution function $f(\mathbf{x},\mathbf{p})$ cannot be understood as a distribution function over coordinates and momenta (because of the uncertainty principle!), but its integrals separately define distribution functions over coordinates and momenta. By itself, Wigner's function can even be negative and does not have the meaning of the usual (classical) distribution function.

1.11 Entropy

1.11.1 Gibbs entropy. Entropy and probability

Let us return to the case of classical statistical mechanics and consider the logarithm of the distribution function (with inverse sign):

$$\eta = -\ln \rho(p,q,t). \tag{1.166}$$

[10] Let us stress that the angular brackets here denote averaging (taking the trace) with the full N-particle density matrix!

This function plays a special role, e.g. above we have already seen that it is additive for factorizing distribution functions of independent subsystems, which is analogous to the additivity of entropy in thermodynamics. The average value of this function is called *Gibbs entropy*:

$$S = \langle \eta \rangle = -\int \frac{dpdq}{(2\pi\hbar)^{3N} N!} \rho(p,q,t) \ln \rho(p,q,t). \tag{1.167}$$

Let us calculate this entropy for the microcanonical distribution (1.58) describing a closed system in equilibrium. Substituting into Equation (1.167) the distribution function given by (1.58):

$$\rho(p,q) = \begin{cases} [\mathcal{W}(E,N,V)]^{-1} & \text{for } E \leq H(p,q) \leq E + \Delta E \\ 0 & \text{outside this layer,} \end{cases} \tag{1.168}$$

where

$$\mathcal{W}(E,N,V) = \frac{1}{N!(2\pi\hbar)^{3N}} \int_{E \leq H(p,q) \leq E+\Delta E} dpdq \tag{1.169}$$

we obtain:

$$\begin{aligned} S(E,N,V) &= -\int_{E \leq H(p,q) \leq E+\Delta E} \frac{dpdq}{(2\pi\hbar)^{3N} N!} \frac{1}{\mathcal{W}(E,N,V)} \ln[\mathcal{W}(E,N,V)]^{-1} \\ &= \ln \mathcal{W}(E,N,V). \end{aligned} \tag{1.170}$$

Thus, for the microcanonical distribution Gibbs entropy is equal to the logarithm of the statistical weight, i.e. to the logarithm of the number of quantum states in the energy layer of width ΔE, corresponding to a given macroscopic state of our system[11].

In quantum statistics we may similarly introduce the operator of entropy via the logarithm of the density matrix:

$$\eta = -\ln \rho. \tag{1.171}$$

We have seen above that the statistical operator ρ is Hermitian and positive definite. Accordingly, its logarithm is also Hermitian and the entropy operator is positive: if

[11] Statistical weight $\mathcal{W}(E,N,V) = \exp S(E,N,V)$, by definition, is the number of energy levels in the energy interval ΔE, which characterize the energy distribution width. Dividing ΔE by $\mathcal{W}(E,N,V)$ we obviously get the average distance between the energy levels in the spectrum, in the vicinity of E. Denoting this distance by $D(E)$ we obtain: $D(E) = \Delta E \exp(-S(E))$. In this sense, the value of entropy $S(E)$ determines the density of states in this energy interval. Due to the additivity of entropy we can claim that the average distance between energy levels of a macroscopic system is dropping off exponentially with the growth of the number of particles, so that the spectrum of a macroscopic body is, in fact, continuous [1].

$w_1, w_2, \ldots$ are the eigenvalues of operator ρ, the eigenvalues of operator η are correspondingly $-\ln w_1, -\ln w_2, \ldots$ as the eigenvalues of the function of an operator are equal to the same function of eigenvalues. From $0 \leq w_k \leq 1$ it follows that $-\ln w_k \geq 0$.

The entropy operator is additive: if the operator ρ is a direct product of operators ρ_1 è ρ_2:

$$\rho = \rho_1 \times \rho_2 \tag{1.172}$$

we get

$$\eta = -\ln \rho_1 - \ln \rho_2 = \eta_1 + \eta_2. \tag{1.173}$$

Now we can again introduce the Gibbs entropy as the average logarithm of the density matrix (with a minus sign):

$$S = \langle \eta \rangle = -\langle \ln \rho \rangle = -\mathrm{Sp}\, \rho \ln \rho. \tag{1.174}$$

Entropy is positive definite and in some diagonal representation it can be written as:

$$S = -\sum_k w_k \ln w_k \geq 0. \tag{1.175}$$

Only for the special case where the density matrix corresponds to a pure state, we have $S = 0$ (one of $w_k = 1$, all others are zero). If ρ describes statistically independent ensembles (1.172), we get $S = S_1 + S_2$, where $S_1 = -\mathrm{Sp}\, \rho_1 \ln \rho_1$ and $S_2 = -\mathrm{Sp}\, \rho_2 \ln \rho_2$, so that the Gibbs entropy is additive (as entropy in thermodynamics).

Let us discuss the statistical meaning of entropy. Consider a macroscopic state of the system characterized by E, N, V *and* some additional macroscopic parameters $(x_1, x_2, \ldots, x_n)$, or just x for brevity. Let the statistical weight of the macroscopic state with the *fixed* values of these parameters be $\mathcal{W}(E, N, V, x)$. Then, the probability that this state (E, N, V, x) is realized, due to the equal probability of all states in the microcanonical ensemble is simply given by (sum of probabilities!):

$$w(x) = \frac{\mathcal{W}(E, N, V, x)}{\sum_x \mathcal{W}(E, N, V, x)} = C \exp(S(E, N, V, x)), \tag{1.176}$$

where

$$S(E, N, V, x) = \ln \mathcal{W}(E, N, V, x) \tag{1.177}$$

is the entropy of the state (E, N, V, x).

In many cases the most probable value of x, which we denote by x^*, and the average value $\langle x \rangle$ just coincide, as the probability $w(x)$ possesses a sharp peak at $x = x^*$

(for a large enough system). The most probable value x^* is determined by the maximum of $w(x)$:

$$S(E, N, V, x) = \text{Max} \quad \text{for} \quad x = x^* \tag{1.178}$$

or

$$\frac{\partial S(E, N, V, x_1^*, \ldots, x_n^*)}{\partial x_j^*} = 0 \quad j = 1, 2, \ldots, n. \tag{1.179}$$

It is easy to conclude that

$$w(\Delta x) = C' \exp\{S(E, N, V, x^* + \Delta x) - S(E, N, V, x^*)\} \tag{1.180}$$

determines the probability of the deviations (fluctuations) Δx of parameters x from their most probable (average or equilibrium!) values[12]. This property of entropy gives the foundation of its statistical applications (Boltzmann's principle), it is also the foundation for the theory of fluctuations (Einstein, see Chapter 7).

1.11.2 The law of entropy growth

In thermodynamics, it is shown that, in the state of thermodynamic equilibrium, the entropy of an isolated system can only increase or remain constant. For the equilibrium state we shall show below that Gibbs' definition of entropy is actually equivalent to thermodynamic entropy. However, for nonequilibrium cases, when the distribution function $\rho(p, q, t)$ depends on time, the situation is much more complicated. In fact, we can easily show that for an isolated system the Gibbs entropy does not depend on time at all, thus it just cannot increase. To see this, let the distribution function at $t = 0$ be $\rho(p^0, q^0, 0)$, while at the moment t it is equal to some $\rho(p, q, t)$, where (p, q) belongs to a phase trajectory passing through (p^0, q^0) and moving according to Hamilton's equations. According to Liouville theorem we have (1.45):

$$\rho(p^0, q^0, 0) = \rho(p, q, t). \tag{1.181}$$

Then at time t the Gibbs entropy is equal to:

$$\begin{aligned} S &= -\int \frac{dpdq}{(2\pi\hbar)^{3N} N!} \rho(p, q, t) \ln \rho(p, q, t) \\ &= -\int \frac{dp^0 dq^0}{(2\pi\hbar)^{3N} N!} \rho(p^0, q^0, 0) \ln \rho(p^0, q^0, 0) \end{aligned} \tag{1.182}$$

as, due to the Liouville theorem on the conservation of the phase volume, we have $dpdq = dp^0 dq^0$. Then it is obvious that the Gibbs entropy cannot serve as the

[12] We can write: $w(x^* + \Delta x) = C \exp\{S(x^* + \Delta x)\} = C' \exp\{S(x^* + \Delta x) - S(x^*)\}$, where $C' = C \exp\{S(x^*)\}$ is just the new normalization constant.

general definition of entropy for the arbitrary nonequilibrium state. This is the major paradox directly connected with the principal difficulty of justifying the irreversible thermodynamic behavior by time-reversible equations of motion of classical (and also quantum) mechanics, which lead to active discussions already at the initial stages of development of statistical mechanics (Boltzmann, Zermelo, Poincare, Gibbs).

Using some early ideas of Gibbs, further developed by Paul and Tatiana Ehrenfest, the following heuristic picture can illustrate the statistical sense of entropy growth with time evolution of a mechanical system. Starting with the idea of the inevitable limitations of measurements of coordinates and momenta in the phase space[13] let us introduce the "coarse grained" distribution function related to "microscopic" distribution $\rho(p,q,t)$ by the following relation:

$$\tilde{\rho}(p,q,t) \equiv \rho_i = \frac{1}{\omega_i}\int_{\omega_i} dpdq\rho(p,q,t), \tag{1.183}$$

where the integration (averaging) is performed over some small fixed "cells" in the phase space ω_i, with size determined by the limitations of the measurements mentioned above. Such averaging ("coarse graining") of the distribution function just means that we introduce some "natural" and finite resolution in the phase space – smaller scales are outside the limits of the measurement procedures available to us. For example, we have already noted that there exists an absolute lower boundary for any "cell" ω_i in the phase space, which can not be smaller than $(2\pi\hbar)^{3N}$ due to the uncertainty principle[14]. The "coarse grained" distribution function (1.183) is obviously constant inside the appropriate "cell" ω_i, surrounding the point (p,q). Then, any integral over the whole phase space with our distribution function can be written as:

$$\int dpdq\rho(p,q)\cdots = \sum_i \rho_i\omega_i\cdots = \sum_i \int_{\omega_i} dpdq\rho(p,q)\cdots = \int dpdq\tilde{\rho}(p,q)\cdots. \tag{1.184}$$

Now we shall see that the Gibbs entropy, constructed with help of the "coarse grained" distribution is, in the general case, time dependent and can increase with time. Let us compare the values of the Gibbs entropy calculated with the "coarse grained" distribution function at the moments t and $t = 0$, assuming that at the initial moment the microscopic distribution function just coincides with "coarse grained":

$$\rho(p^0,q^0,0) = \tilde{\rho}(p^0,q^0,0). \tag{1.185}$$

[13] This may be related to finite resolution of the experimental apparatus, sensitivity to initial conditions etc.

[14] In general, the situation with entropy time dependence in quantum statistical mechanics is quite similar to that in classical statistical mechanics and we shall limit ourselves here to classical case only, referring to the discussions of the quantum case in the literature [3].

We have:

$$S_t - S_0 = -\int d\Gamma \tilde{\rho}(p,q,t)\ln\tilde{\rho}(p,q,t) + \int d\Gamma_0 \rho(p^0,q^0,0)\ln\rho(p^0,q^0,0)$$
$$= -\int d\Gamma\,\{\rho(p,q,t)\ln\tilde{\rho}(p,q,t) - \rho(p,q,t)\ln\rho(p,q,t)\}, \qquad (1.186)$$

where we have used Liouville theorem to write $d\Gamma_0 = d\Gamma$ and also (1.181) and removed the tilde over distribution function, which is not under the logarithm, which according to (1.184) is always correct under integration[15].

For two arbitrary normalized distribution functions ρ and ρ', defined in the same phase space, we can prove the following Gibbs inequality[16]:

$$\int d\Gamma \rho \ln\left(\frac{\rho}{\rho'}\right) \geq 0, \qquad (1.187)$$

where equality is achieved only in the case of $\rho = \rho'$. Then, from Equation (1.186) (taking $\tilde{\rho} = \rho'$) we immediately obtain:

$$S_t \geq S_0. \qquad (1.188)$$

Let $\rho(p^0, q^0, 0)$ be some nonequilibrium distribution. Then at the moment t:

$$\rho(p,q,t) \neq \tilde{\rho}(p,q,t) \qquad (1.189)$$

as though $\rho(p,q,t)$ does not change along phase trajectory, but the "cell", ω, surrounding an arbitrary point (p,q) will be "visited" by phase points from other "cells" (chaotically coming and going), and these processes, in the general case, will not compensate each other. This is called "mixing" of phase points. Taking into account Equation (1.189) from Equation (1.188) it follows that:

$$S_t > S_o \qquad (1.190)$$

i.e. entropy, defined with a "coarse grained" distribution function, grows with time. This conclusion is valid if the motion of phase points is "mixing" in the above-mentioned sense. The "mixing" nature of motion in phase space is intimately related to the local instability of phase trajectories, which appears (as a rule!) for non trivial physical systems, even with a pretty small number of particles [12]. This instability leads to exponential growth (in time) of the distance between phase points on different

[15] We have: $\int dpdq\tilde{\rho}(p,q)\ln\tilde{\rho}(p,q) = \sum_i \rho_i\omega_i \ln\rho_i = \sum_i \left[\int_{\omega_i} dpdq\rho(p,q)\ln\rho_i\right] = \int dpdq\rho(p,q)\ln\tilde{\rho}(p,q)$, which was used in Equation (1.186).

[16] This inequality follows from $\ln\left(\frac{\rho}{\rho'}\right) \geq 1 - \frac{\rho'}{\rho}$ $(\rho > 0, \rho' > 0)$, where equality is valid only for $\rho = \rho'$. It is clear from inequality $\ln x \geq 1 - \frac{1}{x}$, valid for $x > 0$ (equality for $x = 1$), where we put $x = \frac{\rho}{\rho'}$. After multiplication by ρ and integration over the phase space we get: $\int \rho \ln\left(\frac{\rho}{\rho'}\right) d\Gamma \geq \int \rho(1 - \frac{\rho'}{\rho})d\Gamma = 0$, where we have used normalization, thus proving Equation (1.187.)

trajectories initially quite close to each other. We shall discuss this situation in more detail, though still on a rather elementary level, in Appendix A.

However, the introduction of the "coarse grained" distribution function can not quite be considered as a satisfactory solution of the problem. The smaller the scale of "coarse graining" (the size of the "cells" ω) the smaller is the entropy growth, and in the limit of $\omega \to 0$ it just goes to zero. At the same time, the growth of physical entropy should not depend on the scale of "coarse graining". For example, we could have taken $\omega \sim \hbar^{3N}$, in agreement with the requirements of quantum mechanics, but in this case the growth of the entropy would be controlled by the size of Planck's constant $\hbar$. However, this is obviously not so; there is no such relation at all. There are different points of view with respect to this problem. Some researchers believe [3] that "coarse graining" should be performed within two limiting procedures: first we must go to the usual thermodynamic limit of statistical mechanics with the number of particles in the system $N \to \infty$, system volume $V \to \infty$, while the particle density remains finite $N/V = \text{const}$, and *only afterwards* we perform the limit $\omega \to 0$. The modern point of view [12] is that the thermodynamic limit here is irrelevant and the "mixing" of phase points (positive Kolmogorov entropy, see Appendix A) is sufficient to guarantee "correct" physical behavior, even for systems with a rather small number of degrees of freedom $N > 2$. An isolated system, irrespective of the initial conditions, evolves to the equilibrium state, where it can (with equal probability) be discovered in any of its possible microscopic states (ergodic behavior).

Another, probably more physical approach to defining nonequilibrium entropy [1] is based on the unquestionable definition of the entropy of the equilibrium state (1.170). Assume that the system is initially in some not completely equilibrium state and start to analyze its evolution during time intervals Δt. Let us separate the system in some smaller (more or less independent) parts, so small that their relaxation times are also small compared with Δt (relaxation times are usually smaller for smaller systems – an empirical fact!). During the time interval such subsystems Δt can be considered to be in their partial equilibrium states, which are described by their own microcanonical ensembles, when we can use the usual definitions of statistical weight and calculate appropriate (equilibrium) entropies. Then the statistical weight of the complete system is defined as the product $\mathcal{W} = \prod_i \mathcal{W}_i$ of statistical weights of the separate subsystems, and the entropy $S = \sum_i S_i$. In such an approach, the entropy characterizes only some average properties of the body during some finite time interval Δt. Then it is clear that for too small time intervals Δt the notion of entropy just looses its meaning and, in particular, we can not speak about its *instant* value.

More formally, in this approach, we can analyze the entropy growth in the following way. Consider the closed macroscopic system at time t. If we break this system into relatively small parts (subsystems), each will have its own distribution function ρ_i.

The entropy S of the whole system at that moment is equal to:

$$S = -\sum_i \langle \ln \rho_i \rangle = -\Big\langle \ln \prod_i \rho_i \Big\rangle. \tag{1.191}$$

Considering our subsystems as quasi independent, we can introduce the distribution function of the whole system as:

$$\rho = \prod_i \rho_i. \tag{1.192}$$

To obtain the distribution function at some later time t' we have to apply to ρ the mechanical equations of motion for the closed system. Then ρ will evolve at the moment t' to some ρ'. To obtain the distribution function of only i-th part of the system at the moment t' we must integrate ρ' over phase volumes of all subsystems, except the i-th. If we denote this distribution function as ρ_i', then at the moment t' we get:

$$\rho_i' = \int_1 \int_2 \cdots \int_{i-1} \int_{i+1} \cdots d\Gamma_1 d\Gamma_2 \cdots d\Gamma_{i-1} d\Gamma_{i+1} \cdots \rho'. \tag{1.193}$$

Note that in the general case ρ' already cannot be written as a product of all ρ_i'. The entropy at the moment t', according to our definition is:

$$S' = -\sum_i \langle \ln \rho_i' \rangle, \tag{1.194}$$

where the averaging $\langle \cdots \rangle$ is already performed with distribution function ρ'. Let us now use the inequality $\ln x \le x - 1$, valid for $x > 0$. Here, set $x = \frac{\prod_i \rho_i'}{\rho'}$ to obtain:

$$-\ln \rho' + \sum_i \ln \rho_i' \le \frac{\prod_i \rho_i'}{\rho'} - 1. \tag{1.195}$$

Averaging both sides of this inequality with distribution function ρ', we get zero in the r.h.s., as $\int d\Gamma_1 d\Gamma_2 \cdots \prod_i \rho_i' = \prod_i \int d\Gamma_i \rho_i = 1$ due to normalization, while the l.h.s. reduces to $\langle -\ln \rho' \rangle + \sum_i \langle \ln \rho_i' \rangle$. Finally we get:

$$-\langle \ln \rho' \rangle - S' \le 0. \tag{1.196}$$

According to Liouville theorem, the distribution function ρ does not change under mechanical motion, so the value of $-\langle \ln \rho' \rangle$ remains equal to $-\langle \ln \rho \rangle$, which is the initial entropy S. Thus we obtain:

$$S' \ge S, \tag{1.197}$$

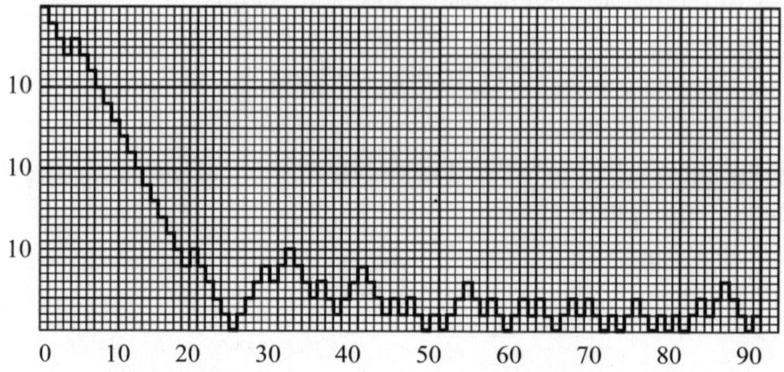

Figure 1.2. Typical realization of Ehrenfest's "H-curve". Ordinate shows the value of $|n_A(s) - n_B(s)| = 2|n_A(s) - R|$.

proving the entropy law: *if the closed system is defined by its macroscopic state at some moment in time, the most probable behavior at some later time is the growth of entropy.*

Mechanical equations of motion are symmetric with respect to the replacement of t by $-t$. If mechanical laws allow some process, e.g. characterized by the growth of entropy, they also must allow just the opposite process, when the system passes through precisely the same configurations in inverse order, so that its entropy diminishes. It may seem that we arrived at a contradiction. However, the formulation of the law of entropy growth used above, does not contradict the time invariance, if we speak only about the *most probable* evolution of some macroscopically defined state. In fact, the above arguments never explicitly used the fact that $t' > t$! A similar discussion will show that $S' \geq S$ also for $t \leq t'$. In other words, the law of entropy growth means only that given the macroscopic state, of all microscopic states forming this macroscopic state, an immensely vast majority will, at a later time, evolve to the state with larger entropy (or the same entropy in case of equilibrium). Thus the entropy law is primarily a *statistical* statement!

To understand this situation better, we may use a heuristic model discussed first by Paul and Tatiana Ehrenfest. Consider $2R$ balls, numbered from 1 to $2R$, and distributed among two boxes A and B. At some discrete moment in time s, a random number generator produces some integer from the interval between 1 and $2R$. Then, the ball with this number is just transferred from one box to another and this procedure continues for many times. Actually, this procedure is simple to realize on any modern PC. Intuitively, it is quite clear what happens. Consider, for simplicity, the initial state when all balls are in the box A. Then, on the first step we necessarily transfer one ball from A to B. On the second step we may return to the initial state, but the probability of this event is $(2R)^{-1}$ and small if $2R$ is big enough. Actually, with much larger probability $1 - (2R)^{-1}$, another ball is transferred to box B. It is clear that until the number of balls n_A in box A is significantly larger than the number of balls n_B in box B, we "almost always" will only observe transitions from A to B. Or, in more detail, let there be $n_A(s)$ balls in box A at time s, while in box B there are $2R - n_A(s)$ balls. At the next moment $s + 1$ the probability of obtaining the ball with the number belonging to box A is $\frac{n_A}{2R}$, while for the ball from box B, the probability, naturally, $\frac{2R-n_A}{2R}$. However, until $n_A > 2R - n_A$, the "relative chance" $\frac{n_A}{2R-n_A}$ of a ball from A to appear, compared to the similar chance of emergence of the ball from B, is obviously larger than 1. Thus, the transition $A \to B$ is more probable and the difference between the number of balls in our boxes diminishes with "time". This tendency persists until we achieve the equality $n_A - (2R - n_A) = 0$, and it becomes weaker as this difference approaches zero. Thus, as the number of balls in both boxes tends to be equal, the probabilities of balls emerging from either A or B become closer to each other, and the result (for further moments in time) becomes less and less clear. The next transfer may lead to further "equalization" of the number of balls in both boxes, but it may also lead to the inverse process. Figure 1.2 shows a typical realization of such an experiment with 40 balls. It is seen that initially the process seems to be irreversible, but close to the "equilibrium state", the difference between the number of balls in our boxes starts to fluctuate, which shows that in fact we are dealing with a reversible process[17]. We cannot say that this difference always diminishes with "time", but we can be absolutely sure that for large values of the number of balls $2R$ it diminishes "almost always", while we are far enough from the "equilibrium". The behavior of entropy in a nonequilibrium many-particle system is precisely the same (with negative sign)!

Ehrenfest's model allows a simple answer to all objections against the statistical mechanical justification of irreversible behavior. According to the principle of microscopic reversibility of mechanical motion, the process, after "time reversal", when the movement of the balls will go in precisely *reverse* order along the same "H-curve", is completely possible. But for large enough values of R such a process is absolutely improbable. The probability of all balls to return "sometime" to a single box is not zero, but it is extremely small (say for $R \sim 10^{22}$!). Precisely the same is the meaning of thermodynamic irreversibility and the law of entropy growth[18].

Thus, the common viewpoint is that an evolving isolated system essentially passes through states corresponding to *more and more probable* distributions. This behavior is overwhelming, due to the factor $\exp(S)$, where in the exponent we have an additive entropy. Thus, the processes in a closed nonequilibrium system develop in such a way, that the system continuously goes from states with lower entropy to states with larger entropy, until the entropy reaches its maximum in the state of statistical equilibrium.

[17] On Figure 1.2 these fluctuations are always positive because the ordinate shows the absolute value of the difference between the number of balls in boxes A and B.

[18] "What, never? No, never! What, never? Well, hardly ever!" to quote Captain Corcoran of *H. M. S. Pinafore* by W. Gilbert and A. Sullivan (1878). This quotation was used in the context of entropy behavior in Ch. 4 of "Statistical Mechanics" by J.Mayer and M.Goeppert-Mayer, Wiley, NY 1940.

Speaking about “most probable” behavior we must take into account that in reality the probability of a transition to a state with larger entropy is immensely larger than the probability of any significant entropy drop, so that such transitions are practically unobservable (up to small fluctuations). This purely statistical interpretation of the entropy growth was first formulated by Boltzmann. “It is doubtful whether the law of increase of entropy thus formulated could be derived on the basis of classical mechanics” [1][19]. In the framework of modern statistical mechanics of non equilibrium systems [3, 16] and physical kinetics [13, 14, 15] it is possible to explicitly demonstrate the entropy growth in a number of concrete statistical models. However, we always need some physical (statistical) assumption to obtain this behavior. We shall return to a brief discussion of these problems later.

[19] Landau made the interesting observation that in quantum mechanics the situation probably changes. Though the Schroedinger equation, by itself, is invariant with respect to time reversal (with simultaneous replacement of ψ by $\psi^{\star}$), quantum mechanics contains some inequivalence of both directions of time. This inequivalence appears due to the importance of the process of interaction of the quantum object with a classical system (e.g. related to the measurement process). If the quantum object undergoes two successive processes of such an interaction, say A and B, the claim that the probability of some outcome of the process B is determined by the result of the process A is justified only if the process A preceded B. Thus it seems that in quantum mechanics there is some physical inequivalence of both directions of time, so that the law of entropy growth may follow from it. However, in this case there should be some inequality containing $\hbar$, justifying the validity of this law. There is no evidence at all that this is true. Similarly, we may mention the possibility to explain irreversible thermodynamic behavior by the experimentally known fact of very weak CP-symmetry violation in the modern physics of elementary particles, which inevitably leads to a weak violation of T-invariance in the processes of elementary particle interactions. Up to now there is no accepted interpretation of this kind.

Chapter 2
Gibbs distribution

2.1 Canonical distribution

Let us consider the from a practical point of view most important task of finding the distribution function of an arbitrary macroscopic body, which is a small part (subsystem) of a much larger closed (adiabatically isolated) system. Suppose that we can consider this large system as consisting of two parts: the body (subsystem) of interest to us and the rest of the system (surrounding the body), which we shall call *thermostat* or *bath* (cf. Figure 2.1). It is assumed that the thermostat is a system with many

Figure 2.1. System (1) in thermostat (bath) (2).

degrees of freedom, which can exchange energy with our subsystem, and that it is so large that its own state is unchanged during such interaction[1].

[1] The following presentation mainly follows [3]. Some points are explained following [1].

Let us start with quantum statistics. Both parts, our subsystem and bath, can be considered as a single, energetically isolated (closed) system with Hamiltonian:

$$H = H_1 + H_2, \tag{2.1}$$

where H_1 is the Hamiltonian of the (sub)system under study and H_2 is the Hamiltonian of the bath (thermostat), which is assumed to be much larger than the system of interest to us. Interaction between our system and the bath is assumed to be very weak but, strictly speaking, finite, as it must be to guarantee the equilibrium state of both parts of the large system (in Hamiltonian (2.1) this interaction is just dropped)[2]. In this case the wave function, corresponding to the Hamiltonian (2.1), is factorized into a product of the wave functions of the bath (system 2) and the body under study (system 1):

$$\psi_{ik}(x, y) = \psi_k(x)\psi_i(y), \tag{2.2}$$

where $\psi_k(x)$ is an eigenfunction of H_1 and $\psi_i(y)$ is an eigenfunction of H_2, while x and y are sets of coordinates of the system and the bath correspondingly.

Energy levels of the whole (composite) system (neglecting small surface interaction effects) are just sums of energy levels of systems (1) and (2):

$$E_{ik} = E_i + E_k, \tag{2.3}$$

where E_k denote the energy levels of the system (1), and E_i denote the energy levels of the bath (2).

The statistical operator (density matrix) of the whole (closed!) system is:

$$\rho(xy; x'y') = \sum_{ik} w_{ik}\psi_{ik}(x, y)\psi^{\star}_{ik}(x', y'), \tag{2.4}$$

where w_{ik} is defined, according to our basic assumption, by the microcanonical distribution (1.58):

$$w(E_{ik}) = \begin{cases} [\mathcal{W}(E)]^{-1} & \text{for } E \le E_{ik} \le E + \Delta E \\ 0 & \text{outside this energy layer.} \end{cases} \tag{2.5}$$

The density matrix of the system under study (1) can be obtained by taking the trace of the statistical operator of the whole (composite) system over the coordinates (variables) of the bath (subsystem (2))[3]:

$$\rho(x, x') = \mathrm{Sp}_2\rho(xy; x'y') = \sum_{ik} w_{ik} \int dy\, \psi_{ik}(x, y)\psi^{\star}_{ik}(x', y). \tag{2.6}$$

[2] For example, the thermal contact of our body with a bath is only through its boundary and can be considered a small surface effect.

[3] This operation is similar to that we used while obtaining e.g. the single particle density matrix from the one for two particles.

From here, using (2.2) and the orthonormality of the wave functions, we immediately obtain:

$$\rho(x,x') = \sum_k w_k \psi_k(x)\psi_k^\star(x'), \tag{2.7}$$

where

$$w_k = \sum_i w_{ik}. \tag{2.8}$$

Now it is clear that to get the probability distribution of quantum states for system (1) we have to simply sum the probability distribution for the whole system over the states of the bath (thermostat):

$$w(E_k) = \sum_i w(E_i + E_k)|_{E_i+E_k=E} = \frac{1}{\mathcal{W}(E)}\sum_i 1|_{E_i=E-E_k}, \tag{2.9}$$

where for brevity we denoted $E_{ik} = E$. It is clear that (2.9) reduces to:

$$w(E_k) = \frac{\mathcal{W}_2(E-E_k)}{\mathcal{W}(E)}, \tag{2.10}$$

where $\mathcal{W}_2(E-E_k)$ is the number of quantum states of the bath with energy $E - E_k$, while $\mathcal{W}(E)$ is the number of states of the whole (composite) system, corresponding to energy E.

Introducing the entropy of the bath $S_2(E)$ and the entropy of the whole system $S(E)$ via (1.170), we rewrite (2.10) as:

$$w(E_k) = \exp\{S_2(E-E_k) - S(E)\}. \tag{2.11}$$

Taking into account that our system (1) is small in comparison with the bath, so that $E_k \ll E$, we can write an expansion:

$$S_2(E-E_k) \approx S_2(E) - \frac{\partial S_2}{\partial E}E_k. \tag{2.12}$$

Substituting (2.12) into (2.11) we get:

$$w(E_k) = A\exp\left(-\frac{E_k}{T}\right), \tag{2.13}$$

where we have introduced the *temperature* T (of the bath!) as:

$$\frac{1}{T} = \frac{\partial S_2(E)}{\partial E} = \frac{\partial \ln \mathcal{W}_2(E)}{\partial E}. \tag{2.14}$$

This definition of the (inverse) temperature coincides with that used in thermodynamics, if we identify our entropy with that of thermodynamics. In Equation (2.13) $A = \exp\{S_2(E) - S(E)\} = \text{const}$, is a constant independent of E_k, i.e. independent of the state of our system under study (1), and this constant can be determined just by a normalization condition. Equation (2.13) is one of the most important expressions of statistical mechanics; it defines the statistical distribution for an arbitrary macroscopic body, which is a relatively small part of some large closed system (essentially, this is probably the most general case of a problem to be solved in reality – there is always some surrounding media for any system of interest!). Equation (2.13) is the called *canonical* Gibbs distribution.

The normalization constant A is determined from $\sum_k w_k = 1$, and using (2.13) we immediately get:

$$\frac{1}{A} \equiv Z = \sum_k e^{-\frac{E_k}{T}}. \tag{2.15}$$

Here we introduced Z, which is usually called the statistical sum or *partition function*. Using this notation, we can rewrite the canonical distribution (2.13) in the following standard form[4]:

$$w(E_k) = Z^{-1} \exp\left(-\frac{E_k}{T}\right). \tag{2.16}$$

The average value of an arbitrary physical variable, described by quantum operator f, can be calculated using the Gibbs distribution as:

$$\langle f \rangle = \sum_k w_k f_{kk} = \frac{\sum_k f_{kk} e^{-\frac{E_k}{T}}}{\sum_k e^{-\frac{E_k}{T}}}, \tag{2.17}$$

where f_{kk} is the diagonal matrix element f calculated with eigenfunctions corresponding to the exact energy levels of the system E_k.

In classical statistics we may proceed in a similar way. Let us consider a small part of an isolated classical system (subsystem), so that we can write a volume element $d\Gamma_0$ of the phase space of the whole (isolated) system as $d\Gamma_0 = d\Gamma' d\Gamma$, where $d\Gamma$ is related to our subsystem, while $d\Gamma'$ relates to the bath (surrounding media). We are interested in the distribution function for the subsystem, and where the bath is in phase space is of no interest to us, so that we just integrate over its variables (coordinates and momenta). Using the equality of the probabilities of all states of microcanonical ensemble (describing the whole closed system, consisting of our subsystem and the bath) we get:

$$dw \sim \mathcal{W}' d\Gamma, \tag{2.18}$$

[4] If we measure the temperature in absolute degrees (K), and not in energy units, as is done in the whole text, we have to replace $T \to k_B T$, where k_B is Boltzmann's constant, $k_B = 1.38\,10^{-16}$erg/K or $k_B = 1.38\,10^{-23}$J/K. In this case we also have to add k_B to our definition of entropy: $S = k_B \ln \mathcal{W}$.

where $\mathcal{W}'$ is the phase space (statistical weight) of the bath. Rewriting this statistical weight via entropy we obtain:

$$\mathcal{W}' \sim \exp\{S'(E_0 - E(p,q))\}, \tag{2.19}$$

where E_0 is the energy of the whole closed system, while $E(p,q)$ is the energy of the subsystem. The last relation takes into account that the energy of the thermostat (bath) is given by: $E' = E_0 - E(p,q)$, because $E_0 = E' + E(p,q)$, if we can neglect interactions between the subsystem and the bath. Now everything is quite easy:

$$dw = \rho(p,q)d\Gamma \sim \exp\{S'(E_0 - E(p,q))\}d\Gamma \tag{2.20}$$

so that

$$\rho(p,q) \sim \exp\{S'(E_0 - E(p,q))\}. \tag{2.21}$$

As above, we can expand:

$$S'(E_0 - E(p,q)) \approx S'(E_0) - E(p,q)\frac{dS'(E_0)}{dE_0} = S'(E_0) - \frac{E(p,q)}{T}, \tag{2.22}$$

where once again we have introduced the temperature of the bath T. Finally we obtain the canonical distribution:

$$\rho(p,q) = Ae^{-\frac{E(p,q)}{T}} \tag{2.23}$$

where $E(p,q)$ is the energy of the body under study (the subsystem in the bath), as a function of the coordinates and momenta of its particles. The normalization constant A is determined by the condition:

$$\int d\Gamma\rho(p,q) = A\int d\Gamma e^{-\frac{E(p,q)}{T}} = 1$$

$$Z = A^{-1} = \int d\Gamma e^{-\frac{E(p,q)}{T}}, \tag{2.24}$$

where Z is called statistical integral or partition function.

Let us return to the quantum case. The density matrix corresponding to the canonical Gibbs distribution can be written as:

$$\rho(x,x') = Z^{-1}\sum_k e^{-\frac{E_k}{T}}\psi_k(x)\psi_k^\star(x'), \tag{2.25}$$

where x is the coordinate set (and probably also spins) of particles (if we work in coordinate representation), and $\psi_k(x)$ are eigenfunctions of Hamiltonian H.

Let us introduce the *operator* $\exp\left(-\frac{H}{T}\right)$. Then we can write down the compact operator expression for the canonical distribution:

$$\rho = Z^{-1}\exp\left(-\frac{H}{T}\right) \tag{2.26}$$

and partition function:

$$Z = \mathrm{Sp}\,\exp\left(-\frac{H}{T}\right). \tag{2.27}$$

This expression for the partition function is very convenient because of the invariance of the trace (Sp) with respect to matrix representations; it is independent of the choice of wave functions $\psi_k(x)$, which may not necessarily be eigenfunctions of H.

Up to now we have spoken about the canonical Gibbs distribution as a statistical distribution for a subsystem inside some large closed system. Note that, in Equation (1.56) above, we in fact already obtained it almost from "nothing", while discussing the role of energy and other additive integrals of motion. This derivation was absolutely correct, but it was relatively obscure and formal from the physical point of view.

It is necessary to stress that the canonical distribution may be successfully applied also to closed systems. In reality, the values of thermodynamic characteristics of the body are independent of whether we consider it as a closed system or a system in some (probably imaginary) thermostat (bath). The difference between an isolated (closed) and an open body is only important, when we analyze the relatively unimportant question of fluctuations in the total energy of this body. The canonical distribution produces some finite value of its average fluctuation, which is a real thing for the body in some surrounding media, while it is fictitious for an isolated body, as its energy is constant by definition and is not fluctuating. At the same time, the canonical distribution is much more convenient in most calculations, than the microcanonical distribution. In fact, it is mostly used in practical tasks, forming the basis of the mathematical apparatus of statistical mechanics.

2.2 Maxwell distribution

As a most simple example of an important application of the canonical distribution, we consider the derivation of Maxwell's distribution function. In the classical case the energy $E(p,q)$ can always be represented as a sum of kinetic and potential energy. Kinetic energy is usually a quadratic form of the momenta of the atoms of the body, while potential energy is given by some function of their coordinates, depending on the interaction law and external fields, if present:

$$E(p,q) = K(p) + U(q) \tag{2.28}$$

so that the probability $dw = \rho(p,q)dpdq$ is written as:

$$dw = Ae^{-\frac{K(p)}{T}}e^{-\frac{U(q)}{T}}dpdq \tag{2.29}$$

i.e. is factorized into the product of the function of momenta and of coordinates. This means that probability distributions for momenta (velocities) and coordinates are independent of each other. Then we can write:

$$dw_p = ae^{-\frac{K(p)}{T}} dp, \tag{2.30}$$

$$dw_q = be^{-\frac{U(q)}{T}} dq. \tag{2.31}$$

Each of these distribution functions can be normalized to unity, which will define the normalization constants a and b.

Let us consider the probability distribution for momenta (velocities) which, within the classical approach is independent of interactions between particles or on external fields, and is in this sense universal. For an atom with mass m we have[5]:

$$dw_p = a \exp\left(-\frac{1}{2mT}(p_x^2 + p_y^2 + p_z^2)\right) dp_x dp_y dp_z \tag{2.32}$$

from which we see that distributions of momentum components are also independent. Using the famous Poisson–Gauss integral[6]:

$$I = \int_{-\infty}^{\infty} dx e^{-\alpha x^2} = \sqrt{\frac{\pi}{\alpha}} \tag{2.33}$$

we find:

$$a \int_{-\infty}^{\infty} dp_x \int_{-\infty}^{\infty} dp_y \int_{-\infty}^{\infty} dp_z \exp\left[-\frac{1}{2mT}(p_x^2 + p_y^2 + p_z^2)\right]$$
$$= a\left(\int_{-\infty}^{\infty} dp e^{-p^2/2mT}\right)^3 = a(2\pi mT)^{3/2}$$

so that:

$$a - (2\pi mT)^{-3/2}. \tag{2.34}$$

Finally, the probability distribution for momenta has the following form:

$$dw_p = \frac{1}{(2\pi mT)^{3/2}} \exp\left(-\frac{p_x^2 + p_y^2 + p_z^2}{2mT}\right) dp_x dp_y dp_z. \tag{2.35}$$

Transforming from momenta to velocities, we can write the similar distribution function for velocities:

$$dw_{\mathrm{v}} = \left(\frac{m}{2\pi T}\right)^{3/2} \exp\left(-\frac{m(v_x^2 + v_y^2 + v_z^2)}{2T}\right) dv_x dv_y dv_z. \tag{2.36}$$

[5] The kinetic energy of the body is the sum of the kinetic energies of the constituent atoms, so that this probability distribution is also factorized into the product of distributions, each of which depends only on the momenta of one atom.

[6] It is easy to see that $I^2 = \int_{-\infty}^{\infty} dx e^{-\alpha x^2} \int_{-\infty}^{\infty} dy e^{-\alpha y^2} = \int_{-\infty}^{\infty} dx \int_{-\infty}^{\infty} dy e^{-\alpha(x^2+y^2)} = 2\pi \int_0^{\infty} d\rho \rho e^{-\alpha\rho^2} = \pi \int_0^{\infty} dz e^{-\alpha z} = \pi/\alpha$, thus proving the Poisson–Gauss expression.

This is the notorious Maxwell's distribution, which is one of the first results of classical statistics. In fact it, is factorized into the product of three independent factors:

$$dw_{v_x} = \sqrt{\frac{m}{2\pi T}} e^{-\frac{mv_x^2}{2T}} dv_x \cdots \tag{2.37}$$

each determining the probability distribution for a separate component of velocity.

Note that the Maxwell distribution is valid also for molecules (e.g., in a molecular gas), independent of the nature of the intra molecular motion of atoms (m in this case is just the molecular mass). It is also valid for Brownian motion of particles in suspensions.

Transforming from Cartesian to spherical coordinates, we obtain:

$$dw_{\mathbf{v}} = \left(\frac{m}{2\pi T}\right)^{3/2} e^{-\frac{mv^2}{2T}} v^2 \sin\theta d\theta d\varphi dv, \tag{2.38}$$

where v is the absolute value of the velocity, while θ nd φ are polar and azimuthal angles, determining the direction of the velocity vector $\mathbf{v}$. Integrating over the angles we find the probability distribution for the absolute values of the velocity:

$$dw_v = 4\pi \left(\frac{m}{2\pi T}\right)^{3/2} e^{-\frac{mv^2}{2T}} v^2 dv. \tag{2.39}$$

As a simple example of the application of the Maxwell distribution, let us calculate the average value of the kinetic energy of an atom. For any of the Cartesian components of the velocity, we have[7]:

$$\langle v_x^2 \rangle = \sqrt{\frac{m}{2\pi T}} \int_{-\infty}^{\infty} dv_x v_x^2 e^{-\frac{mv_x^2}{2T}} = \frac{T}{m}. \tag{2.40}$$

Thus, the average value of the kinetic energy of an atom is equal to $3T/2$, i.e. $3k_BT/2$, if we measure the temperature in absolute degrees. Then the average kinetic energy of all particles of the body in classical statistics is always equal to $3NT/2$, where N is the number of atoms.

2.3 Free energy from Gibbs distribution

According to Equation (1.175) the entropy of a body can be calculated as the average value of the logarithm of the distribution function:

$$S = -\langle \ln w_k \rangle = -\sum_k w_k \ln w_k. \tag{2.41}$$

[7] For the integral of the general form $I_n = \int_0^\infty dx x^n e^{-\alpha x^2}$ we have: $I_n = \frac{1}{2}\alpha^{-\frac{n+1}{2}}\Gamma\left(\frac{n+1}{2}\right)$, where $\Gamma(x)$ is the Γ-function, its values for half-integer values of its argument are well known and can be found in handbooks.

Substituting here the canonical distribution in the form of Equation (2.16), we obtain: $-\langle \ln w_k \rangle = \ln Z + \frac{1}{T}\sum_k w_k E_k = \ln Z + \frac{\langle E \rangle}{T}$, where $\langle E \rangle = \sum_k w_k E_k$ is the average energy. As this average energy $\langle E \rangle$ is precisely the same thing as the energy of the body E in thermodynamics, we can write (2.41) as: $S = \ln Z + \frac{E}{T}$, or using the expression for the free energy in thermodynamics $F = E - TS$:

$$F = -T \ln Z = -T \ln \sum_k e^{-\frac{E_k}{T}}. \tag{2.42}$$

This is the basic relation of equilibrium statistical mechanics, giving an expression for the free energy of an arbitrary system via its statistical sum (partition function). In fact, this fundamental result shows that to calculate the free energy of a body, it is sufficient to know its exact *energy spectrum*. We do not have to know e.g. the wave functions, and finding the spectrum of the Schroedinger equation is much a simpler task, than the solution of the complete quantum mechanical problem, including the determination of the wave functions (eigenvectors).

From Equation (2.42) we can see that the normalization factor in the Gibbs distribution (2.16) is, in fact, expressed via the free energy: $\frac{1}{Z} = e^{\frac{F}{T}}$, so that Equation (2.16) can be written as:

$$w_k = \exp\left(\frac{F - E_k}{T}\right). \tag{2.43}$$

It is the most common way to write the Gibbs distribution.

Similarly, in the classical case, using (1.167), (2.23) and (2.24), we obtain:

$$\rho(p, q) = \exp\left(\frac{F - E(p, q)}{T}\right), \tag{2.44}$$

where

$$F = -T \ln \int d\Gamma \exp\left(-\frac{E(p, q)}{T}\right) \tag{2.45}$$

and $d\Gamma = \frac{dp\,dq}{(2\pi\hbar)^{3N} N!}$. Thus, in the classical approach the statistical sum is just replaced by the statistical integral. Taking into account that $E(p, q)$ here can always be represented by the sum of kinetic $K(p)$ and potential U energies, and kinetic energy is always a quadratic form of momenta, we can perform momentum integration in the statistical integral in its general form (cf. discussion of Maxwell distribution above!). Thus, the problem of the calculation of the statistical integral is reduced to integration over all coordinates in $e^{-\frac{U(q)}{T}}$, which is of course impossible to do exactly.

2.4 Gibbs distribution for systems with varying number of particles

Up to now we implicitly assumed that the number of particles in the system is some predetermined constant. In reality, different subsystems of a large system can exchange particles between them. The number of particles N in a subsystem fluctuates around its average value. In this case the distribution function depends not only on the energy of the quantum state, but also on the number of particles N of the body. In fact, the energy levels E_{kN} themselves are different for different values of N. Let us denote as w_{kN} the probability for the body to be in the k-th state and contain N particles. This probability distribution can be obtained in the same way, as we derived the probability w_k above.

Consider the closed (isolated) system with energy $E^{(0)}$ and number of particles $N^{(0)}$, consisting of two weakly interacting subsystems with energies E' (bath) and E_{kN} (small subsystem) and respective numbers of particles N' (bath) and N (subsystem):

$$E^{(0)} = E_{kN} + E' \quad N^{(0)} = N + N'. \tag{2.46}$$

We assume that the subsystem of interest to us is small in comparison to the bath (particle reservoir), so that:

$$E_{kN} \ll E' \quad N \ll N'. \tag{2.47}$$

As we assume the full composite system to be isolated, it can again be described by the microcanonical distribution. Similar to the derivation of the canonical distribution above, we can find the probability distribution for a small subsystem w_{kN} by summing the microcanonical distribution for the whole system over all states of the bath. In complete analogy with Equation (2.10) we get:

$$w_{kN} = \frac{\mathcal{W}'(E^{(0)} - E_{kN}, N^{(0)} - N)}{\mathcal{W}^{(0)}(E^{(0)}, N^{(0)})}, \tag{2.48}$$

where $\mathcal{W}'$ is the statistical weight of the bath, while $\mathcal{W}^{(0)}$ is the statistical weight of the full (closed) system. Using the definition of entropy we immediately obtain:

$$w_{kN} = \text{Const}\exp\{S'(E^{(0)} - E_{kN}, N^{(0)} - N)\}. \tag{2.49}$$

Now we can again expand S' in powers of E_{kN} and N, restricting ourselves to linear terms only:

$$S'(E^{(0)} - E_{kN}, N^{(0)} - N) \approx S'(E^{(0)}, N^{(0)}) - \left(\frac{\partial S'}{\partial E}\right)_{V,N} E_{kN} - \left(\frac{\partial S'}{\partial N}\right)_{E,V} N + \cdots. \tag{2.50}$$

Then, remembering the thermodynamic relations for the system with a variable number of particles [1]:

$$dE = TdS - PdV + \mu dN; \quad \mu = \left(\frac{\partial E}{\partial N}\right)_{S,V} \tag{2.51}$$

or

$$dS = \frac{dE}{T} + \frac{P}{T}dV - \frac{\mu}{T}dN, \tag{2.52}$$

we obtain:

$$\left(\frac{\partial S}{\partial E}\right)_{V,N} = \frac{1}{T}; \quad \left(\frac{\partial S}{\partial N}\right)_{E,V} = -\frac{\mu}{T}. \tag{2.53}$$

Then we can rewrite the expansion (2.50) as:

$$S'(E^{(0)} - E_{kN}, N^{(0)} - N) \approx S'(E^{(0)}, N^{(0)}) - \frac{E_{kN}}{T} + \frac{\mu N}{T}. \tag{2.54}$$

Notice that both the chemical potential μ and temperature T of the body (subsystem) and the bath (thermostat) just coincide due to the standard conditions of thermodynamic equilibrium.

Finally, we obtain the distribution function:

$$w_{kN} = A \exp\left(\frac{\mu N - E_{kN}}{T}\right). \tag{2.55}$$

The normalization constant A can again be expressed via thermodynamic variables. To see this let us calculate the entropy of the body:

$$S = -\langle \ln w_{kN} \rangle = -\ln A - \frac{\mu}{T}\langle N \rangle + \frac{1}{T}\langle E \rangle \tag{2.56}$$

or

$$T \ln A = \langle E \rangle - TS - \mu \langle N \rangle. \tag{2.57}$$

Identifying $\langle E \rangle$ with energy of the body E in thermodynamics, and $\langle N \rangle$ with the particle number N in thermodynamics, taking into account the thermodynamic relation $E - TS = F$, and introducing the thermodynamic potential Ω as $\Omega = F - \mu N$ [1], we have: $T \ln A = \Omega$, so that Equation (2.55) can be rewritten as:

$$w_{kN} = \exp\left(\frac{\Omega + \mu N - E_{kN}}{T}\right). \tag{2.58}$$

This is the final form of the Gibbs distribution for the system with a variable number of particles, which is called the *grand canonical* distribution.

The usual normalization condition for (2.58) is:

$$\sum_N \sum_k w_{kN} = e^{\frac{\Omega}{T}} \sum_N \left(e^{\frac{\mu N}{T}} \sum_k e^{-\frac{E_{kN}}{T}} \right) = 1. \tag{2.59}$$

From here, we obtain the general expression for the thermodynamic potential Ω in statistical mechanics:

$$\Omega = -T \ln \sum_N \left(e^{\frac{\mu N}{T}} \sum_k e^{-\frac{E_{kN}}{T}} \right), \tag{2.60}$$

where the expression in the right-hand side can be called a grand partition function.

The average number of particles $\langle N \rangle$ in our system is determined by the relation:

$$\langle N \rangle = \sum_N \sum_k N w_{kN} = e^{\frac{\Omega}{T}} \sum_N \left(N e^{\frac{\mu N}{T}} \sum_k e^{-\frac{E_{kN}}{T}} \right), \tag{2.61}$$

which can be considered as a kind of additional "normalization" condition. Actually, this equation implicitly determines the chemical potential μ as function of temperature and *fixed* average particle number $\langle N \rangle$, which is equivalent to the number of particles N in thermodynamics. This is the general recipe to determine μ, which will often be used in future calculations.

Expressions (2.42) and (2.60) determine thermodynamic characteristics for arbitrary systems in equilibrium. The free energy F is determined as a function of T, N and V, while the thermodynamic potential Ω is determined by (2.60) as a function of T, μ and V.

Similar to previous analysis, in classical statistics the grand canonical distribution is written as:

$$dw_N = exp\left(\frac{\Omega + \mu N - E_N(p,q)}{T} \right) \frac{dp^{(N)} dq^{(N)}}{(2\pi\hbar)^{3N} N!} \equiv \rho_N d\Gamma_N. \tag{2.62}$$

The variable N is written here as an index of the distribution function and also of the phase space volume element, to stress that there is a different phase space for each value of N (with its own dimensions $6N$). The expression for the potential Ω is now:

$$\Omega = -T \ln \left\{ \sum_N e^{\frac{\mu N}{T}} \int d\Gamma_N \exp\left(-\frac{E_N(p,q)}{T} \right) \right\}. \tag{2.63}$$

It is clear that in calculations of all statistical (thermodynamic) properties of the body, except fluctuations in the total number of particles, both the canonical and grand canonical Gibbs distributions are equivalent. Neglecting fluctuations in the particle number N we have $\Omega + \mu N = F$ and these distributions just coincide.

The use of one or the other distribution is, in most practical tasks, mostly the question of convenience of calculations. In practice, the microcanonical distribution is most inconvenient, while the most convenient is often the grand canonical distribution.

2.5 Thermodynamic relations from Gibbs distribution

Let us complete the statistical justification of thermodynamics by derivation of its main relations from the Gibbs distribution. Already during our discussion of the role of additive integrals of motion and derivation of Equation (1.56), which is essentially the canonical distribution itself, we noted that the factor β before the energy in Equation (1.56) is the same for all subsystems of the given closed system. Taking into account that in the canonical distribution we have $\beta = -1/T$, we come to the conclusion that this is equivalent to the usual thermodynamic condition for equality of the temperatures for all parts of the system being in the state of thermodynamic equilibrium[8]. It easy to see that for the temperature $T > 0$, otherwise there appears a divergence in the normalization sum $\sum_k w_k$, because the energy levels E_k may be arbitrarily big. All these properties just coincide with the basic properties of temperature in thermodynamics.

Basic thermodynamic relations may be derived in different ways. Let us write down the canonical distribution in operator form as:

$$\rho = e^{\frac{F-H}{T}}. \tag{2.64}$$

Normalization condition $\mathrm{Sp}\,\rho = 1$ can be rewritten as:

$$e^{-\frac{F}{T}} = \mathrm{Sp}\left(e^{-\frac{H}{T}}\right) \tag{2.65}$$

which is in essence the definition of free energy. Differentiating this expression with respect to T we get:

$$\left(\frac{F}{T^2} - \frac{1}{T}\frac{\partial F}{\partial T}\right) e^{-\frac{F}{T}} = \frac{1}{T^2}\mathrm{Sp}\left(He^{-\frac{H}{T}}\right). \tag{2.66}$$

Multiplying this relation by $T^2 e^{\frac{F}{T}}$ and taking into account that $\langle H \rangle = E$, we obtain the basic Gibbs–Helmholtz relation of classical thermodynamics:

$$F = E + T\frac{\partial F}{\partial T}. \tag{2.67}$$

Comparing this expression with the definition of free energy $F = E - TS$, we get:

$$S = -\frac{\partial F}{\partial T} = -\frac{1}{T}(F - \langle H \rangle). \tag{2.68}$$

According to Equation (1.174) we can write down the entropy in operator form as:

$$S = -\mathrm{Sp}\,\rho \ln \rho. \tag{2.69}$$

[8] Equation (1.56) coincides with the canonical distribution (2.43), if we also take $\alpha = F/T$ and consider the system at rest.

The identity of this expression for S with the previous one can be easily seen – according to Equation (2.64) we have $\ln\rho = \frac{1}{T}(F - H)$, the rest is obvious.

Another way to obtain the basic thermodynamic relations is to consider the normalization condition for the Gibbs distribution:

$$\sum_k e^{\frac{F-E_k}{T}} = 1 \tag{2.70}$$

and differentiate it, considering the left-hand side as a function of T and some variables $\lambda_1, \lambda_2, \ldots$, which characterize external conditions for the body under study. These variables may, for example, determine the geometrical form and size of its volume, define external fields etc. Energy levels of the system E_k parametrically depend on $\lambda_1, \lambda_2, \ldots$. After differentiation we obtain (for brevity we write explicitly only one parameter λ)[9]:

$$\sum_k \frac{w_k}{T}\left[dF - \frac{\partial E_k}{\partial\lambda}d\lambda - \frac{F - E_k}{T}dT\right] = 0. \tag{2.71}$$

Then we have:

$$dF\sum_k w_k = d\lambda\sum_k w_k\frac{\partial E_k}{\partial\lambda} + \frac{dT}{T}(F - \sum_k w_k E_k). \tag{2.72}$$

Taking into account $\sum_k w_k = 1$, $\sum_k w_k E_k = \langle E\rangle = E$ and $\sum_k w_k\frac{\partial E_k}{\partial\lambda} = \frac{\partial\langle E_k\rangle}{\partial\lambda}$, as well as $F - E = -TS$ and the relation[10]:

$$\frac{\partial\langle E_k\rangle}{\partial\lambda} = \frac{\partial\langle H\rangle}{\partial\lambda} \tag{2.73}$$

we finally obtain:

$$dF = -SdT + \frac{\partial\langle H\rangle}{\partial\lambda}d\lambda = -SdT + \frac{\partial E}{\partial\lambda}d\lambda \tag{2.74}$$

which represents the general form of the differential of free energy in thermodynamics.

Similarly, from the normalization condition for the grand canonical distribution[11] (2.59) we can obtain the general form of the differential of the thermodynamic poten-

[9] More precisely we write down the full differential on the left-hand side of Equation (2.70): $d\sum_k e^{\frac{F-E_k}{T}} = \sum_k w_k d\left(\frac{F-E_k}{T}\right) = 0$, which gives us Equation (2.71)

[10] If the Hamiltonian H and its eigenvalues E_k depend on the parameter λ, we have: $\frac{\partial E_k}{\partial\lambda} = \left(\frac{\partial H}{\partial\lambda}\right)_{kk}$, so that after the averaging we obtain (2.73).

[11] Note that the grand canonical distribution can also be derived with arguments used in the derivation of Equation (1.56), if we consider the number of particles as N as an additive integral (constant) of motion. Then, for a system at rest we can write: $\ln w_{kN} = \alpha + \beta E_{kN} + \gamma N$, where γ and β are to be the same for all parts of the system at equilibrium. Putting here $\alpha = \Omega/T$, $\beta = -1/T$ and $\gamma = \mu/T$ we obtain the grand canonical distribution. By the way, here we obtained the well known condition of equality of chemical potentials of subsystems at equilibrium with each other.

tial Ω:

$$d\Omega = -SdT - Nd\mu + \frac{\partial \langle H \rangle}{\partial \lambda} d\lambda. \tag{2.75}$$

We assumed above that the external parameters $\lambda_1, \lambda_2, \ldots$ characterize the macroscopic state of the system in equilibrium. These may be the volume (form) of a vessel, the values of external electric or magnetic fields etc. Parameters $\lambda_1, \lambda_2, \ldots$ are also assumed to change very slowly in time, so that during the time of the order of the relaxation time for the system to evolve to equilibrium, these parameters can be considered as practically constant. Then we can suppose that at any moment in time the system is in some equilibrium state, despite the fact that the external parameters change. Such a process of slow change of external parameters may be called quasi-static. If we consider the parameters $\lambda_1, \lambda_2, \ldots$ as generalized coordinates, corresponding generalized forces can be introduced as:

$$\Lambda_i = -\frac{\partial H}{\partial \lambda_i}. \tag{2.76}$$

For a quasi-static process the observed values of the generalized forces can be obtained by averaging over the equilibrium statistical ensemble as:

$$\langle \Lambda_i \rangle = \mathrm{Sp}(\rho \Lambda_i) = -\frac{\partial \langle H \rangle}{\partial \lambda_i}. \tag{2.77}$$

Let us consider some typical examples. If we choose as an external parameter the volume of the system V, the generalized force is pressure:

$$P = -\frac{\partial \langle H \rangle}{\partial V} = -\frac{\partial E}{\partial V}. \tag{2.78}$$

Then Equation (2.74) takes the well known form:

$$dF = -SdT - PdV. \tag{2.79}$$

If we choose as a parameter an external electric field **E**, the generalized force is the polarization (electric dipole moment of the body) **P** and:

$$dF = -SdT - \mathbf{P}d\mathbf{E}; \quad \mathbf{P} = -\frac{\partial \langle H \rangle}{\partial \mathbf{E}}. \tag{2.80}$$

For the case of an external magnetic field **H** the generalized force is the magnetization (magnetic moment) of the body **M** and:

$$dF = -SdT - \mathbf{M}d\mathbf{H}; \quad \mathbf{M} = -\frac{\partial \langle H \rangle}{\partial \mathbf{H}}. \tag{2.81}$$

Thus, we succeeded in the construction of the complete statistical derivation of all basic relations of thermodynamics. Historically, the development of statistical mechanics was directly related to this task.

The last problem to be discussed in relation to the justification of the laws of thermodynamics is Nernst's theorem, sometimes called the third law of thermodynamics. We note from the very beginning that in contrast to the first and the second laws, which directly follow from the Gibbs approach, a similar (in generality) proof of the Nernst's theorem is absent, though for all "reasonable" models of statistical mechanics it is valid. Let us analyze the limiting behavior of the Gibbs distribution

$$w_k = e^{\frac{F-E_k}{T}} \tag{2.82}$$

for temperatures $T \to 0$. Using the expression for the entropy:

$$S = \frac{1}{T}(\langle H \rangle - F), \tag{2.83}$$

we can write $w_k = \exp\{-S + \frac{1}{T}(\langle H \rangle - E_k)\}$, or:

$$w_k = \exp\left\{-S + \frac{\langle H \rangle - E_0}{T} + \frac{E_0 - E_k}{T}\right\}, \tag{2.84}$$

where E_0 is the energy of the ground state of the system, so that $E_k > E_0$ for all $k \neq 0$. Calculating the limit of (2.84) for $T \to 0$, we obtain:

$$\lim_{T\to 0} w_k = w_k(0) = \exp\{-S(0) + C_V(0)\}\delta_{E_k - E_0}, \tag{2.85}$$

where

$$\delta_{E_k-E_0} = \begin{cases} 1 & \text{for } E_k = E_0 \\ 0 & \text{for } E_k \neq E_0. \end{cases} \tag{2.86}$$

In Equation (2.85) $C_V(0) = \left(\frac{\partial \langle H \rangle}{\partial T}\right)_{T=0}$ denotes the specific heat of the body at $T = 0$ and for constant volume. However, from Equation (2.83) it follows (using l'Hôpital's rule) that for $T \to 0$:

$$S(0) = \left(\frac{\partial \langle H \rangle}{\partial T} - \frac{\partial F}{\partial T}\right)_{T\to 0} = C_V(0) + S(0) \tag{2.87}$$

so that $C_V(0) = 0$ (Nernst's theorem). Accordingly Equation (2.85) reduces to:

$$w_k(0) = \exp\{-S(0)\}\delta_{E_k - E_0}, \tag{2.88}$$

which is, in fact, just the microcanonical distribution:

$$w_k(0) = \frac{1}{W_0}\delta_{E_k - E_0}, \tag{2.89}$$

where $\mathcal{W}_0$ is the degeneracy of the ground state. Then the entropy in the ground state at $T = 0$:

$$S(0) = \ln \mathcal{W}_0. \tag{2.90}$$

For the majority of physical systems (like crystals, quantum gases and liquids etc.) the ground state is non-degenerate, so that $\mathcal{W}_0 = 1$, and thus the entropy tends to zero as $T \to 0$. Even for the case of $\mathcal{W}_0 \gg 1$, but for $\lim_{N\to\infty} \frac{1}{N} \ln \mathcal{W}_0 = 0$ (entropy per single particle) we may assume $S(0) = 0$, which is, in fact, the general formulation of Nernst's theorem[12].

Unfortunately, the situation here is not so simple and the physical behavior of systems, described by Nernst's theorem, is not directly related to non-degeneracy of the ground state. Actually it reflects the behavior of an effective behavior of excitation spectra of macroscopic bodies at small energies, and Nernst's theorem manifests itself for temperatures T, which are much larger than the energy difference between the first excited state of the system and its ground state. Above, we have already seen that the energy spectrum of a macroscopic body can be considered as practically continuous, so this energy difference is, in fact, unobservable. This follows even from the simplest estimates. Consider an ideal gas of atoms with mass m, moving in the volume $V = L^3$. Then we can estimate:

$$E_1 - E_0 \sim \frac{\hbar^2}{2m} k_{\min}^2 = \frac{\hbar^2}{2mV^{2/3}} \quad \text{where } k_{\min} = \frac{2\pi}{L} \tag{2.91}$$

and the volume $V \to \infty$. Experimentally, for an ideal gas, manifestations of Nernst's theorem become observable for finite temperatures of the order or below the so called degeneracy temperature $T_0 \sim \frac{\hbar^2}{m}\left(\frac{N}{V}\right)^{2/3}$.

To give the general proof of Nernst's theorem, we have to understand the distribution of energy levels E_k close to the ground state, i.e. to find the general behavior of the statistical weight $\mathcal{W}(E, N, V)$ close to $E = E_0$. Up to now such behavior was only studied for some specific models. The behavior necessary to reproduce Nernst's theorem in all cases, when the weak (low energy) excitations of the system can be represented by an ideal gas of *quasiparticles*. Below, we shall consider only such systems, and the concept of quasiparticles will be of central importance.

This concludes our presentation of the basics of the Gibbs approach to statistical mechanics. The rest of the book will be devoted to applications of this formalism to different concrete problems of the physics of many particle systems.

---

[12] Note that Nernst's theorem is inapplicable to amorphous solids (glasses) or disordered alloys, which are not, in fact, in a state of complete thermodynamic equilibrium, but can be "frozen" (at $T \to 0$) in some of many possible metastable states with quite large or even practically infinite relaxation times.

Chapter 3

Classical ideal gas

3.1 Boltzmann distribution

The simplest model to illustrate the applications of the general principles of statistical mechanics is an ideal gas of noninteracting atoms or molecules[1]. This model played an important role at the early stages of the development of statistical physics[2].

The absence of interactions between the atoms (molecules) of an ideal gas allows us to reduce the quantum mechanical problem of finding the energy levels E_n of a gas as a whole to the problem of finding the energy levels of an isolated atom (molecule). We shall denote these levels as ε_k, where k is the set of quantum numbers, determining the state of an atom (molecule). Because of the absence of interactions the energy levels E_n are just the sums of energies of each of the atoms (molecules). Let us denote as n_k the number of gas particles occupying the quantum state k and calculate its average value $\langle n_k \rangle$ for the important limit of:

$$\langle n_k \rangle \ll 1. \tag{3.1}$$

Physically, this limit corresponds to a strongly diluted gas. Let us apply the canonical Gibbs distribution to gas molecules, considering a single molecule as a subsystem in the bath (of the rest of the molecules). Then it is clear that the probability for the molecule to be in the k-th state, and also the average number $\langle n_k \rangle$ of molecules in this state, will be $\sim e^{-\frac{\varepsilon_k}{T}}$, so that

$$\langle n_k \rangle = a e^{-\frac{\varepsilon_k}{T}}, \tag{3.2}$$

where the coefficient a can be determined by the normalization condition:

$$\sum_k \langle n_k \rangle = N, \tag{3.3}$$

where N is the total number of particles in a gas. The distribution function given by Equation (3.2) is called Boltzmann's distribution.

Let us give another derivation of the Boltzmann distribution, which is based on application of the grand canonical Gibbs distribution to all particles of the gas occupying

---

1 Surely, the existence of some weak interactions (e.g. rare collisions) between atoms or molecules is necessary to reach the equilibrium state. However, during the calculations of the equilibrium thermodynamic properties of an ideal gas we can neglect those from the very beginning.

2 Below we basically follow the presentation of [1].

the same quantum state, which is considered as a subsystem in the bath (of all other particles). In the general expression for the grand canonical distribution (2.58) we now have to set $E = n_k\varepsilon_k$ and $N = n_k$. Adding an index k also to the thermodynamic potential Ω, we obtain:

$$w_{n_k} = e^{\frac{\Omega_k + n_k(\mu - \varepsilon_k)}{T}}. \tag{3.4}$$

In particular, $w_0 = e^{\frac{\Omega_k}{T}}$ is simply the probability of an absence of any particle in this given state. In the limit of interest to us, when $\langle n_k \rangle \ll 1$, the probability $w_0 = e^{\frac{\Omega_k}{T}} \approx 1$, and from Equation (3.4) we obtain:

$$w_1 = e^{\frac{\mu - \varepsilon_k}{T}}. \tag{3.5}$$

As to probabilities of the values of $n_k > 1$, in this approximation they are just zeroes. Thus, in the sum determining $\langle n_k \rangle$, remains only one term:

$$\langle n_k \rangle = \sum_{n_k} w_{n_k} n_k = w_1, \tag{3.6}$$

and we get:

$$\langle n_k \rangle = e^{\frac{\mu - \varepsilon_k}{T}}. \tag{3.7}$$

We see that the coefficient in Equation (3.2) is expressed via the chemical potential of the gas, which is implicitly defined the by normalization condition for the total number of particles (3.3).

3.2 Boltzmann distribution and classical statistics

Previous analysis was based on a quantum approach. Let us consider the same problem in classical statistics. Let dN denote the average number of molecules belonging to an element of the phase space of the molecule $dpdq = dp_1 \cdots dp_r dq_1 \cdots dq_r$ (r is the number of degrees of freedom of the molecule). We can write it as:

$$dN = n(p,q)d\tau \quad d\tau = \frac{dpdq}{(2\pi\hbar)^r}, \tag{3.8}$$

where $n(p,q)$ is probability density in the phase space. Then:

$$n(p,q) = e^{\frac{\mu - \varepsilon(p,q)}{T}}, \tag{3.9}$$

where $\varepsilon(p,q)$ is the energy of the molecule as a function of the coordinates and momenta of its atoms.

For a gas in the absence of any kind of external field this distributions reduces to the Maxwell distribution[3]:

$$dN_p = \frac{N}{V(2\pi mT)^{3/2}} e^{-\frac{p_x^2+p_y^2+p_z^2}{2mT}} dp_x dp_y dp_z, \quad (3.10)$$

$$dN_v = \frac{N}{V}\left(\frac{m}{2\pi T}\right)^{3/2} e^{-\frac{m(v_x^2+v_y^2+v_z^2)}{2T}} dv_x dv_y dv_z, \quad (3.11)$$

where m is the mass of a molecule. Comparing (3.10) and (3.9) we obtain $e^{\frac{\mu}{T}} = \frac{N}{V}(2\pi)^{3/2}\hbar^3(mT)^{-3/2}$, so that the chemical potential of a Boltzmann gas is:

$$\mu = T \ln\left(\frac{N}{V}\frac{(2\pi)^{3/2}\hbar^3}{(mT)^{3/2}}\right). \quad (3.12)$$

This result can also be obtained directly from normalization (3.9) to the total number of particles in a unit volume (density) given by Equation (3.3). In the classical approximation $\varepsilon_k = \frac{p_x^2+p_y^2+p_z^2}{2m}$, so that (3.3) can be written as:

$$\sum_k e^{\frac{\mu-\varepsilon_k}{T}} = N \quad \text{or} \quad e^{\frac{\mu}{T}} \int \frac{d^3p}{(2\pi\hbar)^3} e^{-\frac{p_x^2+p_y^2+p_z^2}{2mT}} = \frac{N}{V} \quad (3.13)$$

which gives (3.12) after calculation of an elementary Gaussian integral:

$$\mu = T\ln\left\{\frac{N}{V}\left(\int \frac{d^3p}{(2\pi\hbar)^3} e^{-\frac{p_x^2+p_y^2+p_z^2}{2mT}}\right)^{-1}\right\} = T\ln\left(\frac{N}{V}\frac{(2\pi)^{3/2}\hbar^3}{(mT)^{3/2}}\right). \quad (3.14)$$

Thus, the chemical potential of the gas is completely determined by the density of the particles and temperature.

Consider now the gas in an external field, when the potential energy of a molecule depends on the coordinates of its center of mass: $U = U(x, y, z)$. A typical example is a gas in a gravitational field. The Maxwell distribution for velocities remains, as was noted above, valid, while the distribution for the center of mass coordinates is given by:

$$dN_r = n_0 e^{-\frac{U(x,y,z)}{T}} dV \quad (3.15)$$

which gives the number of molecules in volume element $dV = dxdydz$. Obviously

$$n(\mathbf{r}) = n_0 e^{-\frac{U(\mathbf{r})}{T}} \quad (3.16)$$

gives the density of particles at the point $\mathbf{r}$. Here n_0 is the density at points, where $U = 0$. Equation (3.16) is sometimes called Boltzmann's law.

[3] In contrast with the form of the Maxwell distribution discussed above, here we introduced an additional factor N/V, which is related to the normalization to particle density used here.

As an example, consider a gas in a homogeneous gravitational field (e.g. on Earth's surface) directed along z-axis, so that $U = mgz$ (g is the free fall acceleration) and for the density distribution of a gas we obtain:

$$n(z) = n_0 e^{-\frac{mgz}{T}}, \tag{3.17}$$

where n_0 is the density at $z = 0$ (at the sea level).

3.3 Nonequilibrium ideal gas

Consider an ideal gas in an arbitrary (in general nonequilibrium) state. Let us assume that all quantum states of a single particle of the gas can be classified into certain groups of levels with energies close to each other, and the number of levels in each group, as well as the number of particles on these levels, are large enough[4]. Let us enumerate these groups of levels by the numbers $j = 1, 2, \ldots$ and let G_j be the number of levels in j-th group, while N_j is the number of particles in these states. The set of numbers N_j completely determines the macroscopic state of the gas, while their arbitrariness, in fact means that we are dealing with an arbitrary, in general, nonequilibrium state of the system.

To calculate the entropy of this macroscopic state, we have to determine its statistical weight $\mathcal{W}$, i.e. the number of microscopic distributions of particles over the levels, which realize such a state. Considering each group of N_j particles as an independent subsystem and denoting its statistical weight by $\mathcal{W}_j$, we can write:

$$\mathcal{W} = \prod_j \mathcal{W}_j. \tag{3.18}$$

Now we have to calculate $\mathcal{W}_j$. In Boltzmann's statistics the average occupation numbers of all quantum states are small in comparison to unity. This means that $N_j \ll G_j$, though N_j are still very large. The smallness of occupation numbers leads to the conclusion that all particles are distributed over different states, independently of each other. Placing each of N_j particles in one of G_j states we obtain in all $G_j^{N_j}$ possible distributions, including physically equivalent ones, which differ only due to permutations of identical particles. Accordingly, we have to divide the total number of possible distributions (configurations) by $N_j!$, so that:

$$\mathcal{W}_j = \frac{G_j^{N_j}}{N_j!}. \tag{3.19}$$

Then the entropy is calculated as:

$$S = \ln \mathcal{W} = \sum_j \ln \mathcal{W}_j = \sum_j (N_j \ln G_j - \ln N_j!). \tag{3.20}$$

[4] This assumption is made just to simplify our analysis and does not restrict its generality.

Using Stirling's asymptotics, valid for $N \gg 1$ [5]:

$$\ln N! \approx N \ln\left(\frac{N}{e}\right) \tag{3.21}$$

we get:

$$S = \sum_j N_j \ln \frac{eG_j}{N_j}. \tag{3.22}$$

This expression determines the entropy of an ideal gas in an arbitrary macroscopic state, defined by the set of numbers N_j. Let us rewrite it, introducing the average numbers $\langle n_j \rangle$ of particles in the j-th group of quantum levels $\langle n_j \rangle = N_j/G_j$. Then:

$$S = \sum_j G_j \langle n_j \rangle \ln \frac{e}{\langle n_j \rangle}. \tag{3.23}$$

Describing particles in a quasi-classic approximation, we can introduce the distribution function in phase space. Dividing the phase space into small elementary volumes $\Delta p^{(j)} \Delta q^{(j)}$, which still contain a large enough number of particles, we can write down the number of quantum states in such a volume as (r is the number of degrees of freedom of a gas molecule, for a one-atom gas $r = 3$):

$$G_j = \frac{\Delta p^{(j)} \Delta q^{(j)}}{(2\pi\hbar)^r} = \Delta\tau^{(j)}. \tag{3.24}$$

The number of particles in these states can be written as $N_j = n(p,q)\Delta\tau^{(j)}$. Substituting these expressions into Equation (3.23), we obtain:

$$S = \int d\tau n(p,q) \ln \frac{e}{n(p,q)}. \tag{3.25}$$

This is the so called Boltzmann's entropy of an ideal gas in an arbitrary (nonequilibrium) state, defined by the *single particle* distribution function $n(p,q)$[6].

What is the connection of the Boltzmann entropy (3.25) with the Gibbs entropy, defined in (1.167)? In the expression for the Gibbs entropy:

$$S = -\int \frac{dpdq}{(2\pi\hbar)^{3N} N!} \rho(p,q,t) \ln \rho(p,q,t) \tag{3.26}$$

---

[5] For $N \gg 1$ the sum $\ln N! = \ln 1 + \ln 2 + \cdots + \ln N$ is approximately expressed as $\int_0^N dx \ln x$, which immediately gives Equation (3.21).

[6] The distribution function $n(p,q)$ can depend on time and this time dependence can be calculated using Boltzmann's *kinetic equation*. For this entropy (3.25) the famous Boltzmann's *H-theorem*, is proved in classical kinetics, describing the time growth of (3.25).

$\rho(p,q)$ denotes the full N-particle distribution function, depending on the coordinates and momenta of all N molecules of gas. For an ideal gas of noninteracting particles this distribution function is obviously factorized (statistical independence – absence of interactions!) into the product of single particle distribution functions for all particles:

$$\rho(p,q) = \frac{N!}{N^N}\prod_{i=1}^{N} n(p_i,q_i), \tag{3.27}$$

where the single particle distribution functions $n(p_i,q_i)$ are normalized as (for one-atom gas, i.e. $r=3$):

$$\int \frac{dp_1 dq_1}{(2\pi\hbar)^3} n(p_1,q_1) = N. \tag{3.28}$$

The factor of $N!/N^N$ in (3.27) is introduced here to insure agreement between this normalization and the one used above for $\rho(p,q)$:

$$\int d\Gamma \rho(p,q) = \left\{\frac{1}{N}\int \frac{dp_1 dq_1}{(2\pi\hbar)^3} n(p_1,q_1)\right\}^N = 1 \quad d\Gamma = \frac{dpdq}{(2\pi\hbar)^{3N} N!} \tag{3.29}$$

Then, using (3.27), (3.21) in (3.26) we get:

$$S = -\int \frac{dp_1 dq_1}{(2\pi\hbar)^3} n(p_1,q_1) \ln \frac{n(p_1,q_1)}{e}, \tag{3.30}$$

which coincides with (3.25).

In the equilibrium state the entropy is to be maximal. This can be used to find the equilibrium distribution function. Let us find $\langle n_j \rangle$, which gives the maximal value of the sum (3.23), with additional demands of the fixed (average) number of particles and average energy of the system:

$$\sum_j N_j = \sum_j G_j \langle n_j \rangle = N, \tag{3.31}$$

$$\sum_j \varepsilon_j N_j = \sum_j \varepsilon_j G_j \langle n_j \rangle = E. \tag{3.32}$$

Using the method of Lagrange multipliers we demand:

$$\frac{\partial}{\partial n_j}(S + \alpha N + \beta E) = 0, \tag{3.33}$$

where α and β are some constants. After differentiation we get:

$$G_j(-\ln\langle n_j \rangle + \alpha + \beta\varepsilon_j) = 0 \tag{3.34}$$

leading to $\ln\langle n_j\rangle = \alpha + \beta\varepsilon_j$, or

$$\langle n_j\rangle = \exp(\alpha + \beta\varepsilon_j). \tag{3.35}$$

We obtained the Boltzmann distribution, where the constants α and β are related to T and μ: $\alpha = \mu/T, \beta = -1/T$. This is clear, in particular, from the possibility to write (3.33) as a relation between differentials: $dS + \alpha dN + \beta dE = 0$, which is to coincide with the well known thermodynamic relation for the differential of energy (for fixed volume): $dE = TdS + \mu dN$.

3.4 Free energy of Boltzmann gas

Let us apply the basic relation of statistical mechanics:

$$F = -T\ln Z = -T\ln\sum_n e^{-\frac{E_n}{T}} \tag{3.36}$$

to the calculation of the free energy of an ideal gas, described by Boltzmann statistics. Energy levels E_n of the whole system (gas) are simply the sums of energies of isolated molecules ε_k, which in the Boltzmann case are all different (because in each quantum state of a gas there is no more than one molecule). Then we can write down $e^{-\frac{E_n}{T}}$ as a product of factors $e^{-\frac{\varepsilon_k}{T}}$ for each molecule and sum over all states of each molecule, which leads to the following expression for the partition function of the gas[7]:

$$Z \sim \left(\sum_k e^{-\frac{\varepsilon_k}{T}}\right)^N. \tag{3.37}$$

This expression is also to be divided by $N!$, taking into account the number of permutations of identical particles (molecules), leading to physically equivalent states (configurations). Then we have:

$$Z = \sum_n e^{-\frac{E_n}{T}} = \frac{1}{N!}\left(\sum_k e^{-\frac{\varepsilon_k}{T}}\right)^N. \tag{3.38}$$

Substituting this expression into (3.36), we get:

$$F = -TN\ln\sum_k e^{-\frac{\varepsilon_k}{T}} + T\ln N! \tag{3.39}$$

[7] We have $e^{-\frac{E_n}{T}} = e^{-\frac{\varepsilon_{k_1}}{T}} e^{-\frac{\varepsilon_{k_2}}{T}} \cdots e^{-\frac{\varepsilon_{k_N}}{T}}$, with N factors in total, with all $k_L (L = 1, 2, \ldots, N)$ different. Calculating now $\sum_{k_1}\sum_{k_2}\cdots\sum_{k_N} \to \left(\sum_k\right)^N$, we get Equation (3.37).

or, using once again $\ln N! \approx N \ln N/e$, we obtain:

$$F = -NT \ln \left\{ \frac{e}{N} \sum_k e^{-\frac{\varepsilon_k}{T}} \right\}. \tag{3.40}$$

In classical statistics we can immediately write:

$$F = -NT \ln \left[\frac{e}{N} \int d\tau e^{-\frac{\varepsilon(p,q)}{T}} \right] \quad d\tau = \frac{d^r p d^r q}{(2\pi\hbar)^r}, \tag{3.41}$$

where r is again the number of degrees of freedom of a gas molecule.

3.5 Equation of state of Boltzmann gas

The energy of a gas molecule can be written as:

$$\varepsilon_k(p_x, p_y, p_z) = \frac{p_x^2 + p_y^2 + p_z^2}{2m} + \varepsilon'_k, \tag{3.42}$$

where the first term is the kinetic energy of molecular motion, while ε'_k denote internal energy levels of the molecule (corresponding e.g. to the rotation of the molecule, atomic oscillations near equilibrium positions, energy levels of atoms etc.). Here it is important to note that ε'_k do not depend on the momenta (velocities) and coordinates of the center of mass of the molecule. Then, the sum under ln in Equation (3.40) is equal to[8]:

$$\sum_k \frac{1}{(2\pi\hbar)^3} e^{-\frac{\varepsilon'_k}{T}} \int_V dV \int_{-\infty}^{\infty} dp_x \int_{-\infty}^{\infty} dp_y \int_{-\infty}^{\infty} dp_z e^{-\frac{p_x^2+p_y^2+p_z^2}{2mT}}$$
$$= V \left(\frac{mT}{2\pi\hbar^2} \right)^{3/2} \sum_k e^{-\frac{\varepsilon'_k}{T}}. \tag{3.43}$$

Then the free energy of the gas is written as:

$$F = -NT \ln \left[\frac{eV}{N} \left(\frac{mT}{2\pi\hbar^2} \right)^{3/2} \sum_k e^{-\frac{\varepsilon'_k}{T}} \right] = -NT \ln \left[\frac{eV}{N} \left(\frac{mT}{2\pi\hbar^2} \right)^{3/2} Z' \right], \tag{3.44}$$

where we have introduced an "internal" partition function of a molecule $Z' = \sum_k e^{-\frac{\varepsilon'_k}{T}}$. This sum cannot be calculated in general form, it depends on the values of the internal energy levels of the molecules, i.e. on the type of gas. However, it is

[8] Integral over dV here is related to integration over coordinates of the center of mass of the molecule and reduces to the total volume occupied by gas V.

important to note that it is some function of temperature only, so that Equation (3.44) gives the complete dependence of the free energy on the volume. This volume dependence can be written explicitly by rewriting Equation (3.44) as:

$$F = -NT \ln \frac{eV}{N} + Nf(T); \quad f(T) = -T \ln \left(\frac{mT}{2\pi\hbar^2} \right)^{3/2} Z'. \tag{3.45}$$

Then for the gas pressure we immediately obtain:

$$P = -\frac{\partial F}{\partial V} = \frac{NT}{V} \quad \text{or} \quad PV = NT \tag{3.46}$$

i.e. an equation of state of an ideal gas. If we measure the temperature in absolute degrees, we have to write:

$$PV = Nk_B T = RT. \tag{3.47}$$

For one gram-molecule (mole) of gas $N = 6.023 \ 10^{23}$ (Avogadro number), $R = 8.314 \ 10^7$ erg/K, $k_B = 1.3804 \ 10^{-16}$ erg/K.

From F we can find other thermodynamic potentials. For example, the Gibbs thermodynamic potential:

$$\Phi = F + PV = E - TS + PV = W - TS = -NT \ln \frac{eV}{N} + Nf(T) + PV, \tag{3.48}$$

where W is the enthalpy. Expressing V via P and T using the equation of state (3.46), to rewrite Φ as a function of P and T (remember that $d\Phi = -SdT + VdP$) and introducing a new function of temperature as: $\chi(T) = f(T) - T \ln T$, we obtain:

$$\Phi = NT \ln P + N\chi(T). \tag{3.49}$$

The entropy of the gas (remember that $dF = -SdT - PdV$):

$$S = -\frac{\partial F}{\partial T} = N \ln \frac{eV}{N} - Nf'(T) \tag{3.50}$$

or, as a function of P and T:

$$S = -\frac{\partial \Phi}{\partial T} = -N \ln P - N\chi'(T). \tag{3.51}$$

The internal energy of the gas:

$$E = F + TS = Nf(T) - NTf'(T) \tag{3.52}$$

and is a function of temperature only. The same is valid for the enthalpy $W = E + PV = E + NT$. The physical reason is simple – molecules of an ideal gas do not interact, so that the change of the average intermolecular distance during the change of volume does not influence the energy. Due to this behavior of E and W, both types of specific heat $C_v = \left(\frac{\partial E}{\partial T}\right)_V$ and $C_p = \left(\frac{\partial W}{\partial T}\right)_P$ also depend only on T. Writing the specific heat per molecule we introduce $C_v = Nc_v$ and $C_p = Nc_p$. For an ideal gas $W - E = NT$, so that the difference $c_p - c_v$ is universal:

$$c_p - c_v = 1 \quad \text{or} \quad c_p - c_v = k_B \tag{3.53}$$

or $C_P - C_V = R$ per mole.

3.6 Ideal gas with constant specific heat

From experiments it is known that in a wide interval of high enough temperatures the specific heat of gases is a constant, independent of T. The physical reasons for such behavior will become clear later, while now we shall show that, under the assumption of temperature independence of the specific heat, the thermodynamic characteristics of a gas can be calculated in general form. More precisely, in this case we can determine the general form of an unknown function of temperature $f(T)$, introduced above in Equation (3.45), expressing it via constants to be determined from experiments. In this case we do not have to calculate the "internal" partition function Z'. Simply differentiating Equation (3.52) for the internal energy with respect to the temperature we find:

$$c_v = -Tf''(T). \tag{3.54}$$

Assuming specific heat to be a constant defined by experiments, we can integrate Equation (3.54) twice to obtain:

$$f(T) = -c_v T \ln T - \zeta T + \varepsilon_0, \tag{3.55}$$

where ζ and ε_0 are two constants of integration. Then, from Equation (3.45) we get the free energy in the form:

$$F = N\varepsilon_0 - NT \ln \frac{eV}{N} - Nc_v T \ln T - N\zeta T. \tag{3.56}$$

The constant ζ is called the chemical constant of a gas and for any concrete gas it is to be determined experimentally. Now using Equation (3.52) we obtain the internal energy as a linear function of temperature:

$$E = N\varepsilon_0 + Nc_v T. \tag{3.57}$$

The Gibbs thermodynamic potential is obtained by adding $PV = NT$ to Equation (3.56), and we have to express the volume of gas via pressure and temperature. Thus we obtain:

$$\Phi = N\varepsilon_0 + NT \ln P - Nc_p T \ln T - N\zeta T. \tag{3.58}$$

Enthalpy $W = E + PV$ is equal to:

$$W = N\varepsilon_0 + Nc_pT. \tag{3.59}$$

Differentiating (3.56) and (3.58) with respect to T, we obtain the entropy expressed via T and V or T and P respectively:

$$S = -\left(\frac{\partial F}{\partial T}\right)_V = -N\ln\frac{eV}{N} + Nc_v\ln T + (\zeta + c_v)N, \tag{3.60}$$

$$S = -\left(\frac{\partial \Phi}{\partial T}\right)_P = -N\ln P + Nc_p\ln T + (\zeta + c_p)N. \tag{3.61}$$

From these expressions, we can obtain the relation between the volume, temperature and pressure of an ideal gas (with constant specific heat) during its adiabatic expansion or compression. During adiabatic processes the entropy remains constant and from Equation (3.61) we have: $-N\ln P + Nc_p\ln T = \text{const}$, so that $T^{c_p}/P = \text{const}$, or using $c_p - c_v = 1$:

$$T^\gamma P^{1-\gamma} = \text{const}, \tag{3.62}$$

where $\gamma = c_p/c_v$. Using the equation of state $PV = NT$, we obtain the relations between T and V and also between P and V:

$$TV^{\gamma-1} = \text{const} \quad PV^\gamma = \text{const}. \tag{3.63}$$

3.7 Equipartition theorem

Let us consider the problem of the calculation of thermodynamic properties of gases from the point of view of classical statistical mechanics. A gas molecule is essentially some configuration of atoms, performing small oscillations near respective equilibrium positions, corresponding to the minimum of potential energy. Obviously, this potential energy can be represented as some quadratic form of the atomic coordinates:

$$U = \varepsilon_0 + \sum_{i,k=1}^{r_{osc}} a_{ik}q_iq_k, \tag{3.64}$$

where ε_0 is the potential energy of the atoms at their equilibrium positions and r_{osc} is the number of vibrational degrees of freedom.

The number r_{osc} can be determined from a very simple analysis, starting with the number of atoms in the molecule n. We know that an n-atomic molecule possess $3n$ degrees of freedom in total. Three of these correspond to free translations of the molecule in space as a whole, and another three – to its rotations as a whole. The rest of the degrees of freedom correspond to atomic oscillations, so that $r_{osc} = 3n - 6$. If all atoms are placed along a straight line (like e.g. in two-atomic molecule), we have only two rotational degrees of freedom, in this case $r_{osc} = 3n - 5$. For a one-atom gas $n = 1$ and there are no oscillations (and rotations) at all, one atom can move only along three directions in space and we have only translational degrees of freedom.

The full energy $\varepsilon(p, q)$ of a molecule is the sum of potential and kinetic energies. Kinetic energy is always a quadratic function of all momenta, the number of these momenta is equal to the total number of degrees of freedom $3n$. Thus this energy can be written as $\varepsilon(p, q) = \varepsilon_0 + f_{II}(p, q)$, where $f_{II}(p, q)$ is some quadratic function of both coordinates and momenta, and the total number of variables in this function is $l = 6n - 6$ (for the general three-dimensional molecule) or $l = 6n - 5$ for a linear molecule. For a one-atom gas $l = 3$ and the coordinates simply do not enter the expression for energy.

As a result for the free energy of a gas, from Equation (3.41) we have:

$$F = -NT \ln \frac{ee^{-\frac{\varepsilon_0}{T}}}{N} \int d\tau e^{-\frac{f_{II}(p,q)}{T}}. \tag{3.65}$$

Let us here make the transformation $p = p'\sqrt{T}$, $q = q'\sqrt{T}$ for all l variables of $f_{II}(p, q)$. Due to the quadratic nature of $f_{II}(p, q)$ we obtain:

$$f_{II}(p, q) = T f_{II}(p', q') \tag{3.66}$$

and T in the exponent under the integral just disappears. A similar transformation in differentials entering $d\tau$ produces the factor $T^{l/2}$, which is moved outside the integral. Integration over the coordinates of the oscillators q is done over the possible values of the atomic oscillations within the molecule. However, due to fast convergence (quadratic function in the exponent) integration over p' and q' can be extended to the infinite interval from $-\infty$ to ∞, so that our integral is reduced to some constant, independent of temperature. Taking into account that integration over the coordinates of the center of mass of the molecule simply gives the total volume V of the gas, we obtain for the free energy the following expression:

$$F = -NT \ln \frac{AVe^{-\frac{\varepsilon_0}{T}} T^{l/2}}{N} \qquad A = \text{const}. \tag{3.67}$$

Then:

$$F = N\varepsilon_0 - NT \ln \frac{eV}{N} - N\frac{l}{2} T \ln T - NT \ln A \tag{3.68}$$

which coincides with Equation (3.56), if we put:

$$c_v = \frac{l}{2} \tag{3.69}$$

and $\zeta = \ln A$. Accordingly:

$$c_p = c_v + 1 = \frac{l+2}{2}. \tag{3.70}$$

Thus the specific heat of a classical ideal gas is a constant, and for each degree of freedom of a molecule $\varepsilon(p,q)$ we get the same contribution of $1/2$ in specific heat c_v (or $k_B/2$ in standard units). It corresponds to the similar $T/2$ ($k_BT/2$ if we measure T in absolute degrees) contribution to the energy of the gas. This rule is called the *equipartition* law or theorem and is a quite general statement of classical statistical mechanics. In particular it is easily generalized also for the case of condensed matter[9]. Taking into account that each of the translational and rotational degrees of freedom enter $\varepsilon(p,q)$ only through respective momenta, we can say that each of these degrees of freedom contributes $1/2$ to the specific heat. For each of the oscillators we have a contribution of two degrees of freedom into $\varepsilon(p,q)$ (coordinate and momentum) and its contribution to the specific heat is 1.

3.8 One-atom ideal gas

Let us consider an ideal gas of single atoms (not molecules). Complete knowledge of the free energy of such a gas requires the calculation of an "internal" partition function Z' introduced in Equation (3.44):

$$Z' = \sum_k e^{-\frac{\varepsilon_k}{T}}, \tag{3.71}$$

where ε_k are the internal energy levels of an atom. These levels may be degenerate, in this case the respective term enters the sum g_k times, where g_k is degeneracy of corresponding level. Then:

$$Z' = \sum_k g_k e^{-\frac{\varepsilon_k}{T}}. \tag{3.72}$$

The free energy of the gas, according to Equation (3.44), is given by:

$$F = -NT \ln\left[\frac{eV}{N}\left(\frac{mT}{2\pi\hbar^2}\right)^{3/2} Z'\right]. \tag{3.73}$$

From quantum mechanics it is known that in atoms the ground state level and first excited level (neglecting superfine splitting) are separated by an energy of the order

[9] As temperature lowers, significant deviations from this law are observed in experiments. It is obvious that constancy of specific heat contradicts Nernst's theorem. Historically, the violation of the equipartition law was one of the first indications of the inadequacy of the classical treatment, which led to the discovery of quantum mechanics.

of the ionization energy (potential) I_{ion}, which for most atoms lies in the interval of $I_{\text{ion}}/k_B \sim 5-28\ 10^4 K$. Thus, for temperatures $T \ll I_{\text{ion}}$, which are of main interest to us, the gas does not contain a significant number of ionized or even excited atoms. All atoms can be assumed to be in their ground states.

Consider the simplest case of atoms with their orbital or spin momentum in the ground state ($L = S = 0$), for example noble gases[10]. In this case the ground state is non degenerate and "internal" partition function consists of one term: $Z' = e^{-\frac{\varepsilon_0}{T}}$. Then from Equation (3.73) we immediately obtain an expression for the free energy similar to (3.56), with constant specific heat:

$$c_v = 3/2 \tag{3.74}$$

and chemical constant:

$$\zeta = \frac{3}{2} \ln \frac{m}{2\pi\hbar^2}. \tag{3.75}$$

The last expression is called the Sakura-Tetrode formula.

These expressions allow us to find the criterion of applicability of the Boltzmann statistics. Previously we obtained the Boltzmann distribution assuming the smallness of the average occupation numbers:

$$\langle n_k \rangle = e^{\frac{\mu-\varepsilon_k}{T}} \ll 1. \tag{3.76}$$

Obviously, it is instead sufficient to require that:

$$e^{\frac{\mu}{T}} \ll 1. \tag{3.77}$$

From this expression it is clear that the chemical potential of a Boltzmann gas is always negative and large in absolute value. Let us find the chemical potential from its thermodynamic definition $\mu = \Phi/N$, using the expression of the Gibbs thermodynamic potential (3.58), substituting $c_p = c_v + 1 = 5/2$ and ζ from Equation (3.75). We obtain:

$$\mu = T \ln \left[\frac{P}{T^{5/2}} \left(\frac{2\pi\hbar^2}{m} \right)^{3/2} \right] = T \ln \left[\frac{N}{V} \left(\frac{2\pi\hbar^2}{mT} \right)^{3/2} \right] \tag{3.78}$$

which obviously coincides with Equation (3.12), determined in another way (from normalization to the fixed average number of particles). Then from (3.77) and (3.78) we obtain the criterion for validity of the Boltzmann statistics in the following form:

$$\frac{N}{V} \left(\frac{\hbar^2}{mT} \right)^{3/2} \ll 1 \quad \text{or} \quad T \gg \frac{\hbar^2}{m} \left(\frac{N}{V} \right)^{2/3}. \tag{3.79}$$

[10] A detailed discussion of more complicated cases, as well as of molecular gases, can be found in [1, 2].

Boltzmann statistics is valid if the gas is sufficiently diluted and the temperature is high enough. The characteristic temperature (energy) from the right-hand side of Equation (3.79) is called the temperature (energy) of degeneracy. It grows with the growth of gas density. Its physical meaning is easily understood from simple estimates as the average distance between atoms of the gas $a \sim (V/N)^{1/3}$. Quantum indeterminacy of the energy of an atom corresponding to its localization on this length scale is of the order of $E_0 \sim \frac{\hbar^2}{ma^2} \sim \frac{\hbar^2}{m}(N/V)^{2/3}$. Condition $T \gg E_0$ in Equation (3.79) means that we can neglect quantum effects. In contrast, for $T < E_0$ quantum effects become important and we have to move from Boltzmann statistics to the quantum statistics of ideal gases[11].

[11] The expressions for thermodynamic characteristics of gases obtained above are obviously unsatisfactory and contradicting Nernst's theorem; neither entropy nor specific heat tend to zero as $T \to 0$.

Chapter 4
Quantum ideal gases

4.1 Fermi distribution

We have seen above that as the temperature of an ideal gas decreases (at fixed density), Boltzmann statistics become invalid due to the emergence of quantum effects (Cf. Equation (3.79)). It is clear that to describe low temperature (or high density) behavior we need another statistics, appropriate for the cases when the average occupation numbers of different quantum states are not assumed to be small[1]. This statistics is different, depending of the nature (type) of the gas particles. The most fundamental classification of particles in modern quantum theory, based on most general theorems of quantum field theory, is a classification into either Fermions (particles with half-integer spins) and Bosons (particles with integer spin). Wave functions of the system of N identical Fermions are antisymmetric with respect to permutations of particles, while those of Bosons are symmetric.

For the system of particles described by antisymmetric wave functions (Fermions) the Pauli exclusion principle applies and the corresponding statistics is called Fermi (or Fermi-Dirac) statistics. Similar to the derivation of the Boltzmann statistics from the grand canonical ensemble given above (Cf. (3.4)–(3.7)), let us apply the Gibbs distribution to a set of particles, occupying the given quantum state (subsystem in the bath). Let us denote as Ω_k the thermodynamic potential of this set of particles. From Equation (2.60), taking into account that for the gas of noninteracting particles $E_{n_k} = n_k \varepsilon_k$, we obtain:

$$\Omega_k = -T \ln \sum_{n_k} \left(e^{\frac{\mu - \varepsilon_k}{T}} \right)^{n_k}, \tag{4.1}$$

where n_k is the number of particles in k-th quantum state. According to the Pauli principle, in the case of Fermions, this number can be either 0 or 1. Then, in the sum over n_k in (4.1) only two terms remain and we get:

$$\Omega_k = -T \ln \left(1 + e^{\frac{\mu - \varepsilon_k}{T}} \right). \tag{4.2}$$

The average number of particles in the system is equal to minus the derivative of the potential Ω_k with respect to the chemical potential μ, so that:

$$\langle n_k \rangle = -\frac{\partial \Omega_k}{\partial \mu} = \frac{e^{\frac{\mu - \varepsilon_k}{T}}}{1 + e^{\frac{\mu - \varepsilon_k}{T}}} \tag{4.3}$$

[1] Below we follow the analysis of [1].

or:

$$\langle n_k \rangle = \frac{1}{e^{\frac{\varepsilon_k - \mu}{T}} + 1}. \tag{4.4}$$

This is called the Fermi distribution. It is easy to see that we always have $\langle n_k \rangle \leq 1$, and for $e^{\frac{\varepsilon_k - \mu}{T}} \gg 1$ Equation (4.4) reduces to the Boltzmann distribution[2].

The normalization condition for the Fermi distribution can be written as:

$$\sum_k \frac{1}{e^{\frac{\varepsilon_k - \mu}{T}} + 1} = N, \tag{4.5}$$

where N is the total number of particles in the gas. This relation gives an implicit equation determining the chemical potential μ, as a function of T and N.

The thermodynamic potential Ω of the gas as a whole is obviously obtained from Ω_k (4.2) summing it over all quantum states:

$$\Omega = -T \sum_k \ln\left(1 + e^{\frac{\mu - \varepsilon_k}{T}}\right). \tag{4.6}$$

4.2 Bose distribution

Consider now the statistics of an ideal gas of particles with integer spin (Bosons), described by symmetric wave functions, which is called Bose (or Bose–Einstein) statistics.

The occupation numbers of quantum states for Bosons can be arbitrary (unlimited). Similar to (4.1) we have:

$$\Omega_k = -T \ln \sum_{n_k} \left(e^{\frac{\mu - \varepsilon_k}{T}}\right)^{n_k}. \tag{4.7}$$

The series entering here is just a geometric progression, which converges if $e^{\frac{\mu - \varepsilon_k}{T}} < 1$. This condition should be satisfied for arbitrary ε_k, so that

$$\mu < 0 \tag{4.8}$$

i.e. the chemical potential of a Bose gas is always negative. Previously we have seen that for a Boltzmann gas $\mu < 0$ and is large in absolute value. Below we shall see that for a Fermi gas μ may have either sign.

Summing the progression in (4.7) we get:

$$\Omega_k = T \ln\left(1 - e^{\frac{\mu - \varepsilon_k}{T}}\right). \tag{4.9}$$

[2] If we require the validity of this inequality for arbitrary ε_k, it reduces to $e^{\mu/T} \ll 1$, coinciding with the criterion of validity of the Boltzmann statistics given in Equation (3.77).

Now for $\langle n_k \rangle = -\frac{\partial \Omega_k}{\partial \mu}$ we obtain:

$$\langle n_k \rangle = \frac{1}{e^{\frac{\varepsilon_k - \mu}{T}} - 1}, \tag{4.10}$$

which is called the Bose distribution. Again, in case of $e^{\frac{\varepsilon_k - \mu}{T}} \gg 1$ it reduces to the Boltzmann distribution.

The normalization condition is again written as:

$$N = \sum_k \frac{1}{e^{\frac{\varepsilon_k - \mu}{T}} - 1} \tag{4.11}$$

and implicitly defines the chemical potential.

The thermodynamic potential Ω for the whole gas, similar to (4.6), is given by:

$$\Omega = T \sum_k \ln\left(1 - e^{\frac{\mu - \varepsilon_k}{T}}\right). \tag{4.12}$$

4.3 Nonequilibrium Fermi and Bose gases

Let us consider the entropy of Fermi and Bose (ideal) gases in general (nonequilibrium) states. Equilibrium Bose and Fermi distributions will be obtained, requiring the maximal value of entropy in equilibrium. This analysis can be performed similar to the case of a Boltzmann gas. Again we can consider groups of levels close in energy, numbered by $j = 1, 2, \ldots$. Let G_j be the number of states in the j-th group and N_j – the number of particles in these states. The set of numbers N_j completely characterizes the microscopic state of a gas.

In the case of Fermi statistics only one particle can occupy each quantum state, but the numbers N_j are not small, but of the order of G_j. The number of possible distributions of N_j identical particles over G_j states, with no more than one particle in each state, is equal to the number of ways that we can choose N_j from G_j states, i.e. the number of combinations of G_j elements by N_j:

$$\mathcal{W}_j = \frac{G_j!}{N_j!(G_j - N_j)!}. \tag{4.13}$$

Taking the logarithm and using for all three factorials in (4.13) Stirling's asymptotics $\ln N! \approx N \ln(N/e)$, we find entropy as:

$$S = \sum_j \{G_j \ln G_j - N_j \ln N_j - (G_j - N_j)\ln(G_j - N_j)\}. \tag{4.14}$$

Introducing again the average occupation numbers $\langle n_j \rangle = N_j/G_j$ we obtain the following expression for the entropy of a nonequilibrium Fermi gas:

$$S = -\sum_j G_j[\langle n_j \rangle \ln\langle n_j \rangle + (1 - \langle n_j \rangle) \ln(1 - \langle n_j \rangle)]. \tag{4.15}$$

Restricting its maximum with additional conditions:

$$\sum_j N_j = \sum_j G_j \langle n_j \rangle = N; \quad \sum_j \varepsilon_j G_j \langle n_j \rangle = E \tag{4.16}$$

i.e. using the method of Lagrange multipliers, from:

$$\frac{\partial}{\partial \langle n_j \rangle}[S + \alpha N + \beta E] = 0 \tag{4.17}$$

we immediately obtain the Fermi distribution as $\langle n_j \rangle = [e^{\alpha + \beta \varepsilon_j} + 1]^{-1}$, where $\alpha = -\mu/T, \beta = 1/T$.

In the case of Bose statistics, in each quantum state we can place an arbitrary number of particles, so that statistical weight $\mathcal{W}_j$ represents the number of all ways to distribute N_j over G_j states:

$$\mathcal{W}_j = \frac{(G_j + N_j - 1)!}{(G_j - 1)! N_j!}. \tag{4.18}$$

To understand this expression, we note that here we are speaking about e.g. the number of ways to distribute N_j identical balls over G_j boxes. Let us denote the balls by N_j points, while the boxes can be numbered and their borders can be visualized by $G_j - 1$ vertical strokes. In total there are $G_j + N_j - 1$ point and strokes. The number we seek is given by the number of ways to choose $G_j - 1$ places for strokes, i.e. the number of combinations of $N_j + G_j - 1$ elements by $G_j - 1$, which gives us Equation (4.18).

Taking the logarithm and neglecting unity in comparison with large numbers $G_j + N_j$ and G_j, we get:

$$S = \sum_j \left\{(G_j + N_j) \ln(G_j + N_j) - N_j \ln N_j - G_j \ln G_j\right\}. \tag{4.19}$$

Introducing $\langle n_j \rangle$ we can write down the entropy of the nonequilibrium Bose gas as:

$$S = \sum_j G_j[(1 + \langle n_j \rangle) \ln(1 + \langle n_j \rangle) - \langle n_j \rangle \ln\langle n_j \rangle]. \tag{4.20}$$

The equilibrium Bose distribution follows from the restriction of the maximum of this expression, similar to the case of Fermi statistics.

For $N_j \ll G_j$ (4.15), (4.20) naturally reduce to the Boltzmann expression (3.23):

$$S = \sum_j G_j \langle n_j \rangle \ln \frac{e}{\langle n_j \rangle} = \sum_j G_j [\langle n_j \rangle (1 - \ln \langle n_j \rangle)]; \quad \langle n_j \rangle \ll 1. \tag{4.21}$$

In the inverse limit of $N_j \gg G_j$, i.e. $\langle n_j \rangle \gg 1$, the entropy of the Bose gas (4.20) reduces to:

$$S = \sum_j G_j \ln \frac{eN_j}{G_j}, \tag{4.22}$$

with statistical weight (4.18) $\mathcal{W}_j = \frac{N_j^{G_j - 1}}{(G_j - 1)!}$.

4.4 General properties of Fermi and Bose gases

Many physical characteristics of Fermi and Bose gases can be written and calculated in general form. In all expressions below, the upper plus corresponds to Fermi statistics, while the lower minus corresponds to Bose statistics.

The energy of a free (non relativistic) particle can be written as:

$$\varepsilon_p = \frac{1}{2m}(p_x^2 + p_y^2 + p_z^2) = \frac{\mathbf{p}^2}{2m}. \tag{4.23}$$

For a given value of the momentum, the state of a particle is defined also by its spin projection. The number of particles in an element of phase space $dp_x dp_y dp_z dV$ can be obtained by multiplication of the Fermi (Bose) distribution by the number of states in this phase space volume:

$$g d\tau = g \frac{dp_x dp_y dp_z dV}{(2\pi\hbar)^3} \quad g = 2s + 1, \tag{4.24}$$

where s is the spin of the particle. Thus we obtain:

$$dN_p = \frac{g d\tau}{e^{\frac{\varepsilon_p - \mu}{T}} \pm 1}. \tag{4.25}$$

Integrating over dV we get the total volume of the gas V. Then, transforming to spherical coordinates in momentum space ($dp_x dp_y dp_z \to 4\pi p^2 dp$), we obtain the momentum distribution as:

$$dN_p = \frac{gVp^2 dp}{2\pi^2\hbar^3 \left(e^{\frac{\varepsilon_p - \mu}{T}} \pm 1\right)} \tag{4.26}$$

or the distribution of energies:

$$dN_\varepsilon = \frac{gVm^{3/2}}{\sqrt{2}\pi^2\hbar^3}\frac{\sqrt{\varepsilon}d\varepsilon}{e^{\frac{\varepsilon-\mu}{T}}\pm 1} = \frac{\mathcal{N}(\varepsilon)d\varepsilon}{e^{\frac{\varepsilon-\mu}{T}}\pm 1}, \tag{4.27}$$

where we have introduced the rather useful function:

$$\mathcal{N}(\varepsilon) = \frac{gVm^{3/2}}{\sqrt{2}\pi^2\hbar^3}\sqrt{\varepsilon} = gV\frac{mp_\varepsilon}{2\pi^2\hbar^3}; \quad \text{where } p_\varepsilon = \sqrt{2m\varepsilon} \tag{4.28}$$

which is called the density of states of a particle in the energy interval $\varepsilon, \varepsilon + d\varepsilon$. These expressions replace the Maxwell distribution for quantum gases.

Integrating (4.27) over $d\varepsilon$, we obtain:

$$N = \int_0^\infty d\varepsilon \frac{\mathcal{N}(\varepsilon)}{e^{\frac{\varepsilon-\mu}{T}}\pm 1} = \frac{gVm^{3/2}}{\sqrt{2}\pi^2\hbar^3}\int_0^\infty d\varepsilon \frac{\sqrt{\varepsilon}}{e^{\frac{\varepsilon-\mu}{T}}\pm 1}. \tag{4.29}$$

Introducing the dimensionless variable $\varepsilon/T = z$ we can write:

$$\frac{N}{V} = \frac{g(mT)^{3/2}}{\sqrt{2}\pi^2\hbar^3}\int_0^\infty dz \frac{\sqrt{z}}{e^{z-\frac{\mu}{T}}\pm 1} \tag{4.30}$$

which gives an implicit equation for the chemical potential μ as a function of T and the particle density N/V.

Making a similar transformation from summation over quantum states to energy integration of Equations (4.6), (4.12) we get:

$$\Omega = \mp\frac{gVTm^{3/2}}{\sqrt{2}\pi^2\hbar^3}\int_0^\infty d\varepsilon\sqrt{\varepsilon}\ln\left(1 \pm e^{\frac{\mu-\varepsilon}{T}}\right). \tag{4.31}$$

After partial integration we obtain:

$$\Omega = -\frac{2}{3}\frac{gVm^{3/2}}{\sqrt{2}\pi^2\hbar^3}\int_0^\infty d\varepsilon \frac{\varepsilon^{3/2}}{e^{\frac{\varepsilon-\mu}{T}}\pm 1}. \tag{4.32}$$

This expression coincides, up to a factor of $-2/3$, with the total energy of the gas given by:

$$E = \int_0^\infty \varepsilon dN_\varepsilon = \frac{gVm^{3/2}}{\sqrt{2}\pi^2\hbar^3}\int_0^\infty d\varepsilon \frac{\varepsilon^{3/2}}{e^{\frac{\varepsilon-\mu}{T}}\pm 1}. \tag{4.33}$$

From thermodynamics it is known that $\Omega = -PV$, so that Equations (4.32), (4.33) give the generalized equation of state for ideal quantum gases:

$$PV = \frac{2}{3}E. \tag{4.34}$$

In the limit of Boltzmann statistics we have $E = 3NT/2$ (equipartition law) and (4.34) reduces to the classical result: $PV = NT$.

Rewriting (4.32) as (Cf. (4.30)):

$$P = \frac{g\sqrt{2}m^{3/2}T^{5/2}}{3\pi^2\hbar^3}\int_0^\infty dz \frac{z^{3/2}}{e^{z-\frac{\mu}{T}} \pm 1} \tag{4.35}$$

we obtain the equation of state in parametric form (parameter μ!), i.e. the relation between P, V and T for a given value of μ.

Consider small corrections to the classical equation of state. We shall use inequality $e^{\mu/T} \ll 1$ (Boltzmann limit) and expand the integrand in (4.35) in powers of $e^{(\mu/T)-z}$, limiting ourselves to the first two terms of the expansion. Then:

$$\int_0^\infty dz \frac{z^{3/2}}{e^{\frac{\varepsilon-\mu}{T}} \pm 1} \approx \int_0^\infty dz z^{3/2} e^{\frac{\mu}{T}-z}\left(1 \mp e^{\frac{\mu}{T}-z}\right) = \frac{3\sqrt{\pi}}{4}e^{\frac{\mu}{T}}\left(1 \mp \frac{1}{2^{5/2}}e^{\frac{\mu}{T}}\right). \tag{4.36}$$

and (4.35) can be rewritten as:

$$\Omega = -PV = -\frac{gVm^{3/2}T^{5/2}}{(2\pi\hbar^2)^{3/2}}e^{\frac{\mu}{T}}\left(1 \mp \frac{1}{2^{5/2}}e^{\frac{\mu}{T}}\right). \tag{4.37}$$

This expression in fact reduces to:

$$\Omega = \Omega_{\text{Boltz}} \pm \frac{gVm^{3/2}T^{5/2}}{16\pi^{3/2}\hbar^3}e^{\frac{2\mu}{T}}. \tag{4.38}$$

Small corrections to thermodynamic potentials, expressed via the appropriate variables, are equal. Thus, rewriting the correction to Ω_{Boltz} via T and V, using the corresponding classical (Boltzmann) expressions (we drop the technical details), we can write down the free energy of the gas as:

$$F = F_{\text{Boltz}} \pm \frac{\pi^{3/2}}{2g}\frac{N^2\hbar^3}{VT^{1/2}m^{3/2}}. \tag{4.39}$$

From here it is easy to find:

$$PV = NT\left\{1 \pm \frac{\pi^{3/2}}{2g}\frac{N\hbar^3}{V(mT)^{3/2}}\right\}. \tag{4.40}$$

We can see that quantum corrections (tending to zero as $\hbar \to 0$) lead to additional growth of the pressure in a Fermi gas and to the opposite effect in a Bose gas. This reflects the natural tendency of Fermions to "avoid" each other (Pauli exclusion principle!), while for Bosons we have just the opposite behavior.

4.5 Degenerate gas of electrons

Quantum effects generally become important in the low temperature limit (in practice these temperatures may be high enough!). Of prime importance are the low temperature properties of Fermi gas. Keeping in mind the most important applications, we shall discuss below mainly the gas of free electrons, and we put $g = 2(s = 1/2)$.

Let us start form the analysis of the situation at $T = 0$. This is the case of a so called completely degenerate Fermi gas. Each quantum state in a Fermi gas can be occupied by no more than one electron. Thus, in fact, at $T = 0$ they just fill all states with energies from zero (ground state) up to some maximum energy (which is called *Fermi energy*), with a value determined simply by the number of particles (density) in the gas.

The number of quantum states of electrons moving in the volume V, with absolute values of momenta in the interval $p, p + dp$, is equal to:

$$2\frac{4\pi p^2 dpV}{(2\pi\hbar)^3}. \tag{4.41}$$

Electrons fill all states with momenta from zero to a maximum momentum $p = p_F$ (*Fermi momentum*). The total number of electrons in these states is determined by[3]:

$$N = \frac{V}{\pi^2\hbar^3}\int_0^{p_F} p^2 dp = \frac{Vp_F^3}{3\pi^2\hbar^3}. \tag{4.42}$$

Then for the Fermi momentum we obtain:

$$p_F = (3\pi^2)^{1/3}\left(\frac{N}{V}\right)^{1/3}\hbar, \tag{4.43}$$

which grows with the growth of electron density. It is clear that from Equation (4.43) follows a simple estimate $p_F \sim \hbar/a$, where a is an average distance between electrons.

Correspondingly, the Fermi energy is defined as[4]:

$$\varepsilon_F = \frac{p_F^2}{2m} = (3\pi^2)^{2/3}\frac{\hbar^2}{2m}\left(\frac{N}{V}\right)^{2/3} \sim \frac{\hbar^2}{ma^2}. \tag{4.44}$$

---

[3] In fact, here we simply calculate the volume of the *Fermi sphere* $V_F = \frac{4\pi p_F^3}{3}$, while the number of electrons is determined by the number of available states "inside" this sphere as $N = 2V\frac{V_F}{(2\pi\hbar)^3}$, which gives (4.42). The surface of the Fermi sphere is called *Fermi surface*. In metals, where the energy spectrum of electrons may be quite different from that of free electrons, the Fermi surface may also be quite different from the simple spherical shape. Geometry and topology of Fermi surfaces plays a very important role in the theory of metals [19]. The simple estimates presented here are, strictly speaking, applicable only to simple metals (like e.g. Na and K).

[4] Note that the value of the Fermi energy is practically the same as the degeneracy temperature (energy) of the gas introduced above (3.79).

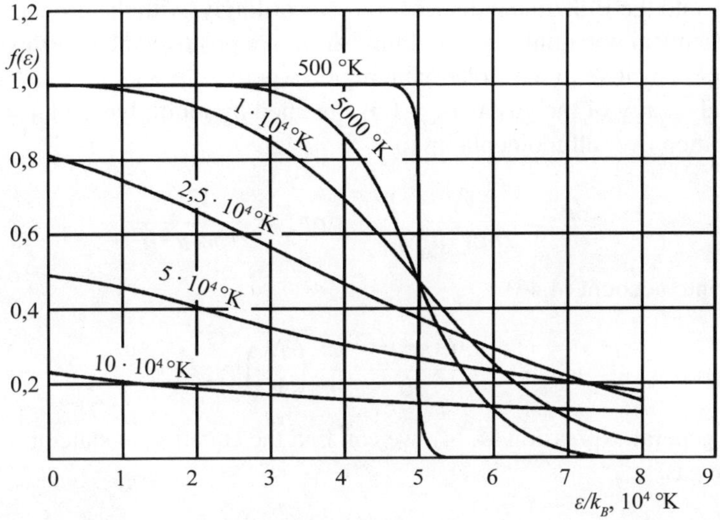

Figure 4.1. Fermi distribution function for different temperatures for $\varepsilon_F / k_B = 50\,000$ K.

Naturally it also grows with the density of the gas $\sim (N/V)^{2/3}$.

Fermi distribution:

$$n_p = \frac{1}{e^{\frac{\varepsilon_p - \mu}{T}} + 1} \tag{4.45}$$

for $T \to 0$ becomes a "Fermi step" function:

$$n_p = \begin{cases} 1 & \text{for } p \leq p_F \\ 0 & \text{for } p > p_F \end{cases} \tag{4.46}$$

or

$$n_\varepsilon = \begin{cases} 1 & \text{for } \varepsilon \leq \mu = \varepsilon_F \\ 0 & \text{for } \varepsilon > \mu = \varepsilon_F . \end{cases} \tag{4.47}$$

The chemical potential of a Fermi gas at $T = 0$ coincides with the Fermi energy:

$$\mu = \varepsilon_F \quad (T = 0). \tag{4.48}$$

At finite temperatures $T \ll \varepsilon_F$ (strongly degenerate gas) the Fermi step is "smeared" in the energy interval $\sim T$ around the Fermi energy (Cf. Figure 4.1). It is easy to see that with the growth of temperature, for $T \gg \varepsilon_F$ the Fermi distribution transforms into the Boltzmann distribution. Accordingly, with the growth of temperature, the chemical potential starts to diminish from a positive value of the order of ε_F and becomes negative in the Boltzmann region where $T \gg \varepsilon_F$.

The total energy of the gas at $T = 0$ is obtained by multiplying (4.41) by $p^2/2m$ and integration over all momenta up to $p = p_F$:

$$E = \frac{V}{2m\pi^2\hbar^3} \int_0^{p_F} dp p^4 = \frac{Vp_F^5}{10m\pi^2\hbar^3} \tag{4.49}$$

or, taking into account (4.43)

$$E = \frac{3(3\pi^2)^{2/3}}{10} \frac{\hbar^2}{m} \left(\frac{N}{V}\right)^{2/3} N. \tag{4.50}$$

Using the general expression (4.34), we can find the equation of state of a completely degenerate gas:

$$P = \frac{(3\pi^2)^{2/3}}{5} \frac{\hbar^2}{m} \left(\frac{N}{V}\right)^{5/3} \tag{4.51}$$

so that at $T = 0$ the pressure of the Fermi gas is $\sim (N/V)^{5/3}$.

In fact all the previous expressions are applicable also for finite but low enough temperatures $T \ll \varepsilon_F$. Corresponding temperature corrections are of the order of $(T/\varepsilon_F)^2$. The Fermi temperature (degeneracy temperature) $T_F \approx \varepsilon_F$ for the gas of electrons with density $N/V \sim 10^{22}\,\text{cm}^{-3}$, typical for metals, and mass $m \sim m_e$, where m_e is the mass of a free electron[5], can be estimated to be in the interval of 10^4–10^5 K. Thus, an electron gas in metals, under normal conditions, is always strongly degenerate. In semiconductors, where the electron density may change within rather wide limits, this is generally not so. Quite often the statistics of current carriers may be Boltzmann's.

To conclude this paragraph, let us make some remarks on the role of interelectron interactions. A degenerate electron gas becomes more "ideal" with the growth of its density. The characteristic kinetic energy of the electrons is of the order of the Fermi energy: $\varepsilon_F \sim \frac{\hbar^2}{m}(N/V)^{2/3} \sim \frac{\hbar^2}{ma^2}$, where a is the interelectron distance (in metals it is practically the same as the interatomic distance). At the same time the characteristic Coulomb repulsion energy $U \sim \frac{e^2}{a}$. Then the dimensionless parameter of perturbation theory over interaction is given by the ratio $\frac{U}{\varepsilon_F} \sim \frac{e^2}{\hbar}\frac{ma}{\hbar} \sim \frac{e^2}{\hbar}\frac{m}{p_F} = \frac{e^2}{\hbar v_F}$, where we have introduced the velocity of electrons on the Fermi surface (Fermi level) $v_F = p_F/m$. Now we see that the smaller a (i.e. for higher densities or Fermi velocity) the smaller is this parameter and interaction effects become weaker. Remember now, that the fine structure constant $\frac{e^2}{\hbar c} \approx \frac{1}{137} \sim 10^{-2}$, where $c \approx 3\ 10^{10}$ cm/sec

[5] Note that in real metals the mass of an electron is not necessarily equal to the mass of a free electron in vacuum.

is the velocity of light in vacuum. In metals (for typical electron densities) it is easy to estimate that $v_F \sim 10^8$ cm/sec. Thus, in real metals the perturbation theory parameter is not small: $\frac{e^2}{\hbar v_F} \sim 1$! Only for electron densities much larger than typical densities in metals, an electron gas can be considered as a nearly free (ideal) gas. So the question arises as to why the nearly free electrons picture is so good to describe many of the electronic properties of metals? The complete solution of this problem is achieved only within the *Fermi-liquid* theory, which will be briefly discussed later.

4.6 Relativistic degenerate electron gas*

Compression of an electron gas leads to the growth of the average electron energy (and Fermi energy ε_F) and sooner or later it becomes comparable to the rest energy mc^2 and even higher. In this situation relativistic effects become important. Let us consider the degenerate ultra-relativistic gas with particle energies much greater than mc^2. In this case the energy spectrum of electrons can be written as:

$$\varepsilon_p = \sqrt{c^2 p^2 + m^2 c^4} \approx cp. \tag{4.52}$$

For the number of quantum states and the Fermi momentum the previous expression remains valid:

$$2\frac{4\pi p^2 dpV}{(2\pi\hbar)^3}, \tag{4.53}$$

$$N = \frac{V}{\pi^2\hbar^3}\int_0^{p_F} p^2 dp = \frac{Vp_F^3}{3\pi^2\hbar^3}, \tag{4.54}$$

$$p_F = (3\pi^2)^{1/3}\left(\frac{N}{V}\right)^{1/3}. \tag{4.55}$$

However, for Fermi energy we have the quite new expression:

$$\varepsilon_F = cp_F = (3\pi^2)^{1/3}\hbar c\left(\frac{N}{V}\right)^{1/3}. \tag{4.56}$$

Correspondingly, the total energy of the gas is:

$$E = \frac{cV}{\pi^2\hbar^3}\int_0^{p_F} dpp^3 = V\frac{cp_F^4}{4\pi^2\hbar^3} \tag{4.57}$$

or

$$E = \frac{3(3\pi^2)^{1/3}}{4}\hbar cN\left(\frac{N}{V}\right)^{1/3}. \tag{4.58}$$

The pressure is obtained by differentiating this expression with respect to volume:

$$P = \frac{E}{3V} = \frac{(3\pi^2)^{1/3}}{4}\hbar c\left(\frac{N}{V}\right)^{4/3} \tag{4.59}$$

and is proportional to the power $4/3$ of density.

The relation

$$PV = \frac{E}{3} \tag{4.60}$$

is valid for ultra-relativistic gases not only at absolute zero $T = 0$, but for arbitrary temperatures. This can be seen as follows. Using $\varepsilon_p = cp$ in Equation (4.6), and going from summation over momenta to integration over energy, we get:

$$\Omega = -\frac{TV}{\pi^2 c^3 \hbar^3}\int_0^\infty d\varepsilon \ln\left(1 + e^{\frac{\mu-\varepsilon}{T}}\right) \tag{4.61}$$

and after partial integration:

$$\Omega = -PV = -\frac{V}{3\pi^2 c^3 \hbar^3}\int_0^\infty d\varepsilon \frac{\varepsilon^3}{e^{\frac{\varepsilon-\mu}{T}} + 1}, \tag{4.62}$$

which reduces to the finite temperature variant of Equation (4.60) Note that the pressure obtained from Equation (4.60) is in fact the highest pressure, which can exist in any macroscopic system [6].

4.7 Specific heat of a degenerate electron gas

At finite temperatures the "Fermi step" is smeared over the interval of the order of $\sim T$. All expressions derived above for $T = 0$ are zeroth order terms of expansion in powers of the small (at low temperatures) parameter T/ε_F. Let us find the corresponding first order corrections. The thermodynamic potential of an electron gas, according to (4.32), can be written as:

$$\Omega = -\frac{4}{3}\frac{Vm^{3/2}}{\sqrt{2}\pi^2\hbar^3}\int_0^\infty d\varepsilon \frac{\varepsilon^{3/2}}{e^{\frac{\varepsilon-\mu}{T}} + 1}. \tag{4.63}$$

Consider the general integral containing the Fermi distribution function:

$$I = \int_0^\infty d\varepsilon \frac{f(\varepsilon)}{e^{\frac{\varepsilon-\mu}{T}} + 1}, \tag{4.64}$$

where $f(\varepsilon)$ is some function (the only limitation is that the integral converges). Equation (4.63) is the specific case of $f(\varepsilon) = \varepsilon^{3/2}$. For the integral (4.64) the following expansion can be derived [1, 2]:

$$I \approx \int_0^\mu d\varepsilon f(\varepsilon) + \frac{\pi^2}{6}T^2 f'(\mu) + \frac{7\pi^4}{360}T^4 f'''(\mu) + \cdots, \tag{4.65}$$

which in fact determines the expansion of all physical characteristics of the form of Equation (4.63) in powers of the small parameter T/ε_F.

Taking here $f(\varepsilon) = \varepsilon^{3/2}$ we write (4.63) as:

$$\Omega = \Omega_0 - VT^2 \frac{\sqrt{2\mu} m^{3/2}}{6\hbar^3}, \tag{4.66}$$

where the first term gives the $T = 0$ contribution. Considering the second term as a small correction to Ω_0 and expressing μ via N and V using the zero-order approximation (4.48) $\mu = \varepsilon_F = (3\pi^2)^{2/3} \frac{\hbar^2}{2m} (N/V)^{2/3}$, we can immediately write the expression for the free energy[6]:

$$F = F_0 - \frac{B}{2} N T^2 \left(\frac{V}{N} \right)^{2/3}, \tag{4.67}$$

where we have introduced the notation $B = (\pi/3)^{2/3} m/\hbar^2$. From this, we find the entropy:

$$S = BNT \left(\frac{V}{N} \right)^{2/3} \tag{4.68}$$

and specific heat:

$$C = T \frac{\partial S}{\partial T} = BNT \left(\frac{V}{N} \right)^{2/3}. \tag{4.69}$$

We see that the specific heat of a degenerate Fermi gas at low temperatures is a linear function of temperature (Pauli specific heat). Using the expression for the density of states (4.28) with $g = 2$ (for electrons), we can easily see that Equation (4.69) can be rewritten as:

$$C = \frac{\pi^2}{3} \nu_F T, \tag{4.70}$$

where we have introduced the density of electronic states at the Fermi surface:

$$\nu_F = \mathcal{N}(\varepsilon = \varepsilon_F) = \frac{m p_F}{\pi^2 \hbar^3} V. \tag{4.71}$$

This is not a simple coincidence. Equation (4.71) is rather simply interpreted in the following way: we have seen that in a degenerate Fermi gas finite temperatures disturb only a narrow energy layer $\sim T$ around the Fermi level. The number of electrons in this layer $\delta N \sim \nu_F T$. Raising the temperature by δT leads to a change in the energy of each of these electrons of $\sim \delta T$. Then the total energy change of the gas

[6] Here we once again use the theorem on small corrections to thermodynamic potentials: $(\delta\Omega)_{T,V,\mu} = (\delta F)_{T,V,N} = (\delta\Phi)_{T,P,N} = (\delta E)_{S,V,N} = (\delta W)_{S,P,N}$.

is $\delta E \sim \nu_F T \delta T$, and the specific heat is $C = \delta E/\delta T = \nu_F T$. This elementary interpretation solves the problem of the contradiction between the classical equipartition law and Nernst's theorem. For $T \to 0$ not all electrons participate in thermal processes, but only those belonging to a narrow energy layer $\sim T$ close to the Fermi level, and the number of such electrons tends to zero as $T \to 0$. The final result (4.70) for specific heat is very important. In fact, it provides one of experimental methods of determination of the density of states at the Fermi level of metals from measurements of electron contributions to the specific heat. In the simplest case of metals, with a spherical Fermi surface, when Equation (4.71) is valid, this also allows experimental determination of the mass of the conduction electrons in a metal, which in the general case does not coincide with that of a free electron.

For completeness, let us write down an expression for the total energy of a degenerate Fermi gas:

$$E = E_0 + \frac{B}{2} N T^2 \left(\frac{V}{N}\right)^{2/3} = E_0 \left[1 + 0.18 \left(\frac{mT}{\hbar^2}\right)^2 \left(\frac{V}{N}\right)^{4/3}\right], \tag{4.72}$$

where E_0 is given by Equation (4.49). From this expression it is easily seen that the relative temperature correction to the energy by parameter $(T/\varepsilon_F)^2$ is small. The specific heat calculated from $C = \frac{\partial E}{\partial T}$ obviously gives the previous result (4.69).

4.8 Magnetism of an electron gas in weak fields

The magnetization of an electron gas in a weak (external) magnetic fields consists of two contributions: *paramagnetic* magnetization, connected with the spin magnetic moment of an electron (Pauli) and *diamagnetic* magnetization, connected with the quantization of orbital motion of an electron in a magnetic field (Landau).

Below we shall analyze only the case of a degenerate electron gas: $T \ll \varepsilon_F$. The magnetic field is considered as weak if $\mu_B H \ll T$, where $\mu_B = \frac{|e|\hbar}{2mc}$ is the Bohr magneton.

Calculations can be conveniently done using the thermodynamic potential Ω, depending on the variables T, V, μ. Then the magnetic moment of the system is defined as:

$$\mathbf{M} = -\left(\frac{\partial \Omega}{\partial \mathbf{H}}\right)_{T,V,\mu}. \tag{4.73}$$

Let us start with the paramagnetic part of the magnetization. The additional energy of the electron, due to the spin interacting with the magnetic field, is given by $\pm \mu_B H$, for two spin projections $\mp 1/2$. Accordingly, in an external field the electron energy $\varepsilon_p = p^2/2m$ is replaced by $\varepsilon_{p\mp} = p^2/2m \pm \mu_B H$. As ε always enters the Fermi distribution function in the combination $\varepsilon - \mu$, the statement equivalent to the previous one is that we have to make the replacement $\mu \to \mu \mp \mu_B H$ in all expressions. Thus, for the thermodynamic potential Ω in a magnetic field we can write:

$$\Omega(\mu) = \frac{1}{2}\Omega_0(\mu + \mu_B H) + \frac{1}{2}\Omega_0(\mu - \mu_B H), \tag{4.74}$$

where $\Omega_0(\mu)$ is the thermodynamic potential in the absence of a magnetic field. The factor $1/2$ is introduced here to account for the change of the number of quantum states for fixed spin projection.

Expanding (4.74) in powers of H we obtain (terms of the first order, obviously cancel each other):

$$\Omega(\mu) = \Omega_0(\mu) + \frac{1}{2}\mu_B^2 H^2 \frac{\partial^2 \Omega_0(\mu)}{\partial \mu^2}. \tag{4.75}$$

Now we get the magnetic moment (4.73) as:

$$\mathbf{M} = -\mu_B^2 \mathbf{H} \frac{\partial^2 \Omega_0(\mu)}{\partial \mu^2}. \tag{4.76}$$

Taking into account that $\frac{\partial \Omega_0}{\partial \mu} = -N$ we get the paramagnetic susceptibility (per volume of the gas):

$$\chi_p = -\frac{\mu_B^2}{V}\frac{\partial^2 \Omega_0(\mu)}{\partial \mu^2} = \frac{\mu_B^2}{V}\left(\frac{\partial N}{\partial \mu}\right)_{T,V}. \tag{4.77}$$

Neglecting small (for $T \ll \varepsilon_F$) temperature effects, i.e. considering the gas as completely degenerate, we have $\mu = \varepsilon_F = (3\pi^2)^{2/3}\frac{\hbar^2}{2m}(N/V)^{2/3}$, and:

$$N = V\frac{(2m\mu)^{3/2}}{3\pi^2\hbar^3}, \tag{4.78}$$

which after differentiation in (4.77) reduces to:

$$\chi_p = \frac{\mu_B^2 (2m)^{3/2}\sqrt{\mu}}{2\pi^2\hbar^3} = \frac{\mu_B^2 m p_F}{\pi^2\hbar^3} \equiv \mu_B^2 \nu_F \tag{4.79}$$

which is called the Pauli paramagnetic susceptibility. Thus, the paramagnetic susceptibility of the degenerate electron gas is independent of temperature (for $T \ll \varepsilon_F$) and is proportional to the electron density of states at the Fermi level. This due to a simple fact – the external magnetic field leads to a difference between the numbers of electrons with spin oriented along and opposite to the direction of magnetic field: $N_\uparrow - N_\downarrow \sim \nu_F \mu_B H$, which leads to the appearance of magnetization along the field $M = \mu_B(N_\uparrow - N_\downarrow) \sim \mu_B^2 \nu_F H$, which gives the susceptibility (4.79)[7].

[7] There are experimental methods allowing direct measurements of only the paramagnetic part of the magnetization (susceptibility) in metals (e.g. Knight shift measurements in NMR), providing information on the value of the density of states, alongside with measurements of electron contributions to specific heat.

Let us turn now to calculations of the diamagnetic part of the susceptibility connected with the orbital motion of electrons. The energy of the orbital motion of an electron in a magnetic field is determined by the Landau levels [5]:

$$\varepsilon_{n,p_z} = \hbar\omega_c(n + \frac{1}{2}) + \frac{p_z^2}{2m} = (2n+1)\mu_B H + \frac{p_z^2}{,}2m \tag{4.80}$$

where $\omega_c = \frac{|e|H}{mc}$ is the cyclotron frequency, $n = 0, 1, 2, \ldots$, p_z is the momentum projection on magnetic field direction. The number of states in the interval dp_z at fixed n is given by [5]:

$$2\frac{V|e|H}{(2\pi\hbar)^2 c} dp_z. \tag{4.81}$$

Then, from Equation (4.6) we get:

$$\Omega = -T\sum_{n=0}^{\infty} 2\frac{V|e|H}{(2\pi\hbar)^2 c}\int_{-\infty}^{\infty} dp_z \ln\left[1 + \exp\left(\frac{\mu - (n+1/2)\hbar\omega_c - p_z^2/2m}{T}\right)\right] \tag{4.82}$$

or

$$\Omega = 2\mu_B H \sum_{n=0}^{\infty} f[\mu - (2n+1)\mu_B H], \tag{4.83}$$

where

$$f(\mu) = -\frac{TmV}{2\pi\hbar^3}\int_{-\infty}^{\infty} dp_z \ln\left[1 + \exp\left(\frac{\mu}{T} - \frac{p_z^2}{2m}\right)\right]. \tag{4.84}$$

Summation over n can be performed using the following formula [1, 2]:

$$\sum_{n=0}^{\infty} F(n + \frac{1}{2}) \approx \int_0^{\infty} dx F(x) + \frac{1}{24}F'(0). \tag{4.85}$$

This expression is valid in case of a small relative change of F during the single step $n \to n+1$. In our case this condition reduces to $\mu_B H \ll T$.

Applying (4.85) to (4.83) and (4.84) we obtain:

$$\begin{aligned}\Omega &= 2\mu_B H \int_0^{\infty} dx f(\mu - 2\mu_B H x) + \frac{2\mu_B H}{24}\frac{\partial f(\mu - 2n\mu_B H)}{\partial n}|_{n=0} \\ &= \int_{-\infty}^{\mu} dx f(x) - \frac{(2\mu_B H)^2}{24}\frac{\partial f(\mu)}{\partial \mu}.\end{aligned} \tag{4.86}$$

The first term here does not contain H and reduces to $\Omega_0(\mu)$ in the absence of a magnetic field. Thus:

$$\Omega = \Omega_0(\mu) - \frac{1}{6}\mu_B^2 H^2 \frac{\partial^2 \Omega_0(\mu)}{\partial \mu^2} \tag{4.87}$$

and similar to paramagnetic case we find diamagnetic susceptibility as:

$$\chi_d = \frac{\mu_B^2}{3V}\frac{\partial^2 \Omega_0(\mu)}{\partial \mu^2} = -\frac{1}{3}\chi_p, \tag{4.88}$$

where the last equality was obtained by comparison with (4.77). We see that the diamagnetic susceptibility (Landau diamagnetism) of an electron gas is equal to $1/3$ of the paramagnetic susceptibility (Pauli paramagnetism) and opposite in sign. The sum of both contributions is positive, so that the electron gas is paramagnetic and its total magnetic susceptibility is equal to:

$$\chi = \chi_p + \chi_d = \frac{2}{3}\chi_p. \tag{4.89}$$

However, it should be noted that these relations between χ_p and χ_d are valid only for the simplest model of free electrons. In real metals the form of the electron spectrum may be quite different from that of free electrons, so that these relations may be significantly changed. Due to this problem, during the discussion of real experiments on magnetic susceptibility of metals, we are always dealing with the complicated problem of separation of paramagnetic and diamagnetic contributions.

Obviously, the total susceptibility can be calculated directly from the single expression, writing the energy levels as $\varepsilon_{n,p_z,\pm} = (2n+1)\mu_B H + p_z^2/2m \mp \mu_B H$, i.e. including the spin splitting into the Landau spectrum. This set of levels can be rewritten as: $\varepsilon_{n,p_z} = 2n\mu_B H + p_z^2/2m$ $(n = 0, 1, 2 \ldots)$, where each value of $n \neq 0$ enters twice, while $n = 0$ enters only once. Similar to the previous analysis we can easily obtain:

$$\Omega = 2\mu_B H \left\{ \frac{1}{2} f(\mu) + \sum_{n=1}^{\infty} f(\mu - 2\mu_B H n) \right\} \tag{4.90}$$

and perform summation using [1, 2]:

$$\frac{1}{2}F(0) + \sum_{n=1}^{\infty} F(n) = \int_0^{\infty} dx F(x) - \frac{1}{12}F'(0). \tag{4.91}$$

Direct calculations lead to the total susceptibility given by Equation (4.89).

4.9 Magnetism of an electron gas in high fields*

Consider now the case of the so called quantizing magnetic field when

$$T < \mu_B H = \hbar\omega_c \ll \varepsilon_F = \mu. \tag{4.92}$$

Under these conditions it is important to take into account the discrete nature of the Landau levels, corresponding to electron motion in the plane orthogonal to the magnetic field[8]. Now we can not separate orbital and spin effects so that during the calculations it is more convenient to use the general expression (4.90). As will be shown below, for $\hbar\omega_c = \mu_B H > T$ the magnetization of an electron gas contains an *oscillating* (as a function of H) part, and the amplitude of these oscillations is not small. We shall drop some details of the calculations, which can be found in [1, 2].

While calculating (4.90) under the conditions of (4.92), we can not use simple summation formulas like (4.91), as the function summed may change rather sharply during the transition from n to $n+1$. Standard approach here is to use the Poisson summation formula[9]:

$$\frac{1}{2}F(0) + \sum_{n=1}^{\infty} F(n) = \int_0^{\infty} dx F(x) + 2\mathrm{Re}\sum_{k=1}^{\infty}\int_0^{\infty} dx e^{2\pi i k x} F(x). \tag{4.93}$$

Then (4.91) can be written as:

$$\Omega = \Omega_0(\mu) + \frac{TmV}{\pi^2\hbar^3}\mathrm{Re}\sum_{k=1}^{\infty} I_k, \tag{4.94}$$

where

$$I_k = -2\mu_B H \int_{-\infty}^{\infty} dp_z \int_0^{\infty} dx e^{2\pi i k x} \ln\left[1 + \exp\left(\frac{\mu}{T} - \frac{p_z^2}{2mT} - \frac{2x\mu_B H}{T}\right)\right]. \tag{4.95}$$

We are only interested in the oscillating (with the change of magnetic field) part of the integrals, which will be denoted as $\tilde{I}_k$. After an appropriate change of variables in

[8] In the classical approximation this motion is a simple cyclotron rotation of an electron around the direction of the field, with angular frequency ω_c. In the quantum case this rotation is described as a quantum oscillator with the same frequency, which leads to the appearance of the first (oscillator like) term in the Landau spectrum (4.80). The second term in (4.80) corresponds to free electron motion along field direction.

[9] The Poisson formula is obtained from the equality: $\sum_{n=-\infty}^{\infty}\delta(x-n) = \sum_{k=-\infty}^{\infty} e^{2\pi i k x}$. The sum of δ-functions in the left-hand side is a periodic function with period 1, while the sum in the right-hand side is the Fourier expansion of this function. Multiplying this equality by an arbitrary function $F(x)$ and integrating over x from 0 to ∞, we obtain the Poisson formula. We only have to take into account that the term of the sum, corresponding to $n = 0$, is equal to $\int_0^{\infty} dx F(x)\delta(x) = F(0)/2$.

(4.95) we obtain:

$$\tilde{I}_k = -\int_{-\infty}^{\infty} dp_z \int_0^{\infty} d\varepsilon \ln\left[1 + \exp\left(\frac{\mu - \varepsilon}{T}\right)\right] \exp\left(\frac{i\pi k\varepsilon}{\mu_B H}\right) \exp\left(-\frac{i\pi k p_z}{2m\mu_B H}\right). \tag{4.96}$$

The integral over p_z can be calculated explicitly [1, 2], so that:

$$\tilde{I}_k = -e^{-i\frac{\pi}{4}} \sqrt{\frac{2m\mu_B H}{k}} \int_0^{\infty} d\varepsilon e^{\frac{i\pi k\varepsilon}{\mu_B H}} \ln\left[1 + e^{\frac{\mu-\varepsilon}{T}}\right]. \tag{4.97}$$

Here we can twice perform partial integration and transform to the variable $\xi = (\varepsilon - \mu)/T$. Dropping the non/oscillating part, we can write [1, 2]:

$$\tilde{I}_k = \frac{\sqrt{2m}(\mu_B H)^{5/2}}{T\pi^2 k^{5/2}} \exp\left(\frac{i\pi k\mu}{\mu_B H} - \frac{i\pi}{4}\right) \int_{-\infty}^{\infty} d\xi \frac{e^{\xi}}{(e^{\xi}+1)^2} \exp\left(\frac{i\pi kT}{\mu_B H}\xi\right). \tag{4.98}$$

For $\mu_B H > T$ the main contribution to the remaining integral comes from the region of $\xi \sim 1$, i.e. the vicinity of the Fermi level $\varepsilon - \mu \sim T$, which allows us to extend the integration to infinity. Practically, the integral is calculated using the formula [1, 2]:

$$\int_{-\infty}^{\infty} d\xi e^{i\alpha\xi} \frac{e^{\xi}}{(e^{\xi}+1)^2} = \frac{\pi\alpha}{sh(\pi\alpha)}. \tag{4.99}$$

Finally we obtain for the oscillating part of the Ω potential:

$$\Omega = \frac{\sqrt{2}(m\mu_B H)^{3/2} T V}{\pi^2 \hbar^3} \sum_{k=1}^{\infty} \frac{\cos\left(\frac{\pi\mu}{\mu_B H}k - \frac{\pi}{4}\right)}{k^{3/2} sh\left(\frac{\pi^2 kT}{\mu_B H}\right)}. \tag{4.100}$$

Calculating the magnetic moment as the derivative of (4.100) with respect to the magnetic field, we only have to differentiate the most oscillating factors of cos in the numerators of the terms of the sum. This gives the Landau result:

$$\tilde{M} = -\frac{\sqrt{2\mu_B} m^{3/2} \mu T V}{\pi\hbar^3 \sqrt{H}} \sum_{k=1}^{\infty} \frac{\sin\left(\frac{\pi\mu}{\mu_B H}k - \frac{\pi}{4}\right)}{\sqrt{k} sh\left(\frac{\pi^2 kT}{\mu_B H}\right)}. \tag{4.101}$$

This expression is oscillating as a function of the inverse magnetic field $1/H$. The period over $1/H$ is given by:

$$\Delta\left(\frac{1}{H}\right) = \frac{2\mu_B}{\mu} \tag{4.102}$$

and is independent of temperature. Here we have $\Delta(1/H)H \sim \mu_B H/\mu \ll 1$, so that the oscillations are of high "frequency". Such oscillations of the magnetic moment in an external magnetic field are observed in metals at low enough temperatures and "clean" enough samples, and are called de Haas–van Alphen effect. For $\mu_B H \sim T$ the amplitude of the oscillating magnetic moment is given by $\tilde{M} \sim V\mu H^{1/2}(m\mu_B)^{3/2}\hbar^{-3}$. The monotonous part of the magnetization M is determined by the susceptibility (4.89) calculated above, so that $M \sim V\mu^{1/2}Hm^{3/2}\mu_B^2\hbar^{-3}$. Then $\tilde{M}/M \sim (\mu/\mu_B H)^{1/2} \gg 1$ and the amplitude of the oscillating part is *large* in comparison to monotonous part. For $\mu_B H \ll T$ this amplitude drops exponentially as $\exp(-\pi^2 T/\mu_B H)$ and becomes negligible.

Expression (4.102) for the period of the oscillations can be rewritten as:

$$\Delta\left(\frac{1}{H}\right) = \frac{|e|\hbar}{mc}\frac{1}{\varepsilon_F} = \frac{2|e|\hbar}{c}\frac{\pi}{\pi p_F^2} = \frac{2\pi|e|\hbar}{cS_F}, \tag{4.103}$$

where $S_F = \pi p_F^2$ is an area of maximal "cross section" of the spherical Fermi surface of free electrons. It turns out that this last expression is also valid for metals with arbitrary Fermi surfaces if S_F is understood as an area of any extremal cross section of the Fermi surface with complicated topology [19]. In a realistic case there may be several such cross sections, so that there appear several periods of de Haas–van Alphen oscillations. Experimental study of these oscillations allows us to determine extremal cross sections of the Fermi surface of a real metal, which helps in determining its form and topology.

The De Haas–van Alphen effect is the first of a number of oscillatory effects in metals in quantizing magnetic fields at low temperatures, e.g. there are similar oscillations of electrical resistivity (Shubnikov–de Haas effect). All of these effects are related to the Landau quantization of the electron spectrum in a magnetic field (4.80), and the "passing" of discrete Landau levels (of transverse motion of electrons) through the Fermi level with the change of external magnetic field [19]. Experimental observation of these effects is a powerful method to study the geometry of Fermi surfaces in real metals.

4.10 Degenerate Bose gas

At low temperatures the properties of a Bose gas are radically different from the properties of a Fermi gas. At $T = 0$ *all* particles of Bose gas occupy the state with lowest energy (ground state) $\varepsilon = 0$, there are no limitations due to the Pauli exclusion principle. Let us consider the equation for the total number of particles, determining the chemical potential (4.30) for the Bose case:

$$\frac{N}{V} = \frac{g(mT)^{3/2}}{\sqrt{2}\pi^2\hbar^3}\int_0^\infty dz\frac{\sqrt{z}}{e^{z-\frac{\mu}{T}} - 1}. \tag{4.104}$$

If for fixed density N/V of the gas we start lowering the temperature, Equation (4.104) immediately shows, that the chemical potential μ drops in absolute value, remaining negative (in accordance with the general requirements of Bose statistics). However, μ can become zero at some *finite* temperature, which is defined by the relation:

$$\frac{N}{V} = \frac{g(mT)^{3/2}}{\sqrt{2}\pi^2\hbar^3}\int_0^\infty dz\frac{\sqrt{z}}{e^z - 1}. \tag{4.105}$$

The integral here is just a dimensionless constant ≈ 2.315. Then, solving Equation (4.105) with respect to T, we obtain the characteristic temperature T_0 [10]:

$$T_0 = \frac{3.31}{g^{2/3}}\frac{\hbar^2}{m}\left(\frac{N}{V}\right)^{2/3} \tag{4.106}$$

which is called the temperature of *Bose condensation*. The physical meaning of this term, as well as of physical effects appearing below this temperature can be understood from the following arguments. For $T < T_0$ Equation (4.105) does not give negative solutions for μ, while in Bose statistics the chemical potential must be, as was shown above, negative for all temperatures. This contradiction appears because under this conditions we can not use the standard transformation from summation over quantum states in Equation (4.11) to integration over a continuous variable (energy) in Equations (4.30), (4.104). In fact, during such a transformation the first term in the sum over k in Equation (4.11), corresponding to energy level $\varepsilon_k = 0$, is multiplied by $\sqrt{\varepsilon} = 0$ (Cf. the expression for the density of states (4.28)) and just drops out of the equation. But in reality, at T, Bose particles will tend to occupy precisely this lowest energy state, until $T = 0$, when all of them will "condense" in this ground state.

Thus, in reality, the physical behavior at temperatures $T < T_0$ is as follows. Particles with energy $\varepsilon > 0$ are distributed according to ($\mu = 0$!):

$$dN_\varepsilon = \frac{gm^{3/2}V}{\sqrt{2}\pi^2\hbar^3}\frac{\sqrt{\varepsilon}d\varepsilon}{e^{\frac{\varepsilon}{T}} - 1}. \tag{4.107}$$

Accordingly, the total number of particles with energies $\varepsilon > 0$ is equal to:

$$N_{\varepsilon>0} = \int dN_\varepsilon = \frac{gV(mT)^{3/2}}{\sqrt{2}\pi^2\hbar^3}\int_0^\infty dz\frac{\sqrt{z}}{e^z - 1} = N\left(\frac{T}{T_0}\right)^{3/2}. \tag{4.108}$$

The remaining

$$N_{\varepsilon=0} = N\left[1 - \left(\frac{T}{T_0}\right)^{3/2}\right] \tag{4.109}$$

[10] Note that, similar to the Fermi temperature, this expression is of the order of the temperature of gas degeneracy (3.79).

particles are already in the state with the lowest energy $\varepsilon = 0$. This effect of a *macroscopic* number of particles being "condensed" in the ground state is called *Bose condensation*. Let us stress that we are speaking here about "condensation" of particles in momentum space (at $p = 0$), which has nothing to do with the usual gas condensation in real (coordinate) space. Particles in a Bose condensate form a macroscopic quantum state with very peculiar properties.

The total energy of the gas at $T < T_0$ is determined by the particles with $\varepsilon > 0$ (Cf. Equation (4.33) written for $\mu = 0$):

$$E = \frac{gV(mT)^{3/2}T}{\sqrt{2}\pi^2\hbar^3}\int_0^\infty dz\frac{z^{3/2}}{e^z - 1} \approx 0.770NT\left(\frac{T}{T_0}\right)^{3/2} = 0.128g\frac{m^{3/2}T^{5/2}}{\hbar^3}V. \tag{4.110}$$

Then we obtain the specific heat as:

$$C_v = \left(\frac{\partial E}{\partial T}\right)_V = \frac{5E}{2T} \sim T^{3/2}. \tag{4.111}$$

Integrating the specific heat we find for the entropy:

$$S = \int_0^T \frac{C_v}{T}dT = \frac{5E}{3T} \tag{4.112}$$

and the free energy $F = E - TS$:

$$F = -\frac{2}{3}E. \tag{4.113}$$

For the gas pressure we obtain:

$$P = -\left(\frac{\partial F}{\partial V}\right)_T \approx 0.0851g\frac{m^{3/2}T^{5/2}}{\hbar^3}. \tag{4.114}$$

At $T = T_0$ all physical characteristics discussed here are continuous, but it can be shown that the derivative of the specific heat with respect to T has a finite discontinuity (jump) at this point [1, 2]. In this sense, the point of Bose condensation, in fact, is a point of some kind of phase transition. Note however, that the properties of this transition essentially depend on the interaction between particles of the gas, which is neglected here.

During many years the phenomenon of Bose condensation in gases remained just a theoretical result, though its importance was clearly understood, and Bose condensation was in fact observed in such phenomena as superfluidity and superconductivity in condensed matter (where interactions are of prime importance). These will be discussed later, but in recent years Bose condensation was directly observed in unique experiments with ultra cold gases of alkali metals (at temperatures $\sim 10^{-7}$ K in special magnetic traps). Apparently, these systems are well described by the model of a nearly free (ideal) Bose gas, though there interactions are also quite important for the explanation of numerous effects. These studies are at present at the center of interests of modern physics of many particle systems [20].

4.11 Statistics of photons

The most important physical object to study with Bose statistics is electromagnetic radiation at thermodynamic equilibrium (for historic reasons called also "black body" radiation), i.e. a gas of photons. Linearity of the equations of electrodynamics leads to validity of the superposition principle, i.e. the absence of interactions between photons – they form an ideal gas! The spin of the photons $s =$), so this is a Bose gas. In fact, to achieve thermodynamic equilibrium we always have to assume the existence of some small interaction of the photons with matter. The mechanism of this interaction consists of absorption and emission of photons by matter[11]. This leads to an important peculiarity of a photon gas: the number of particles (photons) N is not conserved and should be determined from conditions of thermodynamic equilibrium. Requiring the minimum of free energy (at fixed T and V), we obtain the condition: $\left(\frac{\partial F}{\partial N}\right)_{T,V} = \mu = 0$, so that the chemical potential of a photon gas is zero:

$$\mu = 0. \tag{4.115}$$

The distribution function of the photons over the states with definite momenta $\hbar\mathbf{k}$ and energy $\hbar\omega = \hbar ck$ (and definite polarizations – spin projections of photons) is given by the Bose distribution with $\mu = 0$:

$$n_k = \frac{1}{e^{\frac{\hbar\omega}{T}} - 1}, \tag{4.116}$$

which is called the Planck distribution.

Assuming the volume V to be big enough, we can as usual transform from a discrete to a continuous distribution of photon eigenstates. The number of field oscillators with components of the wave vector $\mathbf{k}$ in intervals $d^3k = dk_x dk_y dk_z$ is equal to $V\frac{d^3k}{(2\pi)^3}$ [6]. Then, the number of oscillators with absolute value of the wave vector in interval $k, k + dk$ is given by:

$$\frac{V}{(2\pi)^3} 4\pi k^2 dk. \tag{4.117}$$

Using $\omega = ck$ and multiplying by 2 (there are two independent directions of polarization), we obtain the number of quantum states of photons with frequencies in the interval $\omega, \omega + d\omega$ as:

$$\frac{V\omega^2 d\omega}{\pi^2 c^3}. \tag{4.118}$$

[11] A good example of such a system is the so called "relict" radiation in the Universe, remaining since the "Big Bang" all over space.

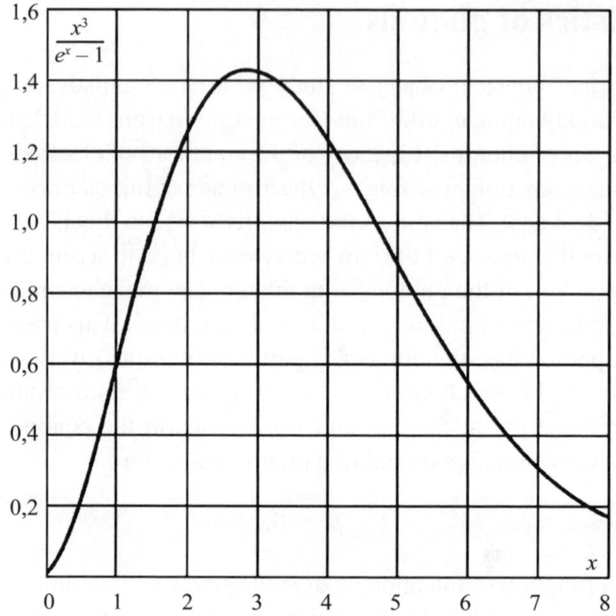

Figure 4.2. Planck function $\frac{x^3}{e^x-1}$; $x = \frac{\hbar\omega}{T}$.

Then the number of photons in this frequency interval is:

$$dN_\omega = \frac{V}{\pi^2 c^3}\frac{\omega^2 d\omega}{e^{\frac{\hbar\omega}{T}} - 1}. \tag{4.119}$$

Multiplying by $\hbar\omega$ we obtain the energy contained in this part of the spectrum:

$$dE_\omega = \frac{V\hbar}{\pi^2 c^3}\frac{\omega^3 d\omega}{e^{\frac{\hbar\omega}{T}} - 1}, \tag{4.120}$$

which is Planck's law. The corresponding graph is presented in Figure 4.2. Expressing everything in terms of the wavelength $\lambda = \frac{2\pi c}{\omega}$, we have:

$$dE_\lambda = \frac{16\pi^2 c\hbar V}{\lambda^5}\frac{d\lambda}{e^{\frac{2\pi\hbar c}{T\lambda}} - 1}. \tag{4.121}$$

For small frequencies $\hbar\omega \ll T$ we obtain from (4.120) the Rayleigh–Jeans law:

$$dE_\omega = V\frac{T}{\pi^2 c^3}\omega^2 d\omega. \tag{4.122}$$

Here, there is no dependence on $\hbar$, as this is a classical limit, and this result can be obtained by multiplying (4.118) by T, i.e. applying the equipartition law to each of the field oscillators[12]. In the inverse limit of $\hbar\omega \gg T$ (quantum limit) from (4.120) we get Wien's formula:

$$dE_\omega = V\frac{\hbar}{\pi^2 c^3}\omega^3 e^{-\frac{\hbar\omega}{T}}\, d\omega. \tag{4.123}$$

The spectral density of the energy distribution of a photon gas $dE_\omega/d\omega$ has a maximum at $\omega = \omega_m$, defined by the condition:

$$\frac{\hbar\omega_m}{T} \approx 2.822. \tag{4.124}$$

Thus, an increase in temperature leads to a shift of the maximum of the energy distribution to higher energies (frequencies) proportional to T (Wien's displacement law)[13].

Let us calculate the thermodynamic properties of a photon gas. For $\mu = 0$ the free energy $F = \Phi - PV = N\mu + \Omega$. Then, putting $\mu = 0$ and transforming from summation over k to integration over ω in (4.12), we obtain:

$$F = T\frac{V}{\pi^2 c^3}\int_0^\infty d\omega\omega^2 \ln\left(1 - e^{-\frac{\hbar\omega}{T}}\right). \tag{4.125}$$

Introducing $x = \hbar\omega/T$ and performing partial integration, we get:

$$F = -V\frac{T^4}{3\pi^2\hbar^3 c^3}\int_0^\infty dx\frac{x^3}{e^x - 1}. \tag{4.126}$$

The integral here is equal to $\pi^4/15$ [1, 2], so that:

$$F = -V\frac{\pi^2 T^4}{45(\hbar c)^3} = -\frac{4\sigma}{3c}VT^4, \tag{4.127}$$

where the coefficient σ (the Stefan–Boltzmann constant) is equal to:

$$\sigma = \frac{\pi^2 k_B^4}{60\hbar^3 c^2} \tag{4.128}$$

if we measure the temperature T in absolute degrees. The entropy of a photon gas is:

$$S = -\frac{\partial F}{\partial T} = \frac{16\sigma}{3c}VT^3. \tag{4.129}$$

---

[12] It is easy to see that the integral (4.122) over all possible frequencies diverges, so that the energy of the photon gas becomes infinite. This is the so called "ultraviolet catastrophe", which historically was one of the strong indications of the shortcomings of classical theory, leading Planck to the introduction of the quanta. Note that Planck suggested his formula (4.120) as the simplest interpolation between (4.122) and the experimentally discovered law (4.123).

[13] For the cosmological "relict" radiation this maximum corresponds to $T \approx 3\,\mathrm{K}$.

For the total energy of the radiation:

$$E = F + TS = \frac{4\sigma}{c} VT^4 = -3F \tag{4.130}$$

which is Boltzmann's law. For the specific heat of the photon gas:

$$C_v = \left(\frac{\partial E}{\partial T}\right)_V = \frac{16\sigma}{c} T^3 \sim T^3. \tag{4.131}$$

The radiation pressure[14] is:

$$P = -\left(\frac{\partial F}{\partial V}\right)_T = \frac{4\sigma}{3c} T^4 \tag{4.132}$$

so that the "equation of state" is:

$$PV = \frac{E}{3} \tag{4.133}$$

characteristic for an (ultra) relativistic gas with $\omega = ck$. The total (average) number of photons at a given temperature is given by:

$$N = \frac{V}{\pi^2 c^3} \int_0^\infty d\omega \frac{\omega^2}{e^{\frac{\hbar\omega}{T}} - 1} = \frac{VT^3}{\pi^2 c^3 \hbar^3} \int_0^\infty dx \frac{x^2}{e^x - 1} \approx 0.244 \left(\frac{T}{\hbar c}\right)^3 V. \tag{4.134}$$

[14] This pressure is very low at normal conditions, but may become enormous for high enough temperature, e.g. in astrophysics. Actually, speaking about "practical" applications of this theoretical expression, we note that the radiation pressure of a photon gas is one of the main driving forces in thermonuclear weapons.

Chapter 5
Condensed matter

5.1 Solid state at low temperature

In crystalline solids atoms oscillate around equilibrium positions, which for a regular crystal lattice. At low temperatures these oscillations are small and can be considered harmonic. Similar situations are characteristic for amorphous solids, where equilibrium positions are disordered in space[1].

Let N denote the number of molecules (atoms) forming a solid, and ν the number of atoms per molecule ($\nu = 1$ if a solid consists of atoms). Then the total number of atoms is equal to $N\nu$. Of total $3N\nu$ degrees of freedom, three correspond to translational and another three to rotational motions of the solid as a whole. The rest $3N\nu - 6$ degrees of freedom correspond to oscillations. Taking into account that $3N\nu$ is an enormous number, we can safely neglect 6 of them and assume that the number of vibrational degrees of freedom is given by $3N\nu$.

Below, we do not take into account electronic degrees of freedom, so that our presentation is related, strictly speaking, only to dielectric solids. In the simplest approximation we can assume that in metals electrons just additively contribute to all thermodynamic quantities.

From a mechanical point of view the system with $3N\nu$ vibrational degrees of freedom can be considered as the set of $3N\nu$ independent oscillators, each corresponding (in harmonic approximation) to a separate *normal oscillator* [11]. From quantum mechanics it is known [5] that the energy of the harmonic oscillator is given by:

$$\varepsilon_n = \hbar\omega\left(n + \frac{1}{2}\right), \tag{5.1}$$

where $\hbar\omega$ is the quantum of oscillation, $n = 0, 1, 2 \ldots$ is the oscillator quantum number. Then the statistical sum of a single oscillator is determined as:

$$Z_{osc} = \sum_{n=0}^{\infty} e^{-\frac{\hbar\omega}{T}(n+1/2)}. \tag{5.2}$$

Let us place the zero of energy at the lowest ($n = 0$) oscillator level, i.e. include the zero-point oscillator energy into a constant ε_0, defining the origin of an energy scale.

[1] Most of the material in this Chapter is based on the presentation of [1, 2].

Then:

$$Z_{\rm osc} = \sum_{n=0}^{\infty} e^{-\frac{\hbar\omega}{T}n} = \frac{1}{1 - e^{-\frac{\hbar\omega}{T}}}, \tag{5.3}$$

and the corresponding free energy of a single oscillator is given by:

$$F_{\rm osc} = T \ln\left(1 - e^{-\frac{\hbar\omega}{T}}\right). \tag{5.4}$$

Then the free energy of a solid can be written as:

$$F = N\varepsilon_0 + T\sum_{\alpha} \ln\left(1 - e^{-\frac{\hbar\omega_\alpha}{T}}\right), \tag{5.5}$$

where the summation is performed over all $3N\nu$ normal oscillators, which are numbered by index α. Here, $N\varepsilon_0$ is the energy of the zero-point oscillations, obviously proportional to the number of the molecules in the solid, while ε_0 is the zero energy of a molecule at $T = 0$.

Consider the limit of low temperatures. At small T, in the sum over α only terms with small $\hbar\omega \sim T$ are relevant. Small frequency vibrations in solids are the usual sound waves. The wavelength of a sound wave is given by $\lambda = u/\omega$, where u is speed of sound. This wavelength is large, compared with lattice constant of a typical crystal (or the average interatomic distance in an amorphous solid): $\lambda \gg a$. The corresponding frequencies $\omega \ll u/a$. To consider the relevant vibrations as sound waves, we have to restrict the temperatures to:

$$T \ll \hbar\frac{u}{a}. \tag{5.6}$$

Let us assume that our solid is isotropic (this is always valid for amorphous solids). In this case, we have to deal with either longitudinal (with velocity u_l) or transversal (with velocity u_t) sound waves, as both can propagate in such a solid. Their frequencies are given by:

$$\omega = u_l k \quad \text{and} \quad \omega = u_t k, \tag{5.7}$$

where $k = |\mathbf{k}|$ is the absolute value of the wave vector.

The number of vibrational modes corresponding to sound waves with absolute value of the wave vector in the interval from k to $k + dk$ and with fixed polarization is given by:

$$V\frac{4\pi k^2 dk}{(2\pi)^3}. \tag{5.8}$$

For longitudinal polarization we have $k = \omega/u_l$, while for the transversal polarization $k = \omega/u_t$, so that in the frequency interval from ω to $\omega + d\omega$ we have the following number of vibrational modes:

$$V\frac{\omega^2 d\omega}{2\pi^2}\left(\frac{1}{u_l^3} + \frac{2}{u_t^3}\right). \tag{5.9}$$

Let us introduce the average speed of sound u via the following relation:

$$\frac{3}{u^3} = \frac{2}{u_t^3} + \frac{1}{u_l^3}. \tag{5.10}$$

Then Equation (5.9) can be written as:

$$V\frac{3\omega^2 d\omega}{2\pi^2 u^3}. \tag{5.11}$$

In this form Equation (5.11) is applicable not only to an amorphous solid, but also to crystals, if we assume that u is a certain average speed of sound in a crystal of given symmetry. Then, using (5.11) we can transform the summation over α in Equation (5.5) into an integration over ω and obtain:

$$F = N\varepsilon_0 + \frac{3VT}{2\pi^2 u^3}\int_0^\infty d\omega\omega^2 \ln\left(1 - e^{-\frac{\hbar\omega}{T}}\right), \tag{5.12}$$

where the integration can be extended to infinity due to fast convergence of the integral at small T. Dropping the contribution $N\varepsilon_0$, we can see that the rest of this expression differs from Equation (4.125) for the free energy of a photon gas only by the replacement of the light speed c by the speed of sound and the factor of 3/2, related to the three polarizations of the sound waves, as opposed to the two polarizations of photons. Now we can conclude that the thermodynamics of a solid is determined by the quanta of sound waves (lattice vibrations), which we shall call *phonons*. Here for the first time we meet the situation where the theoretical description of a many – particle system of (interacting!) atoms (molecules) is reduced to a model of an ideal (noninteracting!) gas of *quasiparticles*.

Now we can just use the expressions obtained above for a photon gas with similar replacements. However, we shall repeat the explicit derivation. We can once again introduce the dimensionless variable $x = \hbar\omega/T$ and perform partial integration in (5.12) to get:

$$F = N\varepsilon_0 - V\frac{T^4}{2\pi^2\hbar^3 u^3}\int_0^\infty dx\frac{x^3}{e^x - 1} = N\varepsilon_0 - V\frac{\pi^2 T^4}{30(\hbar u)^3}. \tag{5.13}$$

The entropy of the system is given by:

$$S = -\frac{\partial F}{\partial T} = V\frac{2\pi^2 T^3}{15(\hbar u)^3}. \tag{5.14}$$

and the energy $E = F + TS$ is:

$$E = N\varepsilon_0 + V\frac{\pi^2 T^4}{10(\hbar u)^3}. \tag{5.15}$$

The specific heat of a solid in this approximation (low temperatures!), is equal to:

$$C = \left(\frac{\partial E}{\partial T}\right) = \frac{2\pi^2}{5(\hbar u)^3}VT^3 \sim T^3. \tag{5.16}$$

Here we can neglect any difference between C_p and C_v as their difference at low temperatures $C_p - C_v \sim T^7$, i.e. is much smaller than the specific heat itself [1, 2].

5.2 Solid state at high temperature

Let us consider now the opposite limit of high temperatures $T \gg \hbar u/a$. In this case we can write:

$$1 - e^{-\frac{\hbar\omega_\alpha}{T}} \approx \frac{\hbar\omega_\alpha}{T} \tag{5.17}$$

so that from Equation (5.5) we obtain:

$$F = N\varepsilon_0 + T\sum_\alpha \ln\frac{\hbar\omega_\alpha}{T} = N\varepsilon_0 - 3N\nu T\ln T + 3N\nu T\ln\hbar\langle\omega\rangle, \tag{5.18}$$

where we have introduced the mean logarithmic frequency of vibrations (phonons) $\langle\omega\rangle$ as:

$$\ln\langle\omega\rangle = \frac{1}{3N\nu}\sum_\alpha \ln\omega_\alpha. \tag{5.19}$$

From Equation (5.18) we find the energy $E = F - T\frac{\partial F}{\partial T}$:

$$E = N\varepsilon_0 + 3N\nu T. \tag{5.20}$$

The case of high temperatures corresponds to the classical analysis of atomic vibrations and Equation (5.20) corresponds to the equipartition theorem – each of the $3N\nu$ vibrational degrees of freedom contributes to the energy T. The specific heat now is given by:

$$C = Nc = 3N\nu, \tag{5.21}$$

where $c = 3\nu$ is the specific heat per one molecule[2]. Thus, at high enough temperatures the specific heat of solids is a constant, independent of temperature and dependent only on the number of atoms. In particular, for all elements ($\nu = 1$) the atomic

[2] Again, here we simply write C, as for solids the difference of C_p and C_v is negligible [1, 2].

high temperature specific heat is the same and equal to 3 (or $3k_B$ in usual units) – Dulong–Petit's law. At normal temperatures this law agrees well with experiments[3].

Using (5.21) we can write free energy as:

$$F = N\varepsilon_0 - NcT \ln T + NcT \ln \hbar \langle \omega \rangle \tag{5.22}$$

$$E = N\varepsilon_0 + NcT. \tag{5.23}$$

Then the entropy of a solid is:

$$S = -\frac{\partial F}{\partial T} = Nc \ln T - Nc \ln \frac{\hbar \langle \omega \rangle}{e}. \tag{5.24}$$

Clearly Equation (5.18) can be directly derived using classical statistics starting from the general expression for the free energy:

$$F = -T \ln \int d\Gamma e^{-\frac{E(p,q)}{T}}. \tag{5.25}$$

Substituting here the oscillator energy:

$$E(p,q) = \frac{1}{2} \sum_\alpha (p_\alpha^2 + \omega_\alpha^2 q_\alpha^2) \tag{5.26}$$

and taking into account $d\Gamma = \frac{1}{(2\pi\hbar)^{3N\nu}} \prod_\alpha dp_\alpha dq_\alpha$, we can see that integral here is factorized into a product of $3N\nu$ identical integrals of the following form:

$$\int_{-\infty}^{\infty} dp_\alpha \int_{-\infty}^{\infty} dq_\alpha \exp\left(-\frac{p_\alpha^2 + \omega_\alpha^2 q_\alpha^2}{2T}\right) = \frac{2\pi T}{\omega_\alpha} \tag{5.27}$$

so that finally we obtain (5.18). Note that the limits of integration here can be extended to infinity due to fast convergence of the integrals, though in reality atoms perform only small oscillations around lattice sites. Accordingly, all areas of integration correspond in fact to physically different microscopic states and there is no need to introduce an additional factor of $N!$ in the denominator of the phase volume.

5.3 Debye theory

Debye proposed a simple, but very effective, interpolation for the specific heat which can be used for arbitrary temperatures. Let us consider a model of a solid, where all vibrational frequencies are distributed according to Equation (5.11), though in reality this expression is valid only for small (sound) frequencies. In fact, the phonon spectrum should be limited from above as the vibrational frequency in a solid cannot

[3] For composite compounds ($\nu > 1$) Dulong–Petit's limit is practically never achieved due melting or chemical decomposition at rather low temperatures.

be larger than some maximal frequency, which can be determined from the condition that the total number of vibrations must be equal to the total number of vibrational degrees of freedom $3N\nu$:

$$\frac{3V}{2\pi^2u^3}\int_0^{\omega_D} d\omega\omega^2 = \frac{V\omega_D^3}{2\pi^2u^3} = 3N\nu. \tag{5.28}$$

Defined thusly, the *Debye frequency* ω_D is equal to[4]:

$$\omega_D = u\left(\frac{6\pi^2N\nu}{V}\right)^{1/3} \sim u/a. \tag{5.29}$$

Accordingly, the frequency distribution or phonon density of states in Debye model is given by:

$$\rho(\omega) = \begin{cases} 9N\nu\frac{\omega^2}{\omega_D^3} & \text{for } \omega \leq \omega_D \\ 0 & \text{for } \omega > \omega_D, \end{cases} \tag{5.30}$$

where we have expressed u via ω_D using (5.29).

Surely, since Debye's work, there has been an enormous progress of solid state physics and nowadays the real phonon density of states is directly measured e.g. by inelastic scattering of neutrons. However, at small frequencies it always reduces to the Debye (sound) dependence $\sim\omega^2$, though at higher frequencies it may become rather complicated (see e.g. Figure 5.1). The limiting frequency always exists, but Equation (5.29) defines it only by the order of magnitude. However, in most cases Debye model produces a rather satisfactory description of the specific heat of real solids. The Debye frequency is usually considered just as a fitting parameter, characterizing the concrete solid, to be determined from experiments.

Again replacing the summation in Equation (5.12) by frequency integration we obtain the free energy of a solid as:

$$F = N\varepsilon_0 + T\frac{9N\nu}{\omega_D^3}\int_0^{\omega_D} d\omega\omega^2 \ln\left(1 - e^{-\frac{\hbar\omega}{T}}\right). \tag{5.31}$$

Let us now introduce the *Debye temperature* as:

$$\theta_D = \hbar\omega_D. \tag{5.32}$$

Then:

$$F = N\varepsilon_0 + 9N\nu T\left(\frac{T}{\theta_D}\right)^3\int_0^{\theta_D/T} dzz^2 \ln\left(1 - e^{-z}\right), \tag{5.33}$$

[4] The existence of such a limiting frequency is crucial for phonon statistics and is the major difference with the statistics of photons. For photons there is no such maximal frequency – the electromagnetic field is the system with an infinite number of degrees of freedom and in Minkowski space-time no minimal length (similar to lattice constant a) exists (at least at the present level of our knowledge).

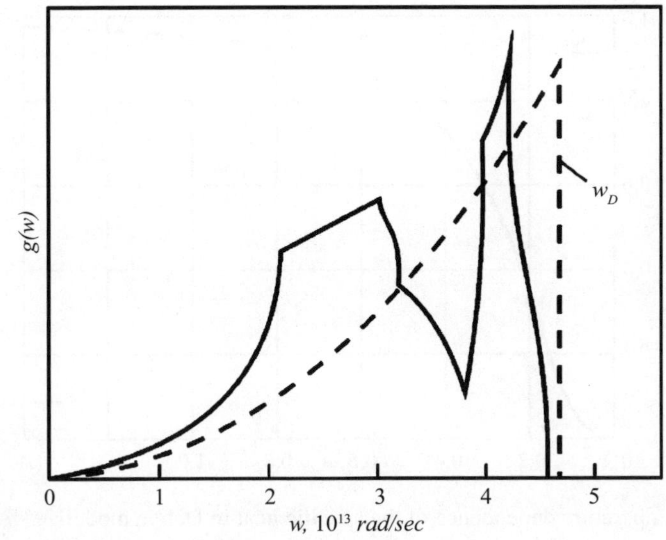

Figure 5.1. Phonon density of states in copper determined from neutron scattering experiments. The dashed line corresponds to the Debye model which is fixed by demanding equality of the areas under this line and the experimental density of states. Debye temperature $\theta_D = 340\,\mathrm{K}$.

where we have introduced the dimensionless variable $z = \frac{\hbar\omega}{T}$. Performing partial integration and introducing the Debye function:

$$D(x) = \frac{3}{x^3}\int_0^x dz \frac{z^3}{e^z - 1} \tag{5.34}$$

we can write (5.33) as:

$$F = N\varepsilon_0 + N\nu T\left\{3\ln\left(1 - e^{-\frac{\theta_D}{T}}\right) - D\left(\frac{\theta_D}{T}\right)\right\}. \tag{5.35}$$

Then the energy $E = F - T\frac{\partial F}{\partial T}$ is given by:

$$E = N\varepsilon_0 + 3N\nu T D\left(\frac{\theta_D}{T}\right) \tag{5.36}$$

and the specific heat is:

$$C = 3N\nu\left\{D\left(\frac{\theta_D}{T}\right) - \frac{\theta_D}{T}D'\left(\frac{\theta_D}{T}\right)\right\}. \tag{5.37}$$

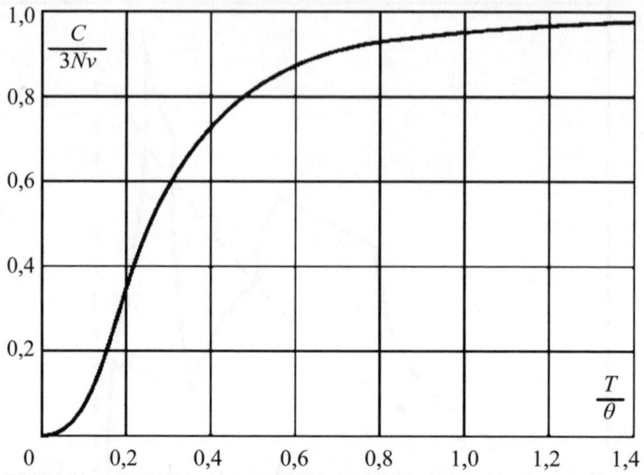

Figure 5.2. Temperature dependence of the specific heat in Debye model.

In Figure 5.2 we show the dependence of $\frac{C}{3N\nu}$ on $\frac{T}{\theta_D}$.

For $T \ll \theta_D$ we have $\frac{\theta_D}{T} \gg 1$, so we can replace the upper limit of integration by infinity and the integral is equal to $\frac{\pi^4}{15}$ and

$$D(x) \approx \frac{\pi^4}{5x^3}. \tag{5.38}$$

Then, from Equation (5.37) we get:

$$C \approx \frac{12N\nu\pi^4}{5}\left(\frac{T}{\theta_D}\right)^3 \tag{5.39}$$

which coincides with (5.16).

For $T \gg \theta_D$ we have $x \ll 1$ and in first approximation we can put $D(x) \approx 1$, so that Equation (5.37) gives $C = 3N\nu$, i.e. the Dulong–Petit law.

Note that the actual form of the Debye function $D(x)$ shows that the border between the different temperature limits is defined by the comparison of T and $\theta_D/4$ – the specific heat is approximately constant for $T \gg \theta_D/4$ and it behaves as $\sim T^3$ for $T \ll \theta_D/4$. In metals, for temperatures $T \ll \theta_D/4$ we can also observe the contribution linear in T to the specific heat from free electrons given by Equation (4.70), which is rather small and is rapidly "masked" by the lattice contribution at higher temperatures. To separate electronic and lattice contributions to the specific heat it is convenient to plot experimental data for the specific heat of metals at low temperatures as the dependence of the ratio C/T on T^2. In a metal we have at low temperatures $C = \gamma T + \beta T^3$, so that $\frac{C}{T} = \gamma + \beta T^2$, and the value of C/T at $T \to 0$ actually determines

Table 5.1. Debye temperatures for some real systems (K).

Pb	Na	KB	Ag	NaCl	Ga	Cu	Al	Mo	SiO_2	Si	LiF
105	158	180	225	280	320	340	430	450	470	645	732

the coefficient γ, which in fact gives us (according to Equation (4.70)) the value of the electron density of states at the Fermi level[5].

In Table 5.1 we present the values of Debye temperatures, determined experimentally for a number of real solids. Excluding special cases like diamond (where $\theta_D \sim 2000\,\text{K}$), Debye temperatures for the majority of solids are of the order of 10^2 K.

5.4 Quantum Bose liquid

In the general case, interaction between atoms (molecules) in liquids is strong, and calculations of thermodynamic characteristics becomes a very complicated task (as opposed to gases or solids, where interactions or atomic vibrations are small, allowing an analytical approach). However, theoretical analysis simplifies in the case of so called quantum liquids, which are close to the ground state at nearly zero temperatures. In reality, there is only one such liquid, which does not crystallize up to the absolute zero – that is liquid Helium. Most importantly, a quantum liquid is also formed by conduction electrons in metals. There are some other more exotic examples of quantum liquids, e.g. nuclear matter, neutron stars etc. Many properties of these systems are quite unusual, including such spectacular phenomena as superfluidity and superconductivity. The theory of quantum liquids is of prime importance and is one of the major areas of the modern theory of many-particle systems.

We have seen that calculation of the thermodynamic properties requires knowledge of the energy spectrum (levels) of the body. For the case of the system of strongly interacting particles, such as a quantum liquid, we have to deal with the energy levels of a liquid as a whole, not of separate atoms forming a liquid. At low temperatures, while calculating the partition function, it is sufficient to take into account only the lowest energy levels (excitations) just above the ground state, which leads to great simplifications.

The basic idea of Landau is that lowest energy levels of a quantum liquid can be reduced to some kind of elementary excitations or *quasiparticles*, with a well defined energy spectrum. In a spatially homogeneous (translationally invariant) liquid these excitations can be characterized by momentum (or quasi-momentum in a crystal). Until the number of quasiparticles is low enough (at low temperatures) we can neglect

[5] Note that in amorphous (insulating) glasses a specific heat linear in T is also sometimes observed, due to the contribution of so called tunneling states (two-level systems). However, we shall not discuss this here as this material is outside the scope of our presentation.

their interactions and assume that, in first approximation, these excitations form an ideal gas[6].

One of the possible types of energy spectrum of weak excitations of a quantum liquid is the Bose-like spectrum, where elementary excitations can appear and disappear one by one. The angular momentum of any quantum system (in our case quantum liquid) can change only by integer multiples of $\hbar$. Thus, elementary excitations appearing one by one, necessarily can possess only an integer angular momentum (spin) and obey Bose statistics. The liquid, consisting of Bose particles must have an energy spectrum of this kind. A typical example is liquid He^4 (while He^3 forms a Fermi liquid).

[6] Let us stress that the concept of quasiparticles is quite nontrivial. Its final justification appeared only within the modern theory of many-particle systems, based on Green's functions and the quantum field theory approach (see Chapter 11 below). Only within this approach we can derive precise criteria for the existence of quasiparticles in concrete systems. In some cases (e.g. in so called strongly correlated systems) the quasiparticle concept breaks down and a much more complicated description is required.

The major characteristic of quasiparticles is dispersion (spectrum), i.e. the dependence of their energy on momentum. In a Bose liquid, elementary excitations with small momenta **p** (large wavelengths $\hbar/p$) correspond to the usual sound waves with dispersion:

$$\varepsilon(p) = up, \tag{5.40}$$

where u is the speed of sound. These excitations (quasiparticles) are called phonons. The knowledge of the spectrum $\varepsilon(p)$ at small p allows us to calculate the thermodynamic characteristics of a liquid at very small temperatures T, when practically all elementary excitations are phonons. Appropriate expressions can be written immediately, using the results obtained above for the thermodynamics of a solid at low temperatures. The only difference is that instead of three independent polarizations (two transverse and one longitudinal) in a solid, we have only one (longitudinal) in a liquid, so that all expressions should be divided by 3. For example, for the free energy of a liquid from Equation (5.13) we obtain:

$$F = F_0 - V\frac{\pi^2 T^4}{90(\hbar u)^3}, \tag{5.41}$$

where F_0 is the free energy of a liquid at $T = 0$. The energy is given by:

$$E = E_0 + V\frac{\pi^2 T^4}{30(\hbar u)^3} \tag{5.42}$$

and the specific heat:

$$C = V\frac{2\pi^2 T^3}{15(\hbar u)^3} \sim T^3. \tag{5.43}$$

With the growth of quasiparticle momentum $\varepsilon(p)$ dependence deviates from a simple linear one, and its behavior becomes dependent on interactions. At large enough momenta $\varepsilon(p)$ dependence ceases to exist, as elementary excitations with large momenta are unstable toward decay into several excitations with smaller momenta.

After a thorough study of experimental data on liquid He^4 Landau has postulated the spectrum of elementary excitations for this system, as shown in Figure 5.3. We can see a characteristic minimum at $p = p_0$ and close to it $\varepsilon(p)$ can be written as:

$$\varepsilon(p) = \Delta + \frac{(p - p_0)^2}{2\tilde{\mu}}. \tag{5.44}$$

Quasiparticles from this part of the spectrum are usually called rotons[7]. Now this form of the spectrum is well confirmed by direct experiments on inelastic neutron

[7] This name is of purely historic origin. In early works Landau assumed the existence of two separate types of quasiparticles in He^4-phonons and rotons, i.e. the existence of two independent branches of the spectrum. Later it was discovered that there is a single branch of the spectrum with phonon and roton parts. Contrary to the initial opinion of Landau it was also discovered that this form of the spectrum is directly related to Bose condensation in He^4.

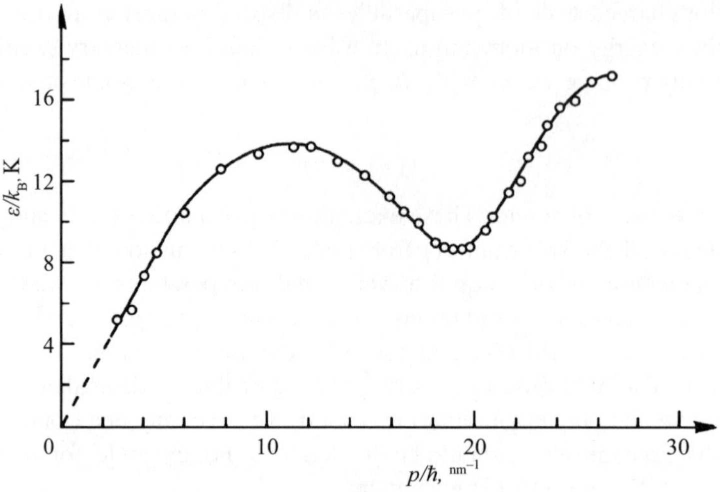

Figure 5.3. Spectrum of elementary excitations in liquid He^4. Points represent experimental data obtained from inelastic neutron scattering.

scattering. Experimental values of the constants for the roton part of the spectrum are:

$$\Delta = 8.5\,\mathrm{K}; \quad \frac{p_0}{\hbar} = 1.9\,10^8\,\mathrm{cm}^{-1}; \quad \tilde{\mu} = 0.16 m_{\mathrm{He}}. \tag{5.45}$$

Note that $p_0 \sim \hbar a^{-1}$, where a is an average interatomic distance in liquid.

As roton energy has a "gap" Δ, at low enough temperatures $T < \Delta$ we are dealing with a dilute gas of rotons, which can be described by Boltzmann statistics. Then, to calculate the "roton" part of the free energy of He^4 we can use Equation (3.41). Substituting $\varepsilon(p)$, which is independent of the coordinates, we immediately obtain:

$$F = -NT \ln \left[\frac{eV}{N} \int \frac{d^3 p}{(2\pi\hbar)^3} e^{-\frac{\varepsilon(p)}{T}} \right]. \tag{5.46}$$

The number of particles N in a roton gas is not fixed and is determined from the requirement for the minimum of F. From the condition $\frac{\partial F}{\partial N} = \mu = 0$ we find the number of rotons as:

$$N_r = \frac{V}{(2\pi\hbar)^3} \int d^3 p e^{-\frac{\varepsilon(p)}{T}}, \tag{5.47}$$

where in the integrand we have just the Boltzmann distribution with $\mu = 0$. Substituting $N = N_r$ from Equation (5.47) into Equation (5.46), we get:

$$F_r = -N_r T \ln e = -TN_r = -\frac{VT}{(2\pi\hbar)^3} \int d^3 p e^{-\frac{\varepsilon(p)}{T}}. \tag{5.48}$$

Taking into account the explicit form of the roton spectrum (5.44) in Equations (5.47) and (5.48), due to $p_0^2 \gg \tilde{\mu}T$ we may take $p^2 \approx p_0^2$ outside integral and perform integration with infinite limits. Then we obtain:

$$N_r = \frac{2(\tilde{\mu}T)^{1/2} p_0^2 V}{(2\pi)^{3/2}\hbar^3} e^{-\frac{\Delta}{T}}; \quad F_r = -TN_r. \tag{5.49}$$

Accordingly, the contribution of the rotons to the entropy and specific heat is:

$$S_r = N_r\left(\frac{3}{2} + \frac{\Delta}{T}\right) \quad C_r = N_r\left[\frac{3}{4} + \frac{\Delta}{T} + \left(\frac{\Delta}{T}\right)^2\right] \tag{5.50}$$

so that the temperature dependence is exponential and for $T < \Delta$ these contributions are small in comparison to the phonon parts determined above. For $T > \Delta$ roton contributions to thermodynamic values may overcome those from phonons. This actually takes place with the rise of temperature.

5.5 Superfluidity

Liquid Helium at temperatures below the so called λ-point $T_\lambda = 2.18\,\text{K}$, acquires the remarkable property of superfluidity – the liquid flows through narrow tubes and channels without friction (viscosity). Superfluidity was discovered by Kapitza in 1938, its theoretical interpretation was given few years later by Landau.

Consider first the case of $T = 0$. Assume that the liquid flows in a tube with a constant velocity $\mathbf{v}$. In the presence of viscosity, friction of the liquid and the tube walls, as well as within the liquid itself, will induce different processes of dissipation of kinetic energy of the flow, so that the flow slows down and finally stops. It is convenient to consider the liquid in a coordinate system moving together with the flow. In this system Helium is at rest in the ground state, while the tube walls move with velocity $(-\mathbf{v})$. In the presence of viscosity (friction), Helium initially at rest should start to move. From a microscopic point of view it is clear that the appearance of this motion should start from some excitation of internal motions within the liquid, i.e. from the appearance of some elementary excitations (quasiparticles).

Consider the situation with the appearance of only one elementary excitation with momentum $\mathbf{p}$ and energy $\varepsilon(p)$. Then, the energy of the liquid E_0 becomes equal to the energy of this excitation $\varepsilon(p)$, while its momentum $\mathbf{P}_0$ becomes equal to $\mathbf{p}$. Let us return to the laboratory coordinate system in which the tube is at rest. Using the well-known Galilean transformations of classical mechanics [11] we obtain for the energy E and momentum $\mathbf{P}$ of the liquid in the laboratory system:

$$\mathbf{P} = \mathbf{P}_0 + M\mathbf{v} \quad E = E_0 + \mathbf{P}_0\mathbf{v} + \frac{Mv^2}{2}, \tag{5.51}$$

where M is the mass of the moving liquid. Substituting now the values $\varepsilon(p)$ and $\mathbf{p}$ for E_0 and $\mathbf{P}_0$ we have:

$$E = \varepsilon(p) + \mathbf{pv} + \frac{Mv^2}{2}. \tag{5.52}$$

The term $\frac{1}{2}Mv^2$ here represents the initial kinetic energy of the liquid flow, while $\varepsilon(p) + \mathbf{pv}$ is now the *change of the liquid energy due to the appearance of a single elementary excitation*. This change should be negative, to diminish the flow energy:

$$\varepsilon(p) + \mathbf{pv} < 0. \tag{5.53}$$

For a given value of $\mathbf{p}$ the left-hand side of Equation (5.53) is minimal for antiparallel $\mathbf{p}$ and $\mathbf{v}$, thus in any case we should have $\varepsilon(p) - pv < 0$, so that:

$$v > \frac{\varepsilon(p)}{p}. \tag{5.54}$$

This inequality is to be satisfied at least for some values of the momentum p of elementary excitation. Thus, to find the final condition for the appearance of elementary excitations in a moving liquid we have to find the *minimum* of $\varepsilon(p)/p$:

$$v_c = \mathrm{Min}\,\frac{\varepsilon(p)}{p}. \tag{5.55}$$

Geometrically, the ratio $\varepsilon(p)/p$ is determined by the slope of a straight line drawn from the origin of the coordinate system in the (ε, p) – plane to some point of the curve $\varepsilon = \varepsilon(p)$. Its minimal value is determined by the point where this line is tangent to the $\varepsilon(p)$ curve. If this minimum is nonzero, then for velocities of the liquid below v_c, determined by Equation (5.55), *no elementary excitations can appear*, so that the flow will be dissipationless (no friction!). This is precisely the case of superfluidity and Equation (5.55) represents Landau's criterion of superfluidity. The value of v_c is called the critical velocity, its existence is confirmed by experiments.

It is easy to see that Landau's spectrum of elementary excitations for He^4 satisfies the criterion of superfluidity. Similarly, this criterion is satisfied by the energy spectrum with a "gap" at $p = 0$. At the same time, the free particle spectrum $\varepsilon(p) = p^2/2m$ obviously does not satisfy this criterion. In essence, it is necessary for the curve $\varepsilon(p)$ not to be tangent to the abscissa at the origin. Thus, superfluidity will appear for almost any spectrum with phonon – like behavior in the long wavelength limit or gap at $p = 0$.

For finite temperatures $T > 0$ the liquid is not in the ground state and there are a number of elementary excitations present. Kinematic arguments given above are still valid, and the motion of the liquid through the tube with velocities satisfying Landau's criterion still does not produce additional excitations. However, we have to clarify the role of the excitations already present due to finite temperatures.

Consider the gas of quasiparticles moving as a whole, relative to the liquid, with velocity $\mathbf{v}$. The distribution function for the gas moving as a whole is obtained from the distribution function $n(\varepsilon)$ at rest by the substitution $\varepsilon \to \varepsilon - \mathbf{pv}$, where $\mathbf{p}$ is the momentum of a quasiparticle[8]. Then, the total momentum of the unit volume of the gas is given by:

$$\mathbf{P} = \int \frac{d^3p}{(2\pi\hbar)^3}\mathbf{p}n(\varepsilon - \mathbf{pv}). \tag{5.56}$$

Let the velocity $\mathbf{v}$ be small, so that we can expand the integrand in powers of $\mathbf{pv} = pv\cos\theta$. The zeroth - order term disappears after integration over the directions of the vector $\mathbf{p}$ (θ angle) and we can write:

$$\mathbf{P} = -\int \frac{d^3p}{(2\pi\hbar)^3}\mathbf{p}(\mathbf{pv})\frac{dn(\varepsilon)}{d\varepsilon}. \tag{5.57}$$

Integrating here again over the directions of the vector $\mathbf{p}$ we get:

$$\mathbf{P} = -\mathbf{v}\frac{4\pi}{3}\frac{1}{(2\pi\hbar)^3}\int dpp^4\frac{dn(\varepsilon)}{d\varepsilon}. \tag{5.58}$$

Substituting here the spectrum of phonons $\varepsilon = up$ and integrating by parts we have:

$$\mathbf{P} = -\mathbf{v}\frac{1}{(2\pi\hbar)^3}\frac{4\pi}{3u}\int_0^\infty dpp^4\frac{dn(p)}{dp} = \mathbf{v}\frac{16\pi}{3u}\frac{1}{(2\pi\hbar)^3}\int_0^\infty dpp^3n(p). \tag{5.59}$$

Here, the integral

$$\frac{1}{(2\pi\hbar)^3}\int_0^\infty dp4\pi p^2upn(p) - \int\frac{d^3p}{(2\pi\hbar)^3}\varepsilon n(\varepsilon) \tag{5.60}$$

reduces to the energy E_{ph} of the unit volume of phonon gas, so that:

$$\mathbf{P} = \mathbf{v}\frac{4E_{ph}}{3u^2}. \tag{5.61}$$

The coefficient before $\mathbf{v}$ here defines the mass density of the liquid transported by the flow of the quasiparticle gas. Nothing can prevent these moving quasiparticles from being scattered by the walls of the tube and exchange momenta as in the usual gas flow. It is clear that this part of the liquid mass will behave as a normal liquid moving

[8] Consider a gas of excitations with respect to the liquid with velocity $\mathbf{v}$. In the coordinate system where the gas is at rest the liquid moves with velocity $-\mathbf{v}$. Then the energy E of liquid in these coordinates is connected with the energy E_0 in the coordinate system where the liquid is at rest, by: $E = E_0 - \mathbf{P}_0\mathbf{v} + \frac{Mv^2}{2}$. Consider an excitation with energy $\varepsilon(p)$, appearing in the rest system of the liquid. Then the additional energy in the rest system of the gas will be given by $\varepsilon - \mathbf{pv}$, which proves our statement.

with friction. However, this is not the whole mass of the liquid, the rest behaves as a superfluid! In fact, after we substitute the expression (5.42) for the energy of the phonon gas into Equation (5.61), we obtain for the phonon part of the *normal* density ρ_n:

$$(\rho_n)_{ph} = \frac{2\pi^2 T^4}{45\hbar^3 u^5}, \tag{5.62}$$

which goes to zero for $T \to 0$, when the whole mass of the liquid becomes superfluid. Now we can say that the total density ρ of He^4 at $T > 0$ consists of normal and superfluid parts (components): $\rho = \rho_n + \rho_s$, though certainly it does not mean that there is any kind of real separation of the liquid into two components. It is important to note that there is no momentum transfer (friction!) between these two parts of the liquid: we have obtained this physical picture at the state of statistical equilibrium in a gas moving with fixed velocity. But any motion in the state of thermodynamic equilibrium is in fact dissipationless.

Above we determined the phonon contribution to ρ_n, to find the roton part we note that rotons can be described by the Boltzmann statistics, so that $\frac{\partial n}{\partial \varepsilon} = -\frac{n}{T}$ and from Equation (5.58) we get:

$$\begin{aligned}(\rho_n)_r &= \frac{4\pi}{3T(2\pi\hbar)^3}\int dpp^4 n(p) = \frac{1}{3T}\int \frac{d^3p}{(2\pi\hbar)^3} p^2 n(p) \\ &\approx \frac{p_0^2}{3T}\frac{N_r}{V} = \frac{2\tilde{\mu}^{1/2} p_0^4}{3(2\pi)^{3/2} T^{1/2}\hbar^3} e^{-\frac{\Delta}{T}},\end{aligned} \tag{5.63}$$

where p_0 is the momentum corresponding to the roton minimum. If we take into account the experimental values for the parameters determining the spectrum of excitations in He^4, it turns out that the roton contribution to ρ_n matches the phonon part at $T \sim 0.6\,\mathrm{K}$ and overcomes it at higher temperatures.

As the temperature T rises, more and more of the liquid becomes normal and $\rho_n \to \rho$ (where ρ is the total density of He^4) for $T \to T_\lambda$ from below. The superfluid density $\rho_s \to 0$ for $T \to T_\lambda$ and $\rho_s = 0$ for $T > T_\lambda$. The value of ρ_n close to the λ-point cannot be calculated precisely, but an approximate estimate for T_λ can be obtained from the condition $(\rho_n)_r \approx \rho$. Using here Equation (5.63) we can obtain $T_\lambda \approx 2.8\,\mathrm{K}$, which is in relatively good agreement with experiments.

Superfluid transition in He^4 is a typical second order phase transition. Such a transition always goes together with the appearance (or disappearance) of some qualitative property (long-range order!). In case of the λ-transition in He^4, this is the appearance (disappearance) of the superfluid component of the liquid. From a microscopic point of view, we can speak about certain properties of the *single-particle* density matrix of our system:

$$\rho(\mathbf{r}, \mathbf{r}') = \int dq \Psi^\star(\mathbf{r}, q)\Psi(\mathbf{r}', q), \tag{5.64}$$

where $\Psi(\mathbf{r}, q)$ is the wave function of the system as a whole, where $\mathbf{r}$ are coordinates of a single particle, while q is the set of all coordinates of the other particles, which are integrated out. For an isotropic medium (liquid) this density matrix depends only on $|\mathbf{r} - \mathbf{r}'|$. In a normal (non-superfluid) state $\rho(\mathbf{r}, \mathbf{r}') \to 0$ for $|\mathbf{r} - \mathbf{r}'| \to \infty$. This is not so in superfluid phase.

Consider the Fourier components of the density matrix:

$$\int d^3(\mathbf{r} - \mathbf{r}') e^{i\mathbf{k}(\mathbf{r}-\mathbf{r}')} \rho(\mathbf{r}, \mathbf{r}'), \tag{5.65}$$

which, up to a constant, coincide with:

$$\int dq \left| \int dV e^{i\mathbf{k}\mathbf{r}} \Psi(\mathbf{r}, q) \right|^2 \tag{5.66}$$

i.e. determine the probability distribution of different values of momentum of a particle $\mathbf{p} = \hbar\mathbf{k}$. If $\rho(\mathbf{r}, \mathbf{r}') \to 0$ for $|\mathbf{r} - \mathbf{r}'| \to \infty$, the probability density in $\mathbf{p}$-space for $\mathbf{p} \to 0$ remains finite. However, if $\rho(\mathbf{r}, \mathbf{r}')$ tends to a finite value $\rho_\infty > 0$ at infinity[9], the integral in (5.65) is equal to $(2\pi)^3 \delta(\mathbf{k}) \rho_\infty$. This corresponds to a finite probability for a particle to have zero momentum. Thus, in the superfluid state (opposite to the case of normal liquid) a *finite* number (fraction) of particles possess zero momentum. This clearly relates superfluidity to Bose condensation. Let us stress that this set of particles should not be identified with the superfluid component of the liquid, discussed above. This obviously will be wrong, as at $T = 0$ all the mass of the liquid is superfluid, though not all particles of the *interacting* system possess zero momentum (cf. below the case of weakly interacting Bose gas).

5.6 Phonons in a Bose liquid*

Let us consider in more detail the origin of the spectrum of elementary excitations of liquid He^4, shown in Figure 5.3. The energy of the liquid can be written as a functional of its (mass) density and hydrodynamic velocity:

$$E[\rho(\mathbf{r}), \mathbf{v}(\mathbf{r})] = \frac{1}{2} \int d\mathbf{r} \rho(\mathbf{r}) \mathbf{v}^2(\mathbf{r}) + E^{(1)}[\rho(\mathbf{r})], \tag{5.67}$$

where $E^{(1)}$ is the part of energy independent of velocity. Consider small oscillations of the density:

$$\rho(\mathbf{r}) = \rho + \delta\rho(\mathbf{r}), \tag{5.68}$$

where ρ is liquid density at equilibrium, while $\delta\rho(\mathbf{r})$ and $\mathbf{v}(\mathbf{r})$ are small deviations, describing oscillations. By definition:

$$\rho = \frac{1}{V} \int d\mathbf{r} \rho(\mathbf{r}) \qquad \int d\mathbf{r} \delta\rho(\mathbf{r}) = 0. \tag{5.69}$$

[9] This is called off-diagonal long-range order (ODLRO).

Limiting ourselves to terms of second order in $\delta\rho$ and v we can replace $\rho(\mathbf{r})$ in the first term in Equation (5.67) by its average value ρ. With the same accuracy $E^{(1)}$ is written as:

$$E^{(1)}[\rho(\mathbf{r})] = E^{(1)}(\rho) + \int d\mathbf{r}\psi(\mathbf{r})\delta\rho(\mathbf{r}) + \frac{1}{2}\int d\mathbf{r}\int d\mathbf{r}'\varphi(\mathbf{r},\mathbf{r}')\delta\rho(\mathbf{r})\delta\rho(\mathbf{r}'). \tag{5.70}$$

The functions $\psi(\mathbf{r})$ and $\varphi(\mathbf{r},\mathbf{r}')$ are determined only by the properties of the unperturbed liquid, which is homogeneous and isotropic, so that $\psi(\mathbf{r}) = \psi = \text{const}$, while $\varphi(\mathbf{r},\mathbf{r}')$ depends only on the distance $|\mathbf{r}-\mathbf{r}'|$: $\varphi(\mathbf{r},\mathbf{r}') = \varphi(|\mathbf{r}-\mathbf{r}'|)$. Then the first order term in the expansion of $E^{(1)}$ given by Equation (5.70) is proportional to $\int dV\delta\rho(\mathbf{r}) = 0$, and finally we obtain:

$$E^{(1)}[\rho(\mathbf{r})] = E^{(1)}(\rho) + \frac{1}{2}\int d\mathbf{r}\int d\mathbf{r}'\varphi(|\mathbf{r}-\mathbf{r}'|)\delta\rho(\mathbf{r})\delta\rho(\mathbf{r}'). \tag{5.71}$$

The velocity $\mathbf{v}$ is related to the density oscillations via the continuity equation:

$$\dot{\rho} + div(\rho\mathbf{v}) = 0, \tag{5.72}$$

which can be written up to first order terms in $\delta\rho$ and $\mathbf{v}$ as:

$$\dot{\delta\rho} + \rho div\mathbf{v} = 0. \tag{5.73}$$

Performing Fourier transformation:

$$\delta\rho(\mathbf{r}) = \frac{1}{V}\sum_{\mathbf{p}}\rho_{\mathbf{p}}e^{i\mathbf{pr}} \quad \mathbf{v}(\mathbf{r}) = \frac{1}{V}\sum_{\mathbf{p}}\mathbf{v}_{\mathbf{p}}e^{i\mathbf{pr}}, \tag{5.74}$$

$$\varphi(\mathbf{r}) = \frac{1}{V}\sum_{\mathbf{p}}\varphi_{\mathbf{p}}e^{i\mathbf{pr}} \tag{5.75}$$

and taking into account the longitudinal nature of liquid oscillations, so that the velocity $\mathbf{v}_{\mathbf{p}}$ in a wave with wave vector $\mathbf{p}$ is directed along $\mathbf{p}$, we can write:

$$\mathbf{v}_{\mathbf{p}} = a_{\mathbf{p}}\mathbf{p}. \tag{5.76}$$

Substituting these expressions into the continuity equation we immediately obtain:

$$\mathbf{v}_{\mathbf{p}} = i\dot{\rho}_{\mathbf{p}}\frac{1}{\rho}\frac{\mathbf{p}}{p^2} \tag{5.77}$$

so that Equation (5.71) is rewritten as:

$$E = E^{(1)}(\rho) + \frac{1}{V}\sum_{\mathbf{p}}\left(\frac{|\dot{\rho}_{\mathbf{p}}|}{2\rho p^2} + \frac{1}{2}\varphi_{\mathbf{p}}|\rho_{\mathbf{p}}^2|\right). \tag{5.78}$$

The first term in Equation (5.78) represents the energy of the unperturbed liquid, the second one reduces to the sum of terms, each having the form of the energy of a harmonic oscillator with frequency ω_p:

$$\omega_p^2 = \rho p^2 \varphi_p, \tag{5.79}$$

where we have taken into account that in an isotropic liquid $\varphi_{\mathbf{p}} = \varphi_p$, i.e. depends only on the absolute value of $|\mathbf{p}|$. In the quantum case the energy of such an oscillator is[10]:

$$\varepsilon(p) = \omega_p \left(n + \frac{1}{2}\right) \quad n = 0, 1, 2 \ldots. \tag{5.80}$$

Thus, the spectrum of our system (liquid) is in fact determined by the spectrum of these oscillators, i.e. by relations (5.79) and (5.80).

To obtain the final solution we have to express φ_p via the characteristics of liquid. In the quantum case the ground state energy does not coincide with $E^{(1)}(\rho)$ (as in classics), we have take into account the zero point oscillator energy $\omega_p/2$. Then the ground state energy of the quantum Bose liquid becomes equal to:

$$E_0 = E^{(1)}(\rho) + \sum_p \frac{1}{2}\omega_p \tag{5.81}$$

or, taking into account Equation (5.78):

$$V\frac{\omega_p}{2} = \frac{1}{2\rho p^2}\langle|\dot{\rho}_p^2|\rangle + \frac{1}{2}\varphi_p\langle|\rho_p|^2\rangle = \varphi_p\langle|\rho_p|^2\rangle, \tag{5.82}$$

where the angular brackets denote averaging over the ground state, and we used the well known result that for a quantum oscillator the average (over the ground state) kinetic energy equals the average potential energy. Then, expressing φ_p in Equation (5.79) via (5.82), we obtain:

$$\varepsilon(p) = \omega_p = V\rho\frac{p^2}{2\langle|\rho_p|^2\rangle} \tag{5.83}$$

or

$$\varepsilon(p) = \frac{p^2}{2mS(p)}, \tag{5.84}$$

where we have introduced:

$$S(p) = \frac{\langle|\rho_p|^2\rangle}{Vm\rho} \tag{5.85}$$

[10] Here we use, for brevity, the system of units, often used by theorists, where $\hbar = 1$ and do not discern momentum and wave vector.

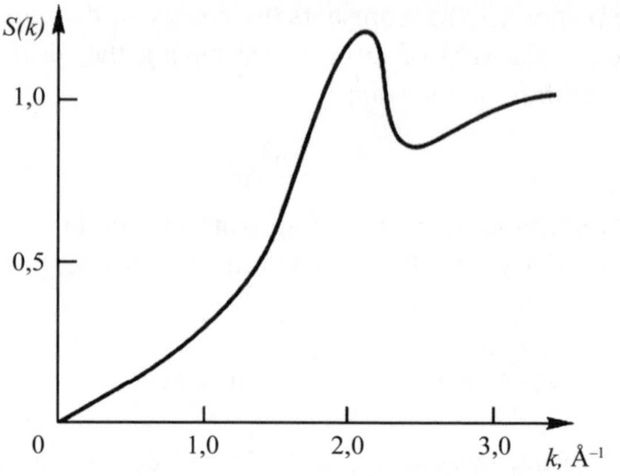

Figure 5.4. Characteristic form of the structure factor of liquid He^4.

– the so called structure factor of the liquid, which is determined by the Fourier transform of the density–density correlation function:

$$S(\mathbf{r}-\mathbf{r}') = \frac{1}{n}\langle[n(\mathbf{r})-n][n(\mathbf{r}')-n]\rangle, \tag{5.86}$$

where $n(\mathbf{r}) = \rho(\mathbf{r})/m$ is (volume) density of the particles at point $\mathbf{r}$, while n is the average density of the particles in the liquid.

Equation (5.84) was first derived by Feynman, and the derivation given above belongs to Pitaevskii. This formula expresses the excitation spectrum via the structure factor of the liquid. The value of $S(p)$ in in the general case can not be derived analytically, but in real liquids it is directly measured in neutron and X-ray scattering experiments.

For small momenta, the excitation spectrum of liquid He^4 is linear over the momentum: $\varepsilon(p) \approx up$, accordingly we have $S(p) \approx p/2mu$. For very large momenta much in excess of the inverse interatomic distance, $p \gg a^{-1}$, we have $S(p) = 1$, which corresponds to $S(r) = \delta(r)$ at small distances. In the intermediate region $p \sim a^{-1}$ the structure factor $S(p)$ is determined from experiments and for the majority of liquids (not only for He^4) it demonstrates the characteristic maximum at $p \sim a^{-1}$ (see Figure 5.4). The presence of this maximum is in fact related to the conservation of the rather strong correlations between atomic positions in the liquid (short range order).

From Feynman's expression (5.84) it becomes clear that for large momenta $p \gg a^{-1}$ the excitation spectrum reduces to the spectrum of free particles: $\varepsilon(p) = p^2/2m$. In the intermediate region of $p \sim a^{-1}$ the presence of the maximum in $S(p)$ leads to the appearance of a roton minimum.

Strictly speaking, this "hydrodynamic" derivation of Feynman's formula is valid only for momenta $p < 1/a$, i.e. when the liquid may be considered as a continuous medium. However, this expression also gives the correct answer for $p \gg 1/a$, i.e. in the free particle limit. It can be considered as a good interpolation also for the region where $p \sim 1/a$, giving a qualitative explanation of the Landau spectrum of He^4.

Note that the spectrum of density oscillations in usual (classical) liquids has a qualitatively similar form, but with a rather strong damping of the oscillations in the region of wave vectors $p \sim 1/a$. The existence of a "roton" minimum in classical liquids is also directly related to the maximum of the structure factor.

5.7 Degenerate interacting Bose gas

Let us consider now the system of interacting Bosons from a microscopic point of view. We shall limit ourselves to the analysis of a weakly interacting Bose gas, which can be described using the rather rigorous approach first proposed by Bogolyubov.

Consider the simple model of a Bose gas with point-like repulsion between the particles and limit ourselves to the case where $T = 0$. The Hamiltonian of the system in second quantization representation can be written as:

$$H = \sum_{\mathbf{p}} \frac{\mathbf{p}^2}{2m} a_{\mathbf{p}}^+ a_{\mathbf{p}} + \frac{v_0}{2V} \sum_{\mathbf{p}_1+\mathbf{p}_2=\mathbf{p}'_1+\mathbf{p}'_2} a_{\mathbf{p}'_1}^+ a_{\mathbf{p}'_2}^+ a_{\mathbf{p}_2} a_{\mathbf{p}_1}, \tag{5.87}$$

where $v_0 > 0$ is the coupling constant of the repulsive interaction and the creation and annihilation operators of Bosons satisfy the commutation relations:

$$a_{\mathbf{p}} a_{\mathbf{p}'}^+ - a_{\mathbf{p}'}^+ a_{\mathbf{p}} = \delta_{\mathbf{p}\mathbf{p}'} \tag{5.88}$$

$$a_{\mathbf{p}} a_{\mathbf{p}'} - a_{\mathbf{p}'} a_{\mathbf{p}} = 0 \quad a_{\mathbf{p}}^+ a_{\mathbf{p}'}^+ - a_{\mathbf{p}'}^+ a_{\mathbf{p}}^+ = 0.$$

In the ground state of an ideal Bose gas all particles are in a Bose *condensate*, i.e. in the state with zero momentum and energy. In terms of occupation numbers $N_{\mathbf{p}=0} = N_0 = N$, where N is the total number of particles in the gas. Accordingly $N_{\mathbf{p}\neq 0} = 0$. In a weakly interacting Bose gas in the ground state and also in weakly excited states $N_{\mathbf{p}\neq 0} \neq 0$, but these occupation numbers are very small compared to the *macroscopic* value of N_0. The fact that $a_0^+ a_0 = N_0 \approx N \gg 1$ means that the expression for the commutator of creation and annihilation operators of condensate particles $a_0 a_0^+ - a_0^+ a_0 = 1$ is small in comparison with a_0 and a_0^+, so that we can neglect unity in the right-hand side and consider these operators as the usual c-numbers[11]:

$$a_0 = a_0^+ = \sqrt{N_0}. \tag{5.89}$$

Then we can accurately separate in the Hamiltonian (5.87) all terms, containing condensate operators and replace them by (5.89). After that, we can build a kind of perturbation theory in powers of the "small" operators $a_{\mathbf{p}}, a_{\mathbf{p}}^+$ with $\mathbf{p} \neq 0$. The main contribution comes from scattering processes (interactions) of condensate particles and particles excited from the condensate (i.e. transitions of particles to and from the condensate), while scattering processes between particles excited "above" the condensate can be just neglected (in first approximation).

The zeroth-order term in the interaction Hamiltonian contains:

$$\frac{v_0}{2V} a_0^+ a_0^+ a_0 a_0 = \frac{v_0}{2V} a_0^4 = \frac{v_0}{2V} N_0^2. \tag{5.90}$$

The terms of first order in $a_{\mathbf{p}}, a_{\mathbf{p}}^+$ with $\mathbf{p} \neq 0$ are absent, as these can not satisfy conservation of momentum, as shown explicitly in Equation (5.87). The second order terms have the form:

$$\frac{v_0}{2V} a_0^2 \sum_{p>0} (a_{\mathbf{p}} a_{-\mathbf{p}} + a_{\mathbf{p}}^+ a_{-\mathbf{p}}^+ + 2a_{\mathbf{p}}^+ a_{\mathbf{p}} + 2a_{-\mathbf{p}}^+ a_{-\mathbf{p}}). \tag{5.91}$$

Limiting ourselves to second order terms we can here replace $a_0^2 = N_0$ by the total particle number N. However, in term (5.90) we have to take into account the more accurate relation:

$$a_0^2 + \sum_{p>0} a_{\mathbf{p}}^+ a_{\mathbf{p}} = N \tag{5.92}$$

and express N_0 via N and $\sum_{\mathbf{p}} a_{\mathbf{p}}^+ a_{\mathbf{p}}$. After explicit calculations, combining (5.90) and (5.91), we obtain:

$$\frac{N^2}{2V} v_0 + \frac{N}{V} v_0 \sum_{p>0} (a_{\mathbf{p}} a_{-\mathbf{p}} + a_{\mathbf{p}}^+ a_{-\mathbf{p}}^+ + a_{\mathbf{p}}^+ a_{\mathbf{p}} + a_{-\mathbf{p}}^+ a_{-\mathbf{p}}). \tag{5.93}$$

[11] More rigorously it is equivalent to an assumption that in the ground state the average values of these operators $\langle a_0 \rangle$ and $\langle a_0^+ \rangle$ are nonzero and equal to $\sqrt{N_0} e^{\pm i\phi}$ (where ϕ is an arbitrary phase of a complex number, which can be assumed here to be just a zero), i.e. there is a *finite* amplitude of creation and annihilation of particles in the condensate. Then the number of particles in the condensate is not conserved, in this sense the ground state of an interacting Bose gas breaks the particle conservation law. Thus, the symmetry of the ground state is *lower* than the symmetry of the Hamiltonian (5.87), which conserves the particle number. This is the first time when we meet the phenomenon of *spontaneous symmetry breaking* and the appearance of *anomalous* averages, breaking the symmetry of Hamiltonian.

Thus, we can rewrite the Hamiltonian (5.87) with the given accuracy as:

$$H = \frac{N^2}{2V} v_0 + \sum_{p>0} \left(\frac{N}{V} v_0 + \frac{\mathbf{p}^2}{2m} \right) (a_{\mathbf{p}}^+ a_{\mathbf{p}} + a_{-\mathbf{p}}^+ a_{-\mathbf{p}}) + \frac{N}{V} v_0 \sum_{p>0} (a_{\mathbf{p}} a_{-\mathbf{p}} + a_{\mathbf{p}}^+ a_{-\mathbf{p}}^+). \tag{5.94}$$

This Hamiltonian is quadratic in the operators $a_{\mathbf{p}}$ and $a_{\mathbf{p}}^+$ and can be *diagonalized* by the so called $u - v$-transformation, first introduced by Bogolyubov. Let us transform $\alpha_{\mathbf{p}}^+$ and $\alpha_{\mathbf{p}}$ to new creation and annihilation operators for Bosons, related to $a_{\mathbf{p}}^+$ and $a_{\mathbf{p}}$ by the linear transformation:

$$\begin{aligned} a_{\mathbf{p}} &= u_p \alpha_{\mathbf{p}} + v_p \alpha_{\mathbf{p}}^+ \\ a_{\mathbf{p}}^+ &= u_p \alpha_{\mathbf{p}}^+ + v_p \alpha_{\mathbf{p}}. \end{aligned} \tag{5.95}$$

New operators should satisfy the usual Bose commutation relations like (5.7), which is guaranteed if the coefficients u_p and v_p satisfy the condition:

$$u_p^2 - v_p^2 = 1. \tag{5.96}$$

Substituting $a_{\mathbf{p}}^+$ and $a_{\mathbf{p}}$ in the form of (5.95) into the Hamiltonian (5.94) we obtain:

$$\begin{aligned} H = &\sum_{p>0} \left\{ \left(\frac{\mathbf{p}^2}{2m} + \frac{N v_0}{V} \right) (u_p^2 + v_p^2) + 2 \frac{N v_0}{V} u_p v_p \right\} (\alpha_{\mathbf{p}}^+ \alpha_{\mathbf{p}} + \alpha_{-\mathbf{p}}^+ \alpha_{-\mathbf{p}}) \\ &+ \sum_{p>0} \left\{ \left(\frac{\mathbf{p}^2}{2m} + \frac{N v_0}{V} \right) 2 u_p v_p + \frac{N v_0}{V} (u_p^2 + v_p^2) \right\} (\alpha_{\mathbf{p}}^+ \alpha_{-\mathbf{p}}^+ + \alpha_{\mathbf{p}} \alpha_{-\mathbf{p}}) \\ &+ \sum_{p>0} \left\{ 2 \left(\frac{\mathbf{p}^2}{2m} + \frac{N v_0}{V} \right) v_p^2 + 2 \frac{N v_0}{V} u_p v_p \right\} + \frac{N^2 v_0}{2V}. \end{aligned} \tag{5.97}$$

To diagonalize this Hamiltonian we have to exclude terms like $\alpha_{\mathbf{p}}^+ \alpha_{-\mathbf{p}}^+$ and $\alpha_{\mathbf{p}} \alpha_{-\mathbf{p}}$, which can be achieved by the requirement:

$$\left(\frac{\mathbf{p}^2}{2m} + \frac{N v_0}{V} \right) 2 u_p v_p + \frac{N v_0}{V} (u_p^2 + v_p^2) = 0, \tag{5.98}$$

which gives the second relation fixing the coefficients u_p and v_p. Solving equations (5.96) and (5.98) we get:

$$u_p = \frac{1}{\sqrt{1 - A_p^2}} \quad v_p = \frac{A_p}{\sqrt{1 - A_p^2}}, \tag{5.99}$$

where

$$A_p = \frac{V}{Nv_0}\left\{\varepsilon(p) - \frac{p^2}{2m} - \frac{Nv_0}{V}\right\} \quad (5.100)$$

$$\varepsilon(p) = \sqrt{\frac{N}{V}\frac{p^2 v_0}{m} + \frac{p^4}{4m^2}}. \quad (5.101)$$

Substituting these coefficients to (5.97) we obtain diagonalized Hamiltonian, having the form of the Hamiltonian of new *noninteracting* quasiparticles, corresponding to operators $\alpha^+_{\mathbf{p}}$ and $\alpha_{\mathbf{p}}$:

$$H = E_0 + \sum_{\mathbf{p}\neq 0} \varepsilon(p)\alpha^+_{\mathbf{p}}\alpha_{\mathbf{p}}, \quad (5.102)$$

where the spectrum of these new quasiparticles $\varepsilon(p)$ (5.101) is radically different from the spectrum of free Bosons due to interaction effects. The ground state energy is given by:

$$E_0 = \frac{N^2}{2V}v_0 + \frac{1}{2}\sum_{p\neq 0}\left[\varepsilon(p) - \frac{p^2}{2m} - \frac{N}{V}v_0\right]. \quad (5.103)$$

At small momenta the quasiparticle energy (5.101) can be written as:

$$\varepsilon(p) = \sqrt{\frac{v_0}{mV_0}}p \equiv up, \quad (5.104)$$

where $V_0 = V/N$ is the volume per particle, while u, which is completely determined by interactions, represents the speed of Bogolyubov's sound. For large momenta, (5.101) reduces to $\varepsilon(p) \approx \frac{p^2}{2m} + \frac{v_0}{V_0}$, i.e. to the spectrum of free particles.

Thus, at small momenta, interactions between Bosons leads to a complete transformation of the spectrum of elementary excitations, which becomes similar to that postulated by Landau, and satisfies the criterion for superfluidity, so that

$$v_c = \left(\frac{\varepsilon(p)}{p}\right)_{p\to 0} = \sqrt{\frac{v_0}{mV_0}} > 0 \quad (5.105)$$

defines the appropriate critical velocity, coinciding in this model with the speed of (Bogolyubov) sound.

From this analysis it becomes clear that the phenomenon of Bose condensation is crucial for the appearance of superfluidity.

5.8 Fermi liquids

A liquid of interacting particles with half-integer spin (Fermi liquid) is characterized by the spectrum of elementary excitations and other properties, which are radically different from those of a Bose liquid. An example of a real Fermi liquid is He^3. Probably, the most common case is the liquid of conduction electrons in metals. More exotic examples are nucleons in atomic nuclei, neutron star matter, etc. We shall see below that the energy spectrum of elementary excitations in a Fermi liquid is somehow similar to that of an ideal Fermi gas, while the role of the interactions reduces to a relatively minor "renormalization" of experimental observables.

The phenomenological theory of Fermi liquids was proposed by Landau. The starting point of this theory is the statement that the classification of energy levels in a Fermi system remains the same after adiabatic "switching" of interaction between particles, as we go from Fermi gas to *normal* Fermi liquid. Elementary excitations (quasiparticles) in a Fermi liquid are in one to one correspondence with free particle excitations of an ideal Fermi gas. Thus, free particles of the gas are replaced by some effective quasiparticles of the liquid, moving in a self-consistent field created by the interactions. The criteria for these quasiparticles to have a well defined momentum will be briefly discussed below. Let n_p be the momentum distribution function of the quasiparticles in a Fermi liquid. The ground state contains no quasiparticle excitations and corresponds to the distribution function of quasiparticles with all states below the Fermi momentum (i.e. for $p < p_F$) occupied. This is equivalent to an assumption of existence of a well defined Fermi surface (sphere) in momentum space. The value of p_F is related to the particle density of the liquid (the number of particles in a unit volume) by the same expression (4.43), as in the Fermi gas[12]:

$$p_F = (3\pi^2)^{1/3} \left(\frac{N}{V}\right)^{1/3} \hbar. \tag{5.106}$$

It must be stressed that the total energy of a liquid E does not reduce to the sum of quasiparticle energies: E is represented by a functional[13] of the distribution function of some general form, which does not reduce to $\int d\tau n_p \varepsilon_p$, as in an ideal gas. At $T = 0$ this functional defines the ground state energy of the Fermi liquid E.

[12] This is directly related to our assumption about the classification of levels in a Fermi liquid and the Pauli principle. In fact, this result can be proved within a modern quantum-field theoretic (microscopic) approach to Fermi liquids, where it is known as Luttinger theorem.

[13] The usual function defines some mapping of one set of numbers into another set of numbers. The functional defines the mapping of a set of *functions* into a set of *numbers*. A typical example of a functional is definite integral: $F[f(x)] = \int_a^b dx f(x)$. Note that the function of a function is again some function, not a functional. Functional (variational) differentiation, as used below is formally defined as follows:

$$\frac{\delta F[f(x)]}{\delta f(y)} = \lim_{\varepsilon \to 0} \frac{F[f(x) + \varepsilon\delta(x-y)] - F[f(x)]}{\varepsilon}. \tag{5.107}$$

For example, for $F[f(x)]$ in the form of the definite integral

$$\frac{\delta F[f(x)]}{\delta f(y)} = \lim_{\varepsilon \to 0} \frac{1}{\varepsilon}\left[\int dx[f(x) + \varepsilon\delta(x-y)] - \int dx f(x)\right] = \int dx \delta(x-y) = 1. \tag{5.108}$$

We can normalize the distribution function as:

$$\int d\tau n_p = \frac{N}{V}, \tag{5.109}$$

where N is the number of particles in the liquid, $d\tau = d^3p/(2\pi\hbar)^3$. The change of E under a small variation of the distribution function can be written as:

$$\frac{\delta E}{V} = \int d\tau \varepsilon_p \delta n_p, \tag{5.110}$$

$$\varepsilon_p = \frac{\delta E}{\delta n_p}. \tag{5.111}$$

The value of ε_p is given by the functional (variational) derivative of E by the distribution function and corresponds to the change of ground state energy of the system due to the addition of a single quasiparticle with momentum $\mathbf{p}$. This energy of a quasiparticle is itself the functional of the distribution function, i.e. the form of ε_p is determined by the distribution of all other quasiparticles in Fermi liquid.

The distribution function of quasiparticles (at equilibrium) has the form of the usual Fermi distribution. This is due to the same classification of energy levels in the liquid as in ideal Fermi gas – the entropy of the liquid is determined by the same combinatorial expression of Equation (4.15), which for the liquid can be written as:

$$S = -\int d\tau [n_p \ln n_p + (1 - n_p)\ln(1 - n_p)]. \tag{5.112}$$

Looking for the maximum (extremum) of this expression with additional conditions of a fixed total number of particles and total energy (similar to our analysis for an ideal gas) we obtain at finite T:

$$n_p = \frac{1}{e^{\frac{\varepsilon_p - \mu}{T}} + 1}. \tag{5.113}$$

However, it should be stressed that ε_p here is some functional of n_p, so that Equation (5.113) gives in fact some complicated implicit definition of n_p. In fact, it can not be found in explicit form[14].

Let us discuss explicitly the spin of the quasiparticles $\vec{\sigma}$. In a homogeneous and isotropic liquid the scalar ε can depend only on scalar arguments, so that $\vec{\sigma}$ can enter the quasiparticle energy (in the absence of an external magnetic field!) only as $\hat{\sigma}^2$ or

[14] Within the microscopic approach to Fermi liquids it was shown by Migdal that the distribution function of the particles (not quasiparticles!) at $T = 0$ contains a finite discontinuity at $\varepsilon_p = \mu$, proving the existence of a Fermi surface in the case of interacting Fermions. The size of this discontinuity in a Fermi liquid < 1, differentiates a liquid from an ideal gas, where it is equal to 1 (see the more detailed discussion below in Chapter 11).

$(\vec{\sigma}\mathbf{p})^2$ (first order term like $\vec{\sigma}\mathbf{p}$ is not allowed, as it is a pseudoscalar due to the axial vector nature of the spin). For spin $s = 1/2$ we have:

$$\vec{\sigma}^2 = \frac{3}{4} \quad (\vec{\sigma}\mathbf{p})^2 = \frac{1}{4}\mathbf{p}^2 \tag{5.114}$$

so that $\boldsymbol{\sigma}$ drops completely and the quasiparticle energy does not depend on spin. Accordingly. all energy levels are twice degenerate and we have to write everywhere $d\tau = 2\frac{d^3p}{(2\pi\hbar)^3}$.

We have attributed to each quasiparticle a well defined momentum. A necessary requirement is that any indeterminacy of this momentum is to be small compared to the value of the momentum itself and also in comparison to the size of the "smearing" region of the distribution function in momentum space (which is defined by small excitation energies or temperatures). The Pauli principle restricts the possible scatterings of quasiparticles precisely to this region and, after the scattering, quasiparticles should arrive also to free (empty) states from this same region. Thus, the probability of quasiparticle scattering is to be proportional to the square of the width Δp of the "smearing" region. This obviously leads to scattering-induced indeterminacy of the quasiparticle momentum of the order of Δp^2. Now it is clear that for small enough Δp the indeterminacy of momentum will be small not only in comparison to the momentum $p \sim p_F$ itself, but also compared to Δp, if we consider it to be small enough. Thus, the quasiparticles in a Fermi liquid the are well defined only *close enough to the Fermi surface* and quasiparticle energy ε_p is also well defined only in this narrow region of energies (or temperatures!). Expanding the quasiparticle energy in a Taylor series in powers of $p - p_F$ we obtain:

$$\xi_p = \varepsilon_p - \mu \approx v_F(|\mathbf{p}| - p_F) \quad \mu = \varepsilon_F, \tag{5.115}$$

where $v_F = \frac{\partial \varepsilon_p}{\partial p}|_{p=p_F}$ is Fermi velocity.

We have already noted above that during quasiparticle creation or annihilation the angular momentum of any quantum system can only change by integer values. If we are dealing with Fermions of spin $s = 1/2$ this means that quasiparticles can be created (annihilated) in pairs. In a Fermi liquid the creation of a quasiparticle with energy above the ground state given by Equation (5.115) takes place via its excitation from the completely filled Fermi sphere to some state above the Fermi surface, with simultaneous creation of a "hole" (of the same energy) below the Fermi surface. Elementary excitation in a Fermi liquid is just this process of quasiparticle – quasihole pair creation. This is quite similar to the case of an ideal Fermi gas, but the major difference is that such excitations are well defined only close enough to the Fermi surface, where scattering (interaction) processes between quasiparticles are strongly suppressed due to Pauli principle limitations.

In an ideal Fermi gas we have $\varepsilon_p = p^2/2m$ and $v_F = p_F/m$. By analogy, in a Fermi liquid we may introduce the value of

$$m^* = \frac{p_F}{v_F} \tag{5.116}$$

which is called the effective mass of a quasiparticle[15]. Then the specific heat of the Fermi liquid is given by the usual "gas-like" expression (4.70), with the simple replacement $m \to m^*$:

$$C = \frac{\pi^2}{3} \nu_F T \quad \nu_F = \frac{m^* p_F}{\pi^2 \hbar^3}. \tag{5.117}$$

To analyze systems with a variable number of particles it is convenient to use the thermodynamic potential $\Omega = F - \mu N$. At $T = 0$ obviously we have $F = E$, so that $\Omega = E - \mu N$. Consider an "excited" state of the system described by the difference:

$$\Omega - \Omega_0 = E - E_0 - \mu(N - N_0), \tag{5.118}$$

where the index 0 denotes the ground state. We can write:

$$N - N_0 = \sum_p \delta n_p = \int d\tau \delta n_p. \tag{5.119}$$

According to Equation (5.111):

$$E[n_p] = E_0 + \sum_p \varepsilon_p \delta n_p + O(\delta n_p^2) \tag{5.120}$$

so that:

$$\Omega - \Omega_0 = \sum_p (\varepsilon_p - \mu)\delta n_p + O(\delta n_p^2). \tag{5.121}$$

We consider only small variations δn_p close to the Fermi surface, i.e. in a narrow energy layer $\sim\delta$ around it, so that $\varepsilon_p - \mu \sim \delta$. But $\delta n_p \sim \delta$ itself, so that $\Omega - \Omega_0 \sim \delta^2$, and in an expansion of Equation (5.121) we have to keep all terms of the order of $\sim\delta^2$. Then we can write:

$$\Omega - \Omega_0 = \sum_p (\varepsilon_p - \mu)\delta n_p + \frac{1}{2}\sum_{pp'} f(\mathbf{p}, \mathbf{p}')\delta n_p \delta n_{p'} + O(\delta_p^3), \tag{5.122}$$

[15] For example, in liquid He^3 it is known from experiments that $m^* \approx 2.4 m_{He^3}$, $p_F/\hbar \approx 0.8\ 10^8\ \mathrm{cm}^{-1}$. The region where quasiparticles (i.e. the concept of Fermi liquid itself) are well defined for He^3 is limited to temperatures $T < 0.5\,\mathrm{K}$.

where we have introduced:

$$f(\mathbf{p},\mathbf{p}') = \frac{\delta^2 E}{\delta n_p \delta n_{p'}} \tag{5.123}$$

– the so called Landau function, describing the *interaction* between quasiparticles. In fact, from the definitions of Equations (5.111) and (5.122) we can see that the variation δn_p leads to a change of quasiparticle energy:

$$\delta\varepsilon_{\mathbf{p}} = \int d\tau' f(\mathbf{p},\mathbf{p}')\delta n_{p'} \tag{5.124}$$

which is completely determined by the Landau function. Here is the main difference of the Fermi liquid theory from the model of an ideal fermi gas.

Let us assume that $f(\mathbf{p},\mathbf{p}')$ is a continuous function for $\mathbf{p}$ and $\mathbf{p}'$ close to p_F. In practice it is sufficient to know $f(\mathbf{p},\mathbf{p}')$ only on the Fermi surface itself, i.e. for $|\mathbf{p}| = |\mathbf{p}'| = p_F$. Then $f(\mathbf{p},\mathbf{p}')$ depends only on the mutual orientation of the vectors $\mathbf{p}$ and $\mathbf{p}'$ (angle in between) and on the spins $\boldsymbol{\sigma}, \boldsymbol{\sigma}'$. It is convenient to write $f(\mathbf{p},\mathbf{p}')$, separating explicitly independent parts, corresponding to the parallel or antiparallel orientations of the spins of the quasiparticles:

$$f_{\uparrow\uparrow}(\mathbf{p},\mathbf{p}') = f^s(\mathbf{p},\mathbf{p}') + f^a(\mathbf{p},\mathbf{p}'), \tag{5.125}$$

$$f_{\uparrow\downarrow}(\mathbf{p},\mathbf{p}') = f^s(\mathbf{p},\mathbf{p}') - f^a(\mathbf{p},\mathbf{p}'). \tag{5.126}$$

We can say that the antisymmetric part $f^a(\mathbf{p},\mathbf{p}')$ is due to some exchange interaction $2f^a(\mathbf{p},\mathbf{p}')$, which appears only when the spins are parallel. Another representation of the Landau function is also widely used in the literature:

$$f_{\sigma,\sigma'}(\mathbf{p},\mathbf{p}') = \varphi(\mathbf{p},\mathbf{p}') + (\hat{\sigma}\hat{\sigma}')\psi(\mathbf{p},\mathbf{p}'), \tag{5.127}$$

where $\hat{\sigma}$ and $\hat{\sigma}'$ are the spin matrices of two fermions.

Thus, in an isotropic Fermi liquid, the functions $f^a(\mathbf{p},\mathbf{p}')$ and $f^s(\mathbf{p},\mathbf{p}')$ depend only on the angle θ between $\mathbf{p}$ and $\mathbf{p}'$. Then these functions can be represented as expansions over Lagrange polynomials:

$$f^{s(a)}(\mathbf{p},\mathbf{p}') = \sum_{l=0}^{\infty} P_l(\cos\theta) f_l^{s(a)} \tag{5.128}$$

so that both functions $f(\mathbf{p},\mathbf{p}')$ are completely determined by the sets of coefficients f_l^s and f_l^a, which are called the Fermi liquid constants. It is convenient to introduce the dimensionless constants $F_l^{s,(a)}$ via:

$$\nu_F f_l^{s,(a)} = \frac{m^* p_F}{\pi^2\hbar^3} f_l^{s,(a)} \equiv F_l^{s,(a)}. \tag{5.129}$$

The values of these constants determine the renormalization of a number of physical characteristics of the Fermi liquid and at least some of them can be determined from experiments. In most cases only a few constants are important. In particular, the following relation between "bare" mass of the particle and effective mass of the quasiparticle can be derived using Galilean invariance [2, 21]:

$$\frac{1}{m} = \frac{1}{m^*} + \frac{p_F}{(2\pi\hbar)^3} 4\pi \int d\cos\theta \cos\theta f(\mathbf{p}, \mathbf{p}'). \tag{5.130}$$

Using (5.128), (5.129) and the properties of Lagrange polynomials we can get:

$$\frac{m^*}{m} = 1 + \frac{F_1^s}{3}. \tag{5.131}$$

From here, it is obvious that $F_1^s > -3$. Similarly, taking into account the interaction with the external magnetic field (see also below), we can derive the spin (paramagnetic) susceptibility of our Fermi liquid as [2, 21]:

$$\chi_P = \mu_B^2 \frac{m^* p_F}{\pi^2 \hbar^3} \frac{1}{1 + F_0^a}, \tag{5.132}$$

which differs from the similar Fermi gas expression (4.79) by the replacement $m \to m^*$ and the Fermi liquid renormalization $1 + F_0^a$.

5.9 Electron liquid in metals*

In our previous discussion we implicitly assumed that a Fermi liquid consists of neutral particles (e.g. like He^3), so that interaction is short range. For a Fermi liquid of electrons in metals, long-range Coulomb interaction becomes important. In case of long-range interactions the basic Fermi liquid theory relation (5.124) becomes, strictly speaking, invalid. However, a certain generalization of the standard Fermi liquid approach for the case of charged Fermi liquids, proposed by Silin, correctly takes into account the Coulomb interaction and reduces the theory to a form quite similar to that of neutral Fermi liquid theory.

Note, first of all, that for the general case of a local in time relation we can write the generalization of Equation (5.124) in the following form:

$$\delta\varepsilon(\mathbf{p}, \mathbf{r}) = \mathrm{Sp}_{\sigma'} \int d\mathbf{r}' \int \frac{d^3 p'}{(2\pi\hbar)^3} F(\mathbf{p}, \mathbf{p}'; \mathbf{r}, \mathbf{r}') \delta n(\mathbf{p}', \mathbf{r}'), \tag{5.133}$$

where we have introduced an explicit dependence on the coordinates, necessary for the analysis of spatially inhomogeneous perturbations and taken Sp over spin[16]. Function $F(\mathbf{p}, \mathbf{p}'; \mathbf{r}, \mathbf{r}')$ here represents the second variational derivative of the ground state

[16] The distribution function of the quasiparticles here is understood to be in the Wigner representation, to account for coordinate dependence.

energy of the Fermi liquid and also depends no only on momenta $\mathbf{p}$, $\mathbf{p}'$ and spins, but also on the coordinates $\mathbf{r}$ and $\mathbf{r}'$. In the simplest case (self-consistent field in Hartree approximation), neglecting exchange effects, for particles interacting via potential $U(|\mathbf{r}-\mathbf{r}'|)$, we have:

$$F_H(\mathbf{p},\mathbf{p}';\mathbf{r},\mathbf{r}') = U(|\mathbf{r}-\mathbf{r}'|). \tag{5.134}$$

This expression neglects the so called correlation effects, while the difference $F - F_H$ by definition is determined by these effects, including the most important effects of exchange correlations. It is important to note that characteristic distances for correlation effects are of the order of electron wavelength at the Fermi level, i.e. of the order of the average distance between the particles (electrons) $(N/V)^{-1/3} \sim 10^{-8}$ cm (in metals). Thus, for the most interesting case, when the characteristic scale of a change of the distribution of the quasiparticles is significantly larger than the correlation range, we may assume:

$$F(\mathbf{p},\mathbf{p}';\mathbf{r},\mathbf{r}') - F_H(\mathbf{p},\mathbf{p}';\mathbf{r},\mathbf{r}') \approx \delta(\mathbf{r}-\mathbf{r}')f(\mathbf{p},\mathbf{p}'). \tag{5.135}$$

Then Equation (5.133) can be rewritten as:

$$\begin{aligned}\delta\varepsilon(\mathbf{p},\mathbf{r}) = \mathrm{Sp}_{\sigma'}\int d\mathbf{r}'\int\frac{d^3p'}{(2\pi\hbar)^3}U(|\mathbf{r}-\mathbf{r}'|)\delta n(\mathbf{p}',\mathbf{r}')\\ + \mathrm{Sp}_{\sigma'}\int\frac{d^3p'}{(2\pi\hbar)^3}f(\mathbf{p},\mathbf{p}')\delta n(\mathbf{p}',\mathbf{r}).\end{aligned} \tag{5.136}$$

For electrons in metals $U(r) = e^2/r$. In equilibrium, when the distribution of particles does not depend on the coordinates, a spatially nonlocal coupling in the first term of (5.136) is irrelevant and the properties of the system of charged particles are, in some sense, similar to those considered above for a neutral Fermi liquid. Note, however, that the first term in (5.136), taken literally, diverges in the case of spatially homogeneous distributions. This divergence is actually fictitious, as we have to take into account the existence in a metal of a homogeneous background of positive ions, guaranteeing the overall electrical neutrality of the system. For spatially inhomogeneous distributions, this term can be considered as a manifestation of the self-consistent scalar potential $\varphi(\mathbf{r})$:

$$e\varphi(\mathbf{r}) = \mathrm{Sp}_{\sigma'}\int d\mathbf{r}'\int\frac{d^3p'}{(2\pi\hbar)^3}\frac{e^2}{|\mathbf{r}-\mathbf{r}'|}\delta n(\mathbf{p}',\mathbf{r}'). \tag{5.137}$$

This potential can be determined by the solution of the Poisson equation:

$$\nabla^2\varphi(\mathbf{r}) = -4\pi e\mathrm{Sp}_{\sigma'}\int\frac{d^3p'}{(2\pi\hbar)^3}\delta n(\mathbf{p}',\mathbf{r}) \tag{5.138}$$

which is an integral part of Landau–Silin theory of charged Fermi liquids.

Let us now take into account the interaction with an external magnetic field $\mathbf{B}$. Then Equation (5.136) for the charged Fermi liquid is rewritten as:

$$\delta\varepsilon(\mathbf{p},\mathbf{r}) = -\mu_B \vec{\sigma} \boldsymbol{B} + e\varphi(\mathbf{r}) + \mathrm{Sp}_{\sigma'} \int \frac{d^3p'}{(2\pi\hbar)^3} f(\mathbf{p},\mathbf{p}')\delta n(\mathbf{p},\mathbf{r}). \tag{5.139}$$

It is important that both $\delta\varepsilon$ and φ are now determined by the system of coupled equations (5.138) and (5.139) in a self-consistent way. In particular, it leads to the phenomenon of the screening of long-range Coulomb forces in a quantum system (metallic Fermi liquid), which will be discussed later in Chapter 11.

Neglecting relativistic effects like spin – orbital coupling, we can again write down the interaction function $f(\mathbf{p},\mathbf{p}')$ as in (5.126) or (5.127). Then again we can introduce the Fermi liquid constants (5.128) and (5.129), which are to be determined from experiments. Finally, for the charged Fermi liquid we can also obtain expressions for the specific heat (5.117), effective mass (5.131) and spin susceptibility (5.132), which are just the same as for the neutral Fermi liquid [21]. Obviously, the values of the Fermi liquid constants in different metals are different and also different from those in liquid He^3, being the characteristics of quasiparticle interactions in a given system (metal). Beside that, in real metals the electronic Fermi liquid may be anisotropic, with a non spherical Fermi surface, due to the effects of the given crystal lattice. This requires the appropriate generalizations of the isotropic model considered here.

Chapter 6

Superconductivity

6.1 Cooper instability

Up to now we have analyzed the so called *normal* Fermi liquid with repulsive interaction between particles. However, as we shall see shortly, the ground state of a Fermi liquid becomes unstable in case of weak (even infinitesimal!) *attraction* between the quasiparticles in the vicinity of the Fermi surface. This instability, discovered by Cooper, leads to the formation of bound states of Fermions (Cooper pairs), i.e. effectively Bosons in a Fermion system. It is fundamental to the understanding of such physical phenomena as superconductivity in metals and superfluidity in liquid He^3.

Below we shall present a simplified analysis of the Cooper instability, which gives an "almost" correct answer [22]. We have already noted that quasiparticles in a Fermi liquid are created in pairs (particle above the Fermi surface and hole below). Close to the Fermi surface, according to Equation (5.115), we can introduce quasiparticle energies as:

$$\begin{aligned} \xi_p &= v_F(|\mathbf{p}| - p_F) \quad \text{(particle)} \\ \xi_p &= v_F(p_F - |\mathbf{p}|) \quad \text{(hole)} \end{aligned} \tag{6.1}$$

so that quasiparticle energy in general can be written as $|\xi_p|$.

Consider the interaction of two particles (or two holes) close to the Fermi surface. The Schroedinger equation for two quasiparticles interacting via potential $U(\mathbf{r}_1, \mathbf{r}_2)$ can be written as[1]:

$$[H_0(\mathbf{r}_1) + H_0(\mathbf{r}_2) + U(\mathbf{r}_1, \mathbf{r}_2)]\psi(\mathbf{r}_1, \mathbf{r}_2) = E\psi(\mathbf{r}_1, \mathbf{r}_2), \tag{6.2}$$

where $H_0(\mathbf{r})$ is the Hamiltonian of a free quasiparticle:

$$H_0(\mathbf{r})\psi_{\mathbf{p}}(\mathbf{r}) = |\xi_p|\psi_{\mathbf{p}}(\mathbf{r}), \tag{6.3}$$

where $\psi_{\mathbf{p}}(\mathbf{r}) = \frac{1}{\sqrt{V}} e^{i\mathbf{pr}/\hbar}$ is the wave function of a free quasiparticle. Let us analyze the possibility of the formation of a bound state two such particles (Cooper pair). In the ground state the momentum of the bound pair should be zero and we assume it have zero spin (singlet state)[2] Thus, the pair is described by the superposition of two

---

[1] This is the point where we actually oversimplify the real many-particle problem – we here analyze two separate quasiparticles on the background of a "rigid" Fermi surface.

[2] We consider here the simplified model with almost point-like attraction of the quasiparticles, and the Pauli principle forbids two Fermions to have the same spin at the same point.

quasiparticles with opposite momenta and spins:

$$\psi(\mathbf{r}_1,\mathbf{r}_2) = \sum_{\mathbf{p}} c_{\mathbf{p}} \psi_{\mathbf{p}\uparrow}(\mathbf{r}_1)\psi_{-\mathbf{p}\downarrow}(\mathbf{r}_2). \tag{6.4}$$

Substituting this expression into Equation (6.2) we obtain the equation for the coefficients $c_{\mathbf{p}}$:

$$2|\xi_{\mathbf{p}}|c_{\mathbf{p}} + \sum_{\mathbf{p}'} U_{\mathbf{p}\mathbf{p}'} c_{\mathbf{p}'} = E c_{\mathbf{p}}, \tag{6.5}$$

where $U_{\mathbf{p}\mathbf{p}'}$ is the matrix element of interaction. Let us assume that this matrix element has the following form:

$$U_{\mathbf{p}\mathbf{p}'} = \begin{cases} -g & \text{for } p_F - \frac{\hbar\omega_D}{v_F} < |\mathbf{p}|, |\mathbf{p}'| < p_F + \frac{\hbar\omega_D}{v_F} \\ 0 & \text{outside this interval.} \end{cases} \tag{6.6}$$

The sign of the coupling constant g corresponds to attraction, while the limitations on the momenta mean that this attraction exists only in rather thin energy layers with a width of $2\hbar\omega_D$ around the Fermi level. The appearance of the Debye frequency here is connected with the well established fact that in most metals the microscopic mechanism of this attraction is due to electron – phonon interaction, and phonons can interact effectively with electrons only in the energy layer $2\hbar\omega_D \ll \varepsilon_F$ near the Fermi surface.

From (6.5) and (6.6) we find the following expression for the coefficient $c_{\mathbf{p}}$:

$$c_{\mathbf{p}} = \frac{gI}{2|\xi_{\mathbf{p}}| - E}, \tag{6.7}$$

where

$$I = \sum_{p'=p_F-\frac{\hbar\omega_D}{v_F}}^{p'=p_F+\frac{\hbar\omega_D}{v_F}} c_{\mathbf{p}'}. \tag{6.8}$$

The bound state of two particles corresponds to the negative value of the energy $E = -2\Delta(\Delta > 0)$. Substituting this into (6.7), and (6.7) into (6.8), we get:

$$\begin{aligned} I &= \frac{1}{2} gI \sum_{p'=p_F-\frac{\hbar\omega_D}{v_F}}^{p'=p_F+\frac{\hbar\omega_D}{v_F}} \frac{1}{|\xi_{p'}| + \Delta} \\ &= \frac{1}{4} gI\nu_F \int_{-\hbar\omega_D}^{\hbar\omega_D} d\xi \frac{1}{|\xi| + \Delta} \approx \frac{1}{2} gI\nu_F \ln\frac{\hbar\omega_D}{\Delta}, \end{aligned} \tag{6.9}$$

where we have transformed the summation over p to integration over $\xi = v_F(p - p_F)$, introducing the density of states at the Fermi level $\nu_F = \frac{mp_F}{\pi^2\hbar^3}$ and taking into account that $\Delta \ll \hbar\omega_D$. The extra coefficient $1/2$ here is due to the summation being done over the states of one of the particles of the pair, with a fixed spin projection, while the expression for the density of states ν_F is written for both spin projections. Accordingly, from (6.9) we obtain the equation for Δ:

$$1 = \frac{1}{2} g\nu_F \ln \frac{\hbar\omega_D}{\Delta}, \tag{6.10}$$

which always (even for infinitesimal values of g) possesses the nontrivial solution:

$$\Delta = \hbar\omega_D \exp\left[-\frac{2}{g\nu_F}\right] \tag{6.11}$$

determining the finite binding energy of the pair. Now we see that our system is unstable to the formation of bound pairs of electrons even in the case of very weak attraction near the Fermi surface. This is called the Cooper instability. Our analysis is slightly inaccurate, as we discussed two separate electrons above the fixed or "rigid" Fermi surface, but it gives the correct order of magnitude estimate of the binding energy. Obviously, Cooper pairs are Bosons and can undergo Bose condensation at low enough temperatures. This is the main physical idea in the explanation of the microscopic nature of superfluidity in Fermi systems (superconductivity in metals).

6.2 Energy spectrum of superconductors

The physical nature of superconductivity in metals is the Cooper pairing of electrons, i.e. the formation of bound states of paired particles, which are close (in momentum space) to the Fermi surface, with equal and opposite momenta and spins. The microscopic mechanism of attractive interaction in traditional superconductors (with a critical temperature for superconducting transition $T_c < 30\,\text{K}$) is, in most cases, attributed to the electron–phonon interaction. The nature of this attraction in high-temperature superconductors (copper oxides, iron pnictides and chalcogenides) with $T_c > 30\,\text{K}$ is up to now not clear; most probably it is connected with the interaction of current carriers (electrons or holes) with antiferromagnetic spin fluctuations. In < superfluid He^3 (where in the temperature region $T < 2.610^{-3} K$ there exist several superfluid phases) this is definitely an exchange by spin fluctuations (paramagnons) among quasiparticles in Helium. A number of other pairing mechanisms were proposed in the literature, e.g. the so called excitonic mechanism. In any case we speak about interaction due to the exchange of some quanta of the collective (Boson) excitations between fermionic quasiparticles. In the following we shall not discuss these microscopic mechanisms of the pairing, but shall limit ourselves to the traditional and simplified model of superconductivity, proposed by Bardeen, Cooper and Schrieffer (BCS model)[3].

Bardeen, Cooper and Schrieffer proposed the following *model* Hamiltonian of a superconductor:

$$H = \sum_{\mathbf{p}\sigma} \xi_p a^+_{\mathbf{p}\sigma} a_{\mathbf{p}\sigma} - \frac{g}{V} \sum_{\mathbf{p}\mathbf{p}'} a^+_{\mathbf{p}'\uparrow} a^+_{-\mathbf{p}'\downarrow} a_{-\mathbf{p}\downarrow} a_{\mathbf{p}\uparrow}, \tag{6.12}$$

where $\xi_p = v_F(|\mathbf{p}| - p_F)$ is the electron energy in a *normal metal* in the vicinity of the Fermi level, $a^+_{\mathbf{p}\sigma}$ and $a_{\mathbf{p}\sigma}$ the creation and annihilation operators of an electron with momentum $\mathbf{p}$ and spin projection σ. The sign of the coupling constant g is taken here corresponding to attraction and it is assumed that this constant is different from zero only in some energy layer around the Fermi surface, as in Equation (6.6). Note that this Hamiltonian is much "reduced" – only electrons with opposite momenta and spins interact, all other interactions are just dropped[4].

To solve the Hamiltonian (6.12) we shall use the method proposed by Bogolyubov. Let us write down the interaction part of the Hamiltonian (6.12):

$$H_{\text{int}} = -\frac{g}{V} \sum_{\mathbf{p}\mathbf{p}'} a^+_{\mathbf{p}'\uparrow} a^+_{-\mathbf{p}'\downarrow} a_{-\mathbf{p}\downarrow} a_{\mathbf{p}\uparrow} \tag{6.13}$$

and make the following approximate replacement of operator part:

$$\begin{aligned} a^+_{\mathbf{p}'\uparrow} a^+_{-\mathbf{p}'\downarrow} a_{-\mathbf{p}\downarrow} a_{\mathbf{p}\uparrow} &\to \langle a^+_{\mathbf{p}'\uparrow} a^+_{-\mathbf{p}'\downarrow} \rangle \langle a_{-\mathbf{p}\downarrow} a_{\mathbf{p}\uparrow} \rangle \\ &+ \langle a^+_{\mathbf{p}'\uparrow} a^+_{-\mathbf{p}'\downarrow} \rangle a_{-\mathbf{p}\downarrow} a_{\mathbf{p}\uparrow} + \langle a_{-\mathbf{p}\downarrow} a_{\mathbf{p}\uparrow} \rangle a^+_{\mathbf{p}'\uparrow} a^+_{-\mathbf{p}'\downarrow}, \end{aligned} \tag{6.14}$$

where angular brackets denote the ground state averaging at $T = 0$ or statistical averaging for $T > 0$, i.e. $\langle \cdots \rangle = Z^{-1}\mathrm{Sp}(e^{-\frac{H}{T}} \cdots)$ (assuming that these averages exist and are nonzero!). This replacement effectively excludes four operator terms in the Hamiltonian, reducing it to the following form, describing interaction with some self-consistent field, determined by these averages:

$$\begin{aligned} H_{\text{int}} = &-\frac{g}{V} \sum_{\mathbf{p}\mathbf{p}'} \left\{ < a^+_{\mathbf{p}'\uparrow} a^+_{-\mathbf{p}'\downarrow} > a_{-\mathbf{p}\downarrow} a_{\mathbf{p}\uparrow} + \langle a_{-\mathbf{p}\downarrow} a_{\mathbf{p}\uparrow} \rangle a^+_{\mathbf{p}'\uparrow} a^+_{-\mathbf{p}'\downarrow} \right\} \\ &- \frac{g}{V} \sum_{\mathbf{p}\mathbf{p}'} \langle a^+_{\mathbf{p}'\uparrow} a^+_{-\mathbf{p}'\downarrow} \rangle \langle a_{-\mathbf{p}\downarrow} a_{\mathbf{p}\uparrow} \rangle. \end{aligned} \tag{6.15}$$

[3] We shall consider only spin singlet pairing (opposite paired spins) of electrons with the pair having zero orbital momentum (s-wave pairing), though in some metals and in superfluid He^3, Cooper pairing takes place in the spin triplet state (parallel paired spins) and not necessarily in the s-wave orbital state. For example in high-temperature copper oxide superconductors d-wave singlet Cooper pairs are well confirmed by many experiments.

[4] As a result of this simplification (separation of the most important interactions) the problem may be studied in detail. BCS theory remains one of the best achievements of modern theoretical physics and its ideas are applied in many other systems (besides metals), energy scales and temperatures. Besides the examples given above, we can mention nucleon pairing in atomic nuclei, superfluidity in neutron star matter and also some models of modern theory of elementary particles.

Finally the total Hamiltonian of the system can be written as[5]:

$$H = \sum_{\mathbf{p}\sigma} \xi_p a^+_{\mathbf{p}\sigma} a_{\mathbf{p}\sigma} + \sum_{\mathbf{p}} \left\{ \Delta^* a_{\mathbf{p}\uparrow} a_{-\mathbf{p}\downarrow} + \Delta a^+_{-\mathbf{p}\downarrow} a^+_{\mathbf{p}\uparrow} \right\} + \frac{1}{g} V |\Delta|^2, \tag{6.16}$$

where we have introduced by definition:

$$\Delta^* = \frac{g}{V} \sum_{\mathbf{p}'} \langle a^+_{\mathbf{p}'\uparrow} a^+_{-\mathbf{p}'\downarrow} \rangle, \tag{6.17}$$

$$\Delta = \frac{g}{V} \sum_{\mathbf{p}'} \langle a_{-\mathbf{p}'\downarrow} a_{\mathbf{p}'\uparrow} \rangle \tag{6.18}$$

the so called *anomalous averages*, directly related to the *order parameter* of the superconducting transition. Combinations of creation and annihilation operators, here taken under the averaging (as well as in Equations (6.14) and (6.15)), are in fact creation and annihilation operators for Cooper pairs (Bosons!) with zero momentum, similar to (5.89). Then, using Bogolyubov's idea applied before to a Bose gas, we can replace this combination of *operators* in Hamiltonian (6.13) by c-numbers, defined by the averages in Equations (6.14), (6.15), or by the directly related (6.17) and (6.18), i.e. assume, that the Cooper pairs undergo Bose condensation at low enough temperatures. Without any limitations we can put here $\Delta^* = \Delta$, i.e. choose the phase of a complex number $\Delta = |\Delta| e^{i\phi}$ (order parameter) equal to zero: $\phi = 0$. In the absence of an external magnetic field this can be done, as the energy of the system does not depend on phase[6]. Note that the existence of anomalous averages of the type (6.18) explicitly breaks the particle conservation law (compare again with Bose gas case!), in a normal metal these averages are obviously zero [18]. The appearance of such averages corresponds to the breaking of this invariance during the phase transition from normal metal to superconductor[7]. Further analysis is aimed to confirm, self-consistently, that such averages are really different from zero at low enough temperatures, corresponding to a phase transition to superconducting state.

Now the Hamiltonian (6.16) is quadratic over the Fermion operators and can be diagonalized by Bogolyubov's $u - v$-transformation. Let us introduce new operators as:

$$b_{\mathbf{p}\downarrow} = u_p a_{\mathbf{p}\downarrow} + v_p a^+_{-\mathbf{p}\uparrow} \quad b_{\mathbf{p}\uparrow} = u_p a_{\mathbf{p}\uparrow} - v_p a^+_{-\mathbf{p}\downarrow}, \tag{6.19}$$

$$b^+_{\mathbf{p}\downarrow} = u_p a^+_{\mathbf{p}\downarrow} + v_p a_{-\mathbf{p}\uparrow} \quad b^+_{\mathbf{p}\uparrow} = u_p a^+_{\mathbf{p}\uparrow} - v_p a_{-\mathbf{p}\downarrow}. \tag{6.20}$$

---

[5] Note the sign change due to the permutation of anticommuting Fermi operators.

[6] This was done before also in (5.89) for the Bose gas model.

[7] Here again we meet the phenomenon of spontaneous symmetry breaking – a new ground state of the system (superconductor) has lower symmetry, than the initial Hamiltonian (6.12). This is typical for any phase transition of second order.

Due to the assumed isotropy of the electronic liquid coefficients u_p and v_p depend only on $|\mathbf{p}|$. The linear transformation (6.20) "intermixes" the operators of quasiparticles with opposite momenta and spins. "Old" operators satisfied the usual Fermion commutation relations

$$\left\{a_{\mathbf{p}\sigma}, a^+_{\mathbf{p}'\sigma'}\right\} = \delta_{\mathbf{pp}'}\delta_{\sigma\sigma'} \quad \{a_{\mathbf{p}\sigma}, a_{\mathbf{p}'\sigma'}\} = \left\{a^+_{\mathbf{p}\sigma}, a^+_{\mathbf{p}'\sigma'}\right\} = 0, \tag{6.21}$$

where figure brackets denote anticommutators. We have to require that the new operators satisfy the same commutation relations:

$$\left\{b_{\mathbf{p}\sigma}, b^+_{\mathbf{p}'\sigma'}\right\} = \delta_{\mathbf{pp}'}\delta_{\sigma\sigma'} \quad \{b_{\mathbf{p}\sigma}, b_{\mathbf{p}'\sigma'}\} = \left\{b^+_{\mathbf{p}\sigma}, b^+_{\mathbf{p}'\sigma'}\right\} = 0 \tag{6.22}$$

so that "new" quasiparticles are also Fermions. It is easy to see that this leads to the following relation between the coefficients u and v:

$$u_p^2 + v_p^2 = 1. \tag{6.23}$$

Inverse transformations have the form:

$$a_{\mathbf{p}\uparrow} = u_p b_{\mathbf{p}\uparrow} + v_p b^+_{-\mathbf{p}\downarrow} \quad a_{\mathbf{p}\downarrow} = u_p b_{\mathbf{p}\downarrow} - v_p b^+_{-\mathbf{p}\uparrow}, \tag{6.24}$$

$$a^+_{\mathbf{p}\uparrow} = u_p b^+_{\mathbf{p}\uparrow} + v_p b_{-\mathbf{p}\downarrow} \quad a^+_{\mathbf{p}\downarrow} = u_p b^+_{\mathbf{p}\downarrow} - v_p b_{-\mathbf{p}\uparrow}. \tag{6.25}$$

Substituting (6.25) into the Hamiltonian (6.16) we obtain:

$$\begin{aligned} H &= 2\sum_p \xi_p v_p^2 - 2\Delta \sum_p u_p v_p + \frac{1}{g} V\Delta^2 \\ &+ \sum_p \left\{[\xi_p(u_p^2 - v_p^2) + 2\Delta u_p v_p](b^+_{\mathbf{p}\uparrow} b_{\mathbf{p}\uparrow} + b^+_{\mathbf{p}\downarrow} b_{\mathbf{p}\downarrow})\right\} \\ &+ \sum_p \left\{[2\xi_p u_p v_p - \Delta(u_p^2 - v_p^2)](b^+_{\mathbf{p}\uparrow} b^+_{-\mathbf{p}\downarrow} + b_{-\mathbf{p}\downarrow} b_{\mathbf{p}\uparrow})\right\}. \end{aligned} \tag{6.26}$$

Now it is seen that if we demand the coefficients u and v to satisfy

$$2\xi_p u_p v_p - \Delta(u_p^2 - v_p^2) = 0 \tag{6.27}$$

the non-diagonal terms in (6.26) vanish. Then we finally obtain the Hamiltonian of the new "free" (!) quasiparticles:

$$H = E_0 + \sum_p \varepsilon(p)[b^+_{\mathbf{p}\uparrow} b_{\mathbf{p}\uparrow} + b^+_{\mathbf{p}\downarrow} b_{\mathbf{p}\downarrow}], \tag{6.28}$$

where

$$E_0 = 2\sum_p [\xi_p v_p^2 - \Delta u_p v_p] + \frac{1}{g} V\Delta^2 \tag{6.29}$$

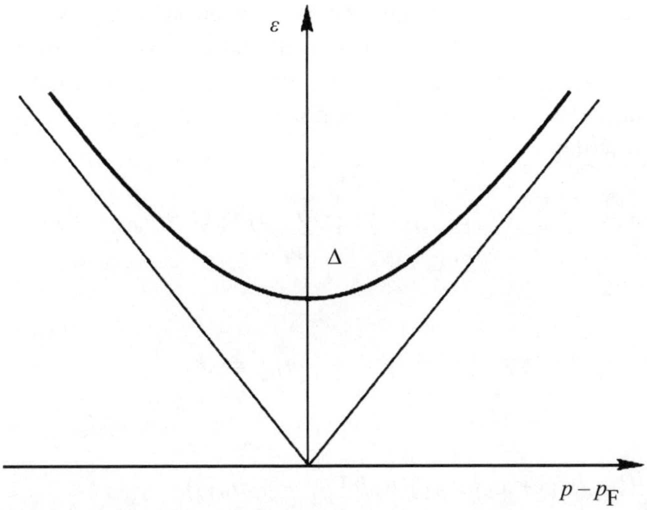

Figure 6.1. Energy spectrum of electrons in BCS theory.

defines the ground state energy, while

$$\varepsilon(p) = \xi_p(u_p^2 - v_p^2) + 2\Delta u_p v_p \tag{6.30}$$

gives the energy of the new quasiparticles. From Equations (6.23) and (6.27) it is easy to obtain explicit expressions for the coefficients u and v:

$$\left.\begin{matrix} u_p^2 \\ v_p^2 \end{matrix}\right\} = \frac{1}{2}\left(1 \pm \frac{\xi_p}{\sqrt{\xi_p^2 + \Delta^2}}\right). \tag{6.31}$$

Then for the spectrum of new quasiparticles from (6.30) we get:

$$\varepsilon(p) = \sqrt{\xi_p^2 + \Delta^2} \tag{6.32}$$

– the BCS spectrum with an *energy gap* of width 2Δ around the Fermi surface! Qualitatively, this spectrum is shown in Figure 6.1. Obviously, this spectrum satisfies the Landau criterion for superfluidity – Min $\frac{\varepsilon(p)}{p} > 0$, i.e. guarantees superconductivity in the system of charged quasiparticles[8].

[8] If there is a current, the whole Fermi surface is shifted in momentum space by some vector $\mathbf{q}$, such that $m\mathbf{v}_s = \hbar\mathbf{q}$, where $\mathbf{v}_s$ is the drift velocity of electrons. Then the energy of an elementary excitation close to the Fermi surface can be written as $\varepsilon(p) \approx \sqrt{\xi_p^2 + \Delta^2} + \mathbf{p}_F\mathbf{v}_s$, where we have taken into account the smallness of the drift velocity (compared to the Fermi velocity), so that $\xi_{\mathbf{p}+\mathbf{q}} \approx \xi_p + \mathbf{v}_F\mathbf{q}$.

Thus, for finite values of Δ (i.e. qualitatively, when there is a Bose condensate of Cooper pairs present) the system becomes a superconductor. However, we still have to show that such a situation is possible, i.e. we have to define conditions when the anomalous averages (6.17), (6.18) become nonzero. Making the $u - v$ transformation in (6.17) we can write:

$$\Delta = \frac{g}{V}\sum_p \langle a^+_{p\uparrow}a^+_{-p\downarrow}\rangle = \frac{g}{V}\sum_p u_p v_p(1 - n_{p\uparrow} - n_{p\downarrow}), \tag{6.33}$$

where

$$n_{p\uparrow} = \langle b^+_{p\uparrow}b_{p\uparrow}\rangle \quad 1 - n_{p\downarrow} = \langle b_{p\downarrow}b^+_{p\downarrow}\rangle. \tag{6.34}$$

In fact:

$$\begin{aligned}\langle a^+_{p\uparrow}a^+_{-p\downarrow}\rangle &= \langle (u_p b^+_{p\uparrow} + v_p b_{-p\downarrow})(u_p b^+_{-p\downarrow} - v_p b_{p\uparrow})\rangle \\ &= u_p^2\langle b^+_{p\uparrow}b^+_{-p\downarrow}\rangle - u_p v_p\langle b^+_{p\uparrow}b_{p\uparrow}\rangle + v_p u_p\langle b_{-p\downarrow}b^+_{-p\downarrow}\rangle - v_p^2\langle b_{-p\downarrow}b_{p\uparrow}\rangle \\ &= u_p v_p(1 - n_{p\uparrow} - n_{p\downarrow})\end{aligned} \tag{6.35}$$

as in the correct ground state we have to satisfy the condition:

$$\langle b^+_{p\uparrow}b^+_{-p\downarrow}\rangle = \langle b_{-p\downarrow}b_{p\uparrow}\rangle = 0,$$

i.e. new quasiparticles should not be spontaneously created or annihilated[9]. Similarly:

$$\langle a_{p\uparrow}a_{-p\downarrow}\rangle = u_p v_p(1 - n_{p\downarrow} - n_{p\uparrow}) = \langle a^+_{p\uparrow}a^+_{-p\downarrow}\rangle. \tag{6.36}$$

Substituting the explicit expression (6.31) for u_p and v_p into (6.33), we obtain:

$$1 = \frac{g}{2V}\sum_p \frac{1 - n_{p\uparrow} - n_{p\downarrow}}{\sqrt{\xi_p^2 + \Delta^2}} \tag{6.37}$$

the fundamental *gap equation* of BCS theory.

For an electron with momentum parallel or antiparallel to $\mathbf{v}_s$ we have $\varepsilon(p) \approx \sqrt{\xi_p^2 + \Delta^2} \pm p_F v_s$. Thus, an energy difference appears $\hbar\omega = 2p_F v_s$ between opposite points on the Fermi surface, so that the excitation spectrum becomes asymmetric. However, until $\hbar\omega = 2p_F v_s < 2\Delta$, the gap in the spectrum persists and for $T = 0$ there are no excited BCS quasiparticles. Accordingly, there is no dissipation of current. For $v_s p_f > \Delta$ the upper and lower quasiparticle bands overlap and the excitation of quasiparticles into the upper band becomes possible even for $T = 0$ and superconductivity vanishes. This leads to the simplest estimate for the critical current of superconductor: $j_c = ev_s^c = \frac{e\Delta}{p_F}$.

[9] Mathematically this follows from the presence of only diagonal elements of the density matrix, corresponding to a diagonalized Hamiltonian (6.28). Accordingly, the averages of the diagonal products of the operators (6.34) are different from zero, while the averages of the nondiagonal products (6.35) are zero.

In the absence of an external magnetic field the occupation numbers are $n_{p\uparrow} = n_{p\downarrow}$ and are defined by the usual Fermi distribution of quasiparticles with spectrum (6.32):

$$n_{p\uparrow} = n_{p\downarrow} = \frac{1}{e^{\frac{\varepsilon(p)}{T}} + 1}. \tag{6.38}$$

Consider first the case of $T = 0$. For $\Delta \neq 0$ there are no (excited) quasiparticles at all, i.e. $n_{p\uparrow} = n_{p\downarrow} = 0$. For $T > 0$ they can be thermally excited in pairs (particles and holes) and appear above (below) the gap. Then in Equation (6.37) we can transform from summation over p to integration and write:

$$1 = \frac{g}{2} \int \frac{d^3p}{(2\pi\hbar)^3} \frac{1 - 2n_p}{\sqrt{\xi_p^2 + \Delta^2}}. \tag{6.39}$$

For $T = 0$ we have:

$$1 = \frac{g}{2} \int \frac{dp}{(2\pi\hbar)^3} \frac{4\pi p^2}{\sqrt{\xi_p^2 + \Delta_0^2}}. \tag{6.40}$$

It is seen immediately that this equation does not have solutions for Δ_0 in case of $g < 0$, i.e. for repulsive interaction, as both sides of this equation have different signs. Remember now that the coupling constant g is nonzero only in a narrow energy layer of width $\sim 2\omega_D$ around the Fermi surface (see Equation (6.6)). Then in (6.40):

$$\int dp p^2 \frac{1}{\sqrt{\Delta_0^2 + v_F^2(p - p_F)^2}} \approx \frac{p_F^2}{v_F} \int_{-\hbar\omega_D}^{\hbar\omega_D} \frac{d\xi_p}{\sqrt{\xi_p^2 + \Delta_0^2}} \approx \frac{2p_F^2}{v_F} \ln \frac{2\hbar\omega_D}{\Delta_0} \tag{6.41}$$

so that Equation (6.40) takes the form:

$$1 = \frac{gmp_F}{2\pi^2\hbar^3} \ln \frac{2\hbar\omega_D}{\Delta_0} \tag{6.42}$$

giving the solution:

$$\Delta_0 = 2\hbar\omega_D \exp\left(-\frac{2}{g\nu_F}\right) \equiv 2\hbar\omega_D \exp\left(-\frac{1}{\lambda_p}\right), \tag{6.43}$$

where $\nu_F = \frac{mp_F}{\pi^2\hbar^3}$ is the electron density of states at the Fermi level and $\lambda_p = g\nu_F/2$ is a dimensionless coupling constant of the pairing interaction. Thus, at $T = 0$ the energy gap Δ_0 is different from zero, formally, even for infinitesimal values of the pairing coupling constant λ_p [10].

---

[10] Note an extra factor of 2 in (6.43), as compared with Equation (6.11), obtained above from a more simple approach. The inaccuracy of Equation (6.11) is connected with the approximation of a separate pair of electrons on the background of a rigid Fermi surface. The correct solution is given by (6.43).

At finite temperatures, setting $\Delta = 0$ in Equation (6.39), we obtain the equation for the critical temperature of the superconducting transition:

$$1 = \frac{g}{2} \int \frac{d^3 p}{(2\pi\hbar)^3} \frac{1 - 2n_p}{|\xi_p|} = \lambda_p \int_{-\hbar\omega_D}^{\hbar\omega_D} d\xi_p \frac{1}{2\xi_p} th \frac{\xi_p}{2T_c}, \tag{6.44}$$

which solution is [22]:

$$T_c = \frac{2\gamma}{\pi} \hbar\omega_D \exp\left(-\frac{1}{\lambda_p}\right), \tag{6.45}$$

where $\gamma \approx 1.78$ is the Euler constant. At this temperature the energy gap goes to zero (cf. below), and the superconductor becomes a normal metal[11].

In Table 6.1 we give the temperatures of the superconducting transition for a number of metals and compounds. In the last row we show the most popular copper oxide high-temperature superconductor. These compounds have been actively studied since 1987. The maximal temperature of the superconducting transition $T_c \sim 135$ K (under pressure up to $\sim$150 K) was observed in $Hg_2Ba_2Ca_2Cu_3O_8$. In 2008 a new class of high-temperature superconductors was discovered, based on iron pnictides and chalcogenides. The highest $T_c = \sim$55 K was observed in this class for the Nd(Sm)FeAsO system. High-temperature superconductors are not described by the simplified version of BCS theory described above, though basic qualitative conclusions are still valid. In fact, in these systems, only on the nature of microscopic mechanism of Cooper pairing, there is no general concensus, though most researchers believe it to be non-phonon, most probably connected with antiferromagnetic spin fluctuations. There are some other differences with the simple BCS approach, e.g. it is well established that in copper oxides the pairing is a singlet, but anisotropic (d-wave pairing). In iron based superconductors the theoretical picture is complicated by their multiple band electronic structure.

In traditional superconductors BCS theory gives a more or less complete description of this phenomenon and there are no doubts as to the electron–phonon nature of Cooper pairing. In Table 6.2 [24] we give the values of λ_p and $\hbar\omega_D$ for a number of superconductors, where the weak coupling BCS model gives a pretty good description[12] As was noted above, in superfluid He^3 Cooper pairing between neutral atoms of Helium takes place at temperatures below 2.6 mK, leading to superfluidity. The

[11] If the microscopic mechanism is not of an electron – phonon nature, the frequency in the pre-exponential factor in this approximation is replaced by the characteristic frequency of Bosons, responsible for the attraction between current carriers. In particular, for so the called excitonic mechanism this is replaced by some energy $\sim E_F \gg \hbar\omega_D$, leading to possible high-temperature superconductivity (Little–Ginzburg). In the real high-temperature superconductors, discovered thus far, we deal here with a characteristic frequency of antiferromagnetic spin fluctuations, while the basic points of BCS theory are conserved.

[12] For the case of strong electron–phonon coupling, BCS theory was generalized by Eliashberg and McMillan, producing more complicated equations but conserving all the main ideas of the BCS approach.

Table 6.1. Temperature of the superconducting transition for a number of metals and compounds (K).

Al	Sn	In	Hg	Pb	Nb	Nb_3Sn	Nb_3Ge	$YBa_2Cu_3O_7$
1.2	3.75	3.4	4.16	7.22	7.78	18.0	23.2	92

Table 6.2. Experimental values of $\hbar\omega_D$, T_c and coupling constant λ_p.

	$\hbar\omega_D$ (K)	T_c (K)	λ_p
Zn	235	0.9	0.18
Cd	164	0.56	0.18
Hg	70	4.16	0.35
Al	375	1.2	0.18
Tl	100	2.4	0.27
Sn	195	3.75	0.25
Pb	96	7.22	0.39

microscopic mechanism of pairing in He^3 is related to the exchange by spin fluctuations (paramagnons). There are several superfluid phases on a rather complicated phase diagram, differing by the type of pairing (orbital and spin momentum of pairs). This leads to an unusual richness of physical phenomena observed in this system [25].

The concept of bound pairs in BCS theory should not be taken too literally. It is more correct to speak about a certain *correlation* between pairs of particles in p-space, leading to a finite probability for particles to have, in fact, the distribution of momenta δp in the region of these correlations corresponding to the binding energy of a pair (gap) $\sim\Delta$, i.e. $\delta p \sim \Delta/v_F$. The appropriate correlation length, given by $\xi \sim \hbar/\delta p \sim \hbar v_F/\Delta$, defines a characteristic scale of distances between correlated particles (the size of a pair). For $T = 0$ this length, also called the coherence length, is equal to:

$$\xi_0 \sim \frac{\hbar v_F}{\Delta_0} \sim \frac{v_F}{\omega_D} \exp\left(\frac{1}{\lambda_p}\right). \tag{6.46}$$

Typically, in metals $\frac{v_F}{\omega_D} \sim \frac{\hbar}{p_F}\frac{\varepsilon_F}{\hbar\omega_D} \gg a$, where a is a characteristic distance between the electrons. Besides that, the exponential factor in (6.46) much exceeds unity, as usually we have $\lambda_p < 1$. From these estimates it is clear that we always have $\xi_0 \gg a$, so that "inside" each pair there is lot of electrons, or, in other words, pairs are much overlapped and lose their individual nature. In high-temperature superconductors, due to much the higher values of T_c (large binding energy of a pair) and a relatively small concentration of current carriers, the size of pairs is not overwhelmingly large in comparison with interparticle distance. These systems belong to a crossover region between very large BCS pairs and "compact" Bosons (BCS-Bose crossover).

In BCS theory, the electrons of a normal metal are transformed into Fermion quasiparticles with the spectrum given by Equation (6.32). Simultaneously, a reconstruction of the ground state takes place. Here, we present (without derivation) the main expressions describing the ground state of a superconductor [24]. This state is described by the following state vector:

$$|BCS\rangle = \prod_p (u_p + v_p a^+_{p\uparrow} a_{-p\downarrow})|0> \tag{6.47}$$

where $|0>$ is the state without electrons (vacuum), satisfying the obvious condition: $a_{p\sigma}|0\rangle = 0$. Equation $u_p^2 + v_p^2 = 1$ guarantees the normalization $\langle BCS|BCS\rangle = 1$. The average number of particles in the BCS ground state is given by:

$$\langle N\rangle = \sum_{p\sigma} \langle BCS|a^+_{p\sigma} a_{p\sigma}|BCS\rangle = 2\sum_p v_p^2 = \frac{V}{(2\pi\hbar)^3}\int d^3p 2v_p^2. \tag{6.48}$$

However, the fluctuation in particle numbers in the BCS state is different from zero, as this ground state (as was noted above) breaks the particle conservation:

$$\langle N^2\rangle - \langle N\rangle^2 = \sum_p 4u_p^2 v_p^2. \tag{6.49}$$

From here it is easily seen that $\langle N^2\rangle - \langle N\rangle^2 \sim V \sim \langle N\rangle$, but the relative fluctuation

$$\frac{\langle N^2\rangle - \langle N\rangle^2}{\langle N\rangle^2} \sim \frac{1}{\langle N\rangle} \tag{6.50}$$

and the relative mean square fluctuation behaves as $1/\sqrt{\langle N\rangle}$ for $\langle N\rangle \to \infty$.

Direct calculations show that the BCS ground state satisfies the condition:

$$b_{p\uparrow}|BCS\rangle = b_{p\downarrow}|BCS\rangle = 0,$$

i.e. is the correct vacuum state for BCS quasiparticles, originating from quasiparticles of a normal metal via the $u - v$-transformation.

6.3 Thermodynamics of superconductors

Consider now finite temperatures $T > 0$. The gap equation (6.39) can be rewritten as:

$$-1 + \frac{g}{2}\int \frac{d^3p}{(2\pi\hbar)^3}\frac{1}{\varepsilon(p)} = g\int \frac{d^3p}{(2\pi\hbar)^3}\frac{n_p}{\varepsilon(p)}, \tag{6.51}$$

where $\varepsilon(p)$ is given by (6.32). Note that the integral in the left-hand side differs here from those in Equation (6.40) only by the replacement of Δ by Δ_0. Then, replacing unity in the left-hand side by the logarithm form of Equation (6.42), we can rewrite the left-hand side of Equation (6.51) as $g\frac{mp_F}{2\pi^2\hbar^3}\ln\frac{\Delta_0}{\Delta}$. In the right-hand side we write explicitly the Fermi function $n_p = [e^{\frac{\varepsilon(p)}{T}} + 1]^{-1}$ and transform to an integration over $d\xi = v_F dp$. Then (6.51) takes the following form:

$$\ln\frac{\Delta_0}{\Delta} = \int_{-\infty}^{\infty} \frac{d\xi}{\sqrt{\xi^2+\Delta^2}\left(e^{\frac{\sqrt{\xi^2+\Delta^2}}{T}}+1\right)} = 2I\left(\frac{\Delta}{T}\right), \tag{6.52}$$

where

$$I(u) = \int_{-\infty}^{\infty} \frac{dx}{\sqrt{x^2+u^2}(\exp\sqrt{x^2+u^2}+1)}. \tag{6.53}$$

This integral can be calculated in limiting cases [1], and we obtain:

$$I(u) = \begin{cases} \left(\frac{\pi}{2u}\right)^{1/2} e^{-u} & \text{for } u \gg 1 \\ \ln\left(\frac{\pi}{\gamma u}\right) + \frac{7\zeta(3)}{8\pi^2}u^2 & \text{for } u \ll 1, \end{cases} \tag{6.54}$$

where $\gamma \approx 1.78$ is the Euler constant, $\zeta(3) \approx 1.202$ is Riemann's ζ-function with argument 3. Substituting these limiting expressions to (6.52), we obtain for low temperatures $T \ll \Delta$:

$$\Delta = \Delta_0\left[1 - \sqrt{\frac{2\pi T}{\Delta_0}}e^{-\frac{\Delta_0}{T}}\right], \tag{6.55}$$

while in the vicinity of the transition to the normal state, where $\Delta \to 0$, we get:

$$\ln\frac{\Delta_0}{\Delta} = \ln\frac{\pi T}{\gamma\Delta} + \frac{7\zeta(3)}{8\pi^2}\frac{\Delta^2}{T^2}. \tag{6.56}$$

From this equation we can wee that the gap Δ becomes zero at the critical temperature:

$$T_c = \frac{\gamma}{\pi}\Delta_0 \approx 0.57\Delta_0, \tag{6.57}$$

which, taking into account (6.43), coincides with (6.45). Note the characteristic BCS ratio, following from these expressions: $\frac{2\Delta_0}{T_c} \approx 3.52$, its experimental verification in traditional superconductors was one of the first confirmations of BCS theory[13].

[13] In many real superconductors, significant deviations from this BCS theory prediction are widely observed. In fact, the "ideal" BCS value of 3.52 the ratio of full width of the energy gap and T_c is characteristic for weakly coupled superconductors (with small values of the pairing coupling constant), in accordance with BCS theory. The observed deviations (mostly growth) of this ratio are typical for strongly coupled superconductors and are well described by the Eliashberg–McMillan approach.

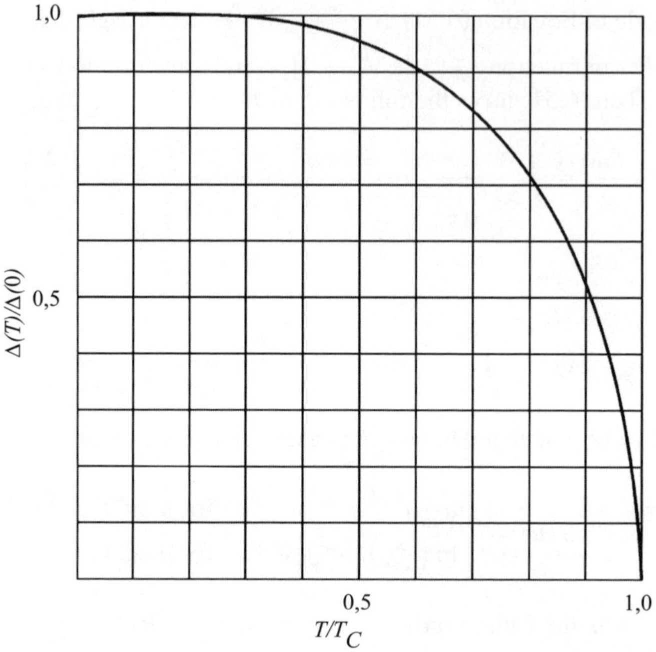

Figure 6.2. Temperature dependence of the gap in BCS theory.

Close to T_c it follows from (6.56) that:

$$\Delta(T) = T_c \left[\frac{8\pi^2}{7\zeta(3)}\left(1 - \frac{T}{T_c}\right)\right]^{1/2} \approx 3.06 T_c \sqrt{1 - \frac{T}{T_c}} \tag{6.58}$$

demonstrating the characteristic square root behavior of the gap, typical for the order parameter of a second order phase transition.

The general the form of the temperature dependence of the gap Δ in BCS theory, following from Equation (6.52), is shown in Figure 6.2. This dependence is also well confirmed by experiments on traditional superconductors with a relatively low transition temperature T_c.

Let us consider some other properties of a superconductor at finite temperatures (dropping the details of the derivation). The difference in free energies of the superconducting and the normal state, close to $T_c (T < T_c)$, following from BCS theory [1, 2], is given by:

$$F_s - F_n = -V \frac{2mp_F T_c^2}{7\zeta(3)\hbar^3}\left(1 - \frac{T}{T_c}\right)^2, \tag{6.59}$$

so that the superconducting state at $T < T_c$ has lower free energy than the normal state. The difference of entropies, following from (6.59) is:

$$S_s - S_n = -\frac{\partial(F_s - F_n)}{\partial T} = -V\frac{4mp_F T_c}{7\zeta(3)\hbar^3}\left(1 - \frac{T}{T_c}\right). \tag{6.60}$$

Accordingly, we obtain the value for the specific heat *discontinuity* at the transition point:

$$C_s - C_n = T\frac{\partial(S_s - S_n)}{\partial T} = V\frac{4mp_F T_c}{7\zeta(3)\hbar^3}. \tag{6.61}$$

Taking into account that $C_n = Vmp_F T/3\hbar^3$ (cf. (4.70)), we obtain:

$$\frac{C_s(T_c)}{C_n(T_c)} = \frac{12}{7\zeta(3)} + 1 \approx 2.43. \tag{6.62}$$

This universal value is also rather well confirmed in measurements of the specific heat on traditional (weakly coupled) superconductors, while strong coupling leads to significant deviations from this prediction of simple BCS theory.

To calculate the specific heat at low temperatures we can use the relation:

$$\delta E = \sum_p \varepsilon(p)(\delta n_{p\uparrow} + \delta n_{p\downarrow}) = 2\sum_p \varepsilon(p)\delta n_p \tag{6.63}$$

for the total quasiparticle energy change due to variation of occupation numbers. Dividing this expression by δT and going from summation to integration, we obtain the specific heat as:

$$C = V\frac{mp_F}{\pi^2\hbar^3}\int_{-\infty}^{\infty} d\xi_p \varepsilon(p)\frac{\partial n_p}{\partial T}. \tag{6.64}$$

For $T \ll \Delta_0$ we can write $n_p \approx e^{-\frac{\varepsilon(p)}{T}}$ and $\varepsilon(p) \approx \Delta_0 + \frac{\xi_p^2}{2\Delta_0}$. Then, simple integration gives:

$$C = V\frac{\sqrt{2}mp_F\Delta_0^{5/2}}{\pi^{3/2}\hbar^3 T^{3/2}}e^{-\frac{\Delta_0}{T}} \tag{6.65}$$

so that at $T \to 0$ the specific heat of the electron gas in a superconductor is exponentially small, due to the existence of a finite gap in the quasiparticle spectrum.

At $T = 0$ it can be shown [1, 2] that the difference between the ground state energies of superconducting and normal states is given by:

$$E_s - E_n = -V\frac{mp_F}{4\pi^2\hbar^3}\Delta_0^2 = -\frac{1}{4}V\nu_F\Delta_0^2. \tag{6.66}$$

The negative sign here corresponds to the instability of the "normal" ground state in case of attraction between quasiparticles and makes the superconducting state the real (stable) ground state of the system. The physical meaning of Equation (6.66) is pretty clear: in an energy layer of width $\sim\Delta_0$ around the Fermi level we have $\sim\nu_F\Delta_0$ quasiparticles, each gaining energy of the order of $\sim\Delta_0$ due to gap formation. The estimate of the total energy gain per one electron is $\sim\Delta^2/\varepsilon_F$.

6.4 Coulomb repulsion*

Up to now, we assumed, that there is an attractive interaction between the electrons, acting in a narrow energy layer of width $2\omega_D$ around the Fermi surface[14]. Such an attraction can exist in metals due to electron–phonon interactions. However, a strong Coulomb repulsion is obviously acting between all electrons in metals, which definitely opposes the formation of Cooper pairs (and thus superconductivity). Let us show how this repulsion can be taken into account in the equations of BCS theory.

In the general case the energy gap of a superconductor, when taking into account different interaction mechanisms, is defined by a rather complicated integral equation. Close to T_c this equation can be linearized over Δ as the gap goes to zero for $T \to T_c$. In the weak coupling approximation we can write the following gap equation close to T_c, which is the direct generalization of Equation (6.44) and determines the critical temperature of the superconducting transition [24]:

$$\Delta(\xi) = \int_{-\infty}^{\infty} d\tau V(\xi,\xi')N(\xi')\frac{1}{2\xi'}th(\frac{\xi'}{2T_c})\Delta(\xi'), \tag{6.67}$$

where $N(\xi)$ is the density of electron states in a normal metal (per one spin projection), and $V(\xi,\xi')$ is the "potential" of an effective interaction between the electrons. We assume that $\Delta(\xi)$ here is some unknown function of the energy of a quasiparticle ξ, which is to be determined depending on the accepted model of the interactions. In our previous discussion Δ was assumed to be a constant and just canceled out, dropping out from Equation (6.44).

Effective electron–electron attraction in superconductors is determined in reality by some balance between attraction due to electron–phonon interaction and Coulomb repulsion. We may assume for the "potential" $V(\xi,\xi')$ the following very crude model[15]:

$$V(\xi,\xi') = -V_c(\xi,\xi') + V_{ph}(\xi,\xi'), \tag{6.68}$$

where

$$V_c(\xi,\xi') = V_c\theta(\varepsilon_F - |\xi|)\theta(\varepsilon_F - |\xi'|), \tag{6.69}$$

$$V_{ph}(\xi,\xi') = V_{ph}\theta(\omega_D - |\xi|)\theta(\omega_D - |\xi'|) \tag{6.70}$$

are the "potentials" of electron–electron and electron–phonon interactions respectively, ω_D is Debye frequency. Constants $V_c > 0$ and $V_{ph} > 0$ describe repulsion and attraction, acting (due to $\varepsilon_F \gg \omega_D$) in significantly different intervals of energy:

[14] In this section we put $\hbar = 1$ and measure frequency ω_D in units of energy.

[15] We assume interelectron repulsion to be short-ranged due to the strong screening of the Coulomb interaction in metals.

electron–phonon attraction acts only on electrons in an energy layer of width $2\omega_D$ close to the Fermi level, while the (screened) Coulomb repulsion acts between all conduction electrons on an energy scale of the order of the Fermi energy ε_F.

After substitution of this expression into Equation (6.67) and simple transformations, using the (presumably) even gap function $\Delta(\xi)$, we get:

$$\Delta(\xi) = [V_{ph}\theta(\omega_D - \xi) - V_c\theta(\varepsilon_F - \xi)] \int_0^{\omega_D} d\xi' N(\xi')\frac{1}{\xi'}th(\frac{\xi'}{2T_c})\Delta(\xi')$$
$$- V_c\theta(\varepsilon_F - \xi) \int_{\omega_D}^{\varepsilon_F} d\xi' N(\xi')\frac{1}{\xi'}th(\frac{\xi'}{2T_c})\Delta(\xi'). \quad (6.71)$$

In rough approximation we can seek a solution of this equation of the form of two "step" functions [24]:

$$\Delta(\xi) = \begin{cases} \Delta_{ph}, & |\xi| < \omega_D, \\ \Delta_c, & \omega_D < |\xi| < \varepsilon_F, \end{cases} \quad (6.72)$$

where Δ_{ph} and Δ_c are some constants, which can be determined (after substitution of (6.72) into (6.71) from the following system of homogeneous linear equations (obtained after substitution of (6.72) into (6.71)):

$$\left\{1 - (V_{ph} - V_c)N(0)K\left(\frac{\omega_D}{2T_c}\right)\right\}\Delta_{ph} + V_cN(0)\left[K\left(\frac{\varepsilon_F}{2T_c}\right) - K\left(\frac{\omega_D}{2T_c}\right)\right]\Delta_c = 0,$$
$$V_cN(0)K\left(\frac{\omega_D}{2T_c}\right)\Delta_{ph} + \left\{1 + V_cN_0(0)\left[K\left(\frac{\varepsilon_F}{2T_c}\right) - K\left(\frac{\omega_D}{2T_c}\right)\right]\right\}\Delta_c = 0, \quad (6.73)$$

where we have replaced the density of states by its constant value $N(0) = \frac{1}{2}\nu_F$ at the fermi level and introduced the notation:

$$K(x) = \int_0^x dx' \frac{1}{x'}th(x'). \quad (6.74)$$

A nontrivial solution of this system exists if the determinant of this system of equations is zero, which gives the equation for T_c:

$$(\lambda - \mu^*)K\left(\frac{\omega_D}{2T_c}\right) = 1,$$
$$\mu^* = \mu\left\{1 + \mu\left[K\left(\frac{\varepsilon_F}{2T_c}\right) - K\left(\frac{\omega_D}{2T_c}\right)\right]\right\}^{-1}, \quad (6.75)$$

where we have introduced μ^* – the so called Coulomb pseudo-potential, $\mu = V_cN_0(0)$ is the dimensionless Coulomb (repulsion) coupling constant, while $\lambda = V_{ph}N_0(0)$ is a dimensionless pairing coupling constant due to electron–phonon interaction.

Due to inequality $\varepsilon_F \gg \omega_D \gg T_c$ the integral in (6.74) can be calculated for $x \gg 1$, so that $K(x) = \ln\left(\frac{4\gamma}{\pi}x\right)$, where γ is again the Euler constant. Then for the critical temperature of the superconducting transition we immediately obtain[16]:

$$T_c = \frac{2\gamma}{\pi}\omega_D \exp\left(-\frac{1}{\lambda - \mu^*}\right), \tag{6.76}$$

which coincides with the BCS expression (6.45), if we write the pairing constant as $\lambda_p = \lambda - \mu^*$. The Coulomb potential μ^* is given here by the following expression:

$$\mu^* \approx \frac{\mu}{1 + \mu \ln \frac{\varepsilon_F}{\omega_D}}. \tag{6.77}$$

From this result we can see that Coulomb repulsion naturally opposes pairing and reduces T_c, diminishing λ_p by μ^*. However, in most metals this effect is largely suppressed due to relatively large (for $\varepsilon_F \gg \omega_D$) value of $\ln(\varepsilon_F/\omega_D)$ (so called Tolmachev's logarithm). In particular, even for $\lambda < \mu$, i.e. when for all energies the total constant of electron–electron interaction is formally repulsive, superconductivity may still persist if $\lambda > \mu^*$.

Using Equation (6.76) we may propose the following ways to raise the critical temperature of the superconducting transition:

1. We may raise the value of ω_D or try to use another (non-phonon) mechanism of pairing via the exchange by some collective excitations with characteristic frequencies larger than ω_D. A typical example is the so called excitonic mechanism, for which ω_D is replaced by an energy of the order of ε_F.

2. Another way is to raise the pairing coupling constant λ_p, either by rising attractive coupling λ, or by reducing the Coulomb pseudo-potential μ^*.

Nearly all attempts to search for high-temperature superconductivity were undertaken along this way. Many theoretical explanations of high transition temperatures observed in real high-temperature superconductors are explicitly or implicitly based on these or similar ideas. In fact, the practical realization of these tasks is pretty complicated. Even on this elementary level it can be seen that the necessary requirements are rather contradictory. For example, raising the pre-exponential ω_D in (6.76) up to the values of the order of ε_F inevitably leads to the appropriate growth of the Coulomb pseudo-potential, due to the diminishing value of Tolmachev's logarithm. On the other hand, raising the effective pairing constant demands the replacement of the weak coupling approximation used in simple BCS theory[17].

Concluding our review of the microscopic theory of superconductivity we note that above we always supposed that Cooper pairing takes place in a singlet state (antiparal-

[16] This important result was obtained by Tolmachev soon after the BCS work.

[17] As an example of the appropriate development of microscopic theory we give here the interpolation formula for T_c, proposed by Allen and Dynes, which is valid for the wide interval of the dimensionless

lel spins), and with zero orbital momentum of the pair (s-wave pairing). In real superconductors the situation may be more complicated. It was shown by experiments that in some systems Cooper pairing takes place in a triplet state (parallel spins), and also in a state with nonzero orbital momentum (He^3, the so called "heavy Fermion" systems etc.). In copper oxides Cooper pairing is a singlet in spin, but a d-wave. In iron pnictides the situation is complicated by a multiple band electronic structure, leading to different superconducting gaps in different bands etc. Obviously, the microscopic description of such systems requires more complicated theories, but the main ideas and qualitative conclusions of BCS theory remain valid.

6.5 Ginzburg–Landau theory

The complete microscopic theory, describing the behavior of superconductors in an external electromagnetic field is too complicated to be discussed here. However, the analysis can be very much simplified if we restrict ourselves to the temperature region of $T \to T_c$, where the phenomenological Ginzburg–Landau (GL) theory can be applied. In fact, GL theory is one of most outstanding physical theories; its main ideas play a major role not only in superconductivity, but in many other branches of theoretical physics (such as e.g. the Standard Model of elementary particles). At the same time, from a phenomenological point of view, GL theory is an impressive example of the use of general the Landau theory of phase transitions of second order [1][18].

In the general Landau theory of phase transitions of the second order, the difference between "non-symmetric" and "symmetric" phases is described by the *order parameter*. For a superconductor the natural choice of the order parameter is the complex energy gap, or more precisely, the anomalous average (6.18), which is proportional to the *condensate wave function* of Cooper pairs. In the general case this order parameter can be inhomogeneous in space. Assuming for simplicity cubic symmetry of the crystal lattice, we note that the superconducting state is characterized by a scalar n_s-

coupling constant of electron–phonon pairing interaction, including the values $\lambda \sim 1$:

$$T_c = \frac{f_1 f_2}{1.20}\omega_{\log} \exp\left\{-\frac{1.04(1+\lambda)}{\lambda - \mu^\star(1+0.62\lambda)}\right\}, \tag{6.78}$$

where

$$\begin{aligned} f_1 &= [1 + (\lambda/\lambda_1)^{3/2}]^{1/3}; \quad & f_2 &= 1 + \frac{[\langle\omega^2\rangle^{1/2}/\omega_{\log} - 1]\lambda^2}{\lambda^2 + \lambda_2^2}, \\ \lambda_1 &= 2.46(1 + 3.8\mu^\star); \quad & \lambda_2 &= 1.82(1 + 6.3\mu^\star)\frac{\langle\omega^2\rangle^{1/2}}{\omega_{\log}}, \end{aligned} \tag{6.79}$$

where $\omega_{\log}$ is the average logarithmic frequency of the phonons, while $\langle\omega^2\rangle$ is the average square of the phonon frequency (the averaging in both cases is over the phonon spectrum). These parameters replace ω_D of BCS theory, the other parameters were defined above.

[18] Note that GL theory can actually be *derived* from microscopic BCS theory, though in fact the GL approach was formulated nearly a decade earlier.

density of superconducting electrons (pairs). Thus, it is convenient to normalize the condensate wave function by the condition $|\Psi|^2 = n_s/2$, and introducing its phase ϕ write it in the form [2]:

$$\Psi = \sqrt{\frac{n_s}{2}} e^{i\phi} \sim \Delta. \tag{6.80}$$

Thus, the order parameter is the complex (two-component) function.

According to the general rules of quantum mechanics we can write down the density of the supercurrent as:

$$\mathbf{j}_s = -\frac{ie\hbar}{2m}(\Psi^\star \nabla \Psi - \Psi \nabla \Psi^\star) = \frac{e\hbar}{2m} n_s \nabla \phi, \tag{6.81}$$

where the last equality is valid for the case of a spatially homogeneous density n_s, while the doubled mass is introduced here just formally, to stress that the supercurrent carriers are Cooper pairs.

The starting point of GL theory is the expression for the free energy of a superconductor as a functional of $\Psi(\mathbf{r})$. Consider first a superconductor in the absence of an external magnetic field. It is obvious that the physical properties should be invariant with respect to a gauge (phase) transformation $\Psi \to \Psi e^{i\alpha}$. This requirement excludes odd power terms in the Landau expansion of the free energy[19].

Thus, the free energy expansion in powers of the order parameter Ψ for a superconductor can be written as[20]:

$$F = F_n + \int dV \left\{ \frac{\hbar^2}{4m} |\nabla \Psi|^2 + a|\Psi|^2 + \frac{b}{2}|\Psi|^4 \right\}. \tag{6.82}$$

Here F_n is the free energy of a normal state, coefficient $b > 0$, and coefficient a is written in usual Landau form:

$$a = \alpha(T - T_c) \quad \alpha > 0 \tag{6.83}$$

so that for $T < T_c$ we have $a < 0$. The coefficient before $|\nabla\Psi|^2$ is taken in the form which leads to the expression (6.81) for the current (see below). Identification of m with the electron mass is of no importance, as well as the definition of n_s.

For the case of the homogeneous order parameter we have:

$$F = F_n + \alpha V(T - T_c)|\Psi|^2 + \frac{bV}{2}|\Psi|^4. \tag{6.84}$$

---

[19] Note that phase invariance in quantum mechanics is responsible for particle conservation. The order parameter itself is not invariant with respect to this transformation. In this sense, as was noted above, in the superconducting state this symmetry is broken. Symmetry breaking takes place at any second order phase transition, so that the condensed phase is always "non-symmetric".

[20] The basic postulate of Landau theory is precisely the possibility to perform such an expansion due to the smallness of the order parameter close to the transition temperature [1].

The value of $|\Psi|^2$ at equilibrium is determined by the minimum of this expression and is given by:

$$|\Psi|^2 = -\frac{a}{b} = \frac{\alpha}{b}(T_c - T) \tag{6.85}$$

for $T < T_c$, and is zero for $T > T_c$. The value of the order parameter $|\Psi|$ goes to zero for $T \to T_c$, according to the square root law, in complete accord with Equation (6.58). The value of $n_s \sim |\Psi|^2 \to 0$ linearly in $T_c - T$.

Substituting (6.85) into (6.84) we obtain:

$$F_s - F_n = -V\frac{\alpha^2}{2b}(T - T_c)^2 \tag{6.86}$$

which is equivalent to Equation (6.59)[21]. Differentiating Equation (6.86) with respect to T, similar to (6.60), we can find the difference between the entropies and the specific heat discontinuity at the transition point:

$$C_s - C_n = V\frac{\alpha^2 T_c}{b} \tag{6.87}$$

which agrees with Equation (6.87).

Close to T_c (6.86) gives a small correction to the free energy, according to thermodynamics it also represents (being expressed via T, P instead of T, V) the difference between the Gibbs thermodynamic potentials $\Phi_s - \Phi_n$. This difference coincides with the value of $-V\frac{B_{ct}^2}{8\pi}$, where B_{ct} is the thermodynamic critical field destroying the superconducting state. Then we easily obtain:

$$B_{ct} = \left(\frac{4\pi a^2}{b}\right)^{1/2} = \left(\frac{4\pi\alpha^2}{b}\right)(T_c - T). \tag{6.88}$$

In the presence of an external magnetic field Equation (6.82) for the free energy can be written as:

$$F = F_n + \int dV\left\{\frac{\mathbf{B}^2}{8\pi} + \frac{\hbar^2}{4m}\left|\left(\nabla - \frac{2ie}{\hbar c}\mathbf{A}\right)\Psi\right|^2 + a|\Psi|^2 + \frac{b}{2}|\Psi|^4\right\}, \tag{6.89}$$

where $\mathbf{B} = \operatorname{rot}\mathbf{A}$. The structure of gradient term here is determined by the gauge invariance of electrodynamics, in particular the coefficient $\frac{2ie}{\hbar c}$ here is expressed via

[21] GL theory was derived from microscopic BCS theory of superconductivity by Gorkov, giving an explicit microscopic expression for the GL coefficients α and b. These expressions can be easily obtained by direct comparison of Equations (6.85), (6.58) with Equations (6.58), (6.86). Thus, for "pure" superconductors (without impurities) we have: $\alpha = \frac{6\pi^2 T_c}{7\zeta(3)\varepsilon_F}$ and $b = \frac{\alpha T_c}{n}$, where $n = \frac{p_F^3}{3\pi^2\hbar^3}$ is the electron density, with T_c given by the BCS expression (6.45).

fundamental constants, in contrast to $\hbar^2/4m$. The presence of $2e$ reflects the charge of a Cooper pair – in GL theory we are dealing with a charged order parameter!

Looking for an extremum of F as a functional of three independent variables $\Psi, \Psi^\star, \mathbf{A}$ [22], we can find a differential equation determining the distribution of Ψ and the magnetic field in the superconductor. Varying (6.89) with respect to $\Psi^\star$ and transforming the integral of $(\nabla - 2ie\mathbf{A}/\hbar c)\nabla\delta\Psi^\star$ by partial integration, we get:

$$\delta F = \int dV \left\{ -\frac{\hbar^2}{4m}(\nabla - \frac{2ie}{\hbar c}\mathbf{A})^2\Psi + a\Psi + b|\Psi|^2\Psi \right\} \delta\Psi^\star + \frac{\hbar^2}{4m}\oint d\mathbf{s}(\nabla\Psi - \frac{2ie}{\hbar c}\mathbf{A}\Psi)\delta\Psi^\star, \tag{6.90}$$

where the second integral is taken over the surface of the superconductor. Demanding $\delta F = 0$, we get the condition for the volume integral being zero for arbitrary $\delta\Psi^\star$, in the form of the following Ginzburg–Landau equation:

$$\frac{1}{4m}\left(-i\hbar\nabla - \frac{2e}{c}\mathbf{A}\right)^2 \Psi + a\Psi + b|\Psi|^2\Psi = 0. \tag{6.91}$$

Variation over Ψ gives the complex conjugate equation for $\Psi^\star$. Variation of (6.89) over $\mathbf{A}$ leads to Maxwell's equation:

$$\operatorname{rot}\mathbf{B} = \frac{4\pi}{c}\mathbf{j}, \tag{6.92}$$

where

$$\mathbf{j} = -\frac{ie\hbar}{2m}(\Psi^\star\nabla\Psi - \Psi\nabla\Psi^\star) - \frac{2e^2}{mc}|\Psi|^2\mathbf{A}. \tag{6.93}$$

Here we have written $\mathbf{j}$ as the superconducting current because in the equilibrium state the normal current is absent.

The boundary condition for these equations is obtained from the condition that the surface integral in (6.90) be zero:

$$\mathbf{n}\left(-i\hbar\nabla\Psi - \frac{2e}{\hbar c}\mathbf{A}\right)\Psi = 0, \tag{6.94}$$

where $\mathbf{n}$ is the unit vector normal to the surface of the superconductor. It leads to the equivalent relation: $\mathbf{nj} = 0$. Equation (6.94) is valid on the boundary of the superconductor with a vacuum (insulator) which, in case of a boundary with a normal metal takes another form. The boundary condition for $\mathbf{B}$ reduces to the continuity of $\mathbf{B}$ at the border.

[22] Complex Ψ consists of independent real and imaginary parts, so it is convenient to consider Ψ and $\Psi^\star$, as independent variables.

In a weak magnetic field we may neglect its influence on $|\Psi|^2$ and set it equal to (6.85). For spatially homogeneous $n_s = 2|\Psi|^2$ from (6.93) we get (cf. (6.81)):

$$\mathbf{j} = \frac{\hbar e}{2m} n_s \left(\nabla \phi - \frac{2e}{\hbar c} \mathbf{A} \right). \tag{6.95}$$

Applying rot to both parts of this equation and using rot $\mathbf{A} = \mathbf{B}$ we obtain London's equation:

$$\operatorname{rot} \mathbf{j} = -\frac{n_s e^2}{mc} \mathbf{B}. \tag{6.96}$$

From Maxwell's equations (6.92) and div $\mathbf{B} = 0$, substituting $\mathbf{j}$ from the first equation into (6.96) and using rot rot $\mathbf{B} =$ grad div $\mathbf{B} - \nabla^2 \mathbf{B} = -\nabla^2 \mathbf{B}$, we can write London's equation as:

$$\nabla^2 \mathbf{B} = \frac{1}{\delta^2} \mathbf{B}, \tag{6.97}$$

where

$$\delta^2 = \frac{mc^2}{4\pi e^2 n_s} \quad \delta = \left(\frac{mc^2 b}{8\pi e^2 |a|} \right)^{1/2} = \left[\frac{mc^2 b}{8\pi e^2 \alpha (T_c - T)} \right]^{1/2}. \tag{6.98}$$

Near the flat surface of the superconductor, taking it as the yz-plane and directing the x-axis into the body of superconductor, we can reduce Equation (6.97) to:

$$\frac{d^2 \mathbf{B}}{dx^2} = \frac{1}{\delta^2} \mathbf{B} \tag{6.99}$$

and immediately get the solution:

$$\mathbf{B}(x) = \mathbf{B}_0 e^{-x/\delta}, \tag{6.100}$$

where the vector $\mathbf{B}_0$ is parallel to the surface. This gives the description of the Meissner effect – the "exclusion" of the external magnetic field from a superconductor. The characteristic length δ is called the penetration depth and it is directly measurable. Its typical values for real superconductors at low temperatures $\delta \sim 10^{-5} - 10^{-6}$ cm. For $T \to T_c$ it diverges according to (6.98), which corresponds to complete penetration of the external magnetic field into a normal metal.

In addition to δ, another characteristic length appears in GL theory – the so called coherence length or correlation length of the order parameter fluctuations $\xi(T)$. Using the standard expressions of the Landau theory of second order phase transitions (see below) this length is expressed via the GL coefficient as follows:

$$\xi(T) = \frac{\hbar}{2(m|a|)^{1/2}} = \frac{\hbar}{2(m\alpha)^{1/2}(T - T_c)^{1/2}} \sim \xi_0 \sqrt{\frac{T_c}{T_c - T}}; \quad \xi_0 \sim \frac{\hbar v_F}{T_c}, \tag{6.101}$$

where in the last estimate we used the microscopic expressions for the GL coefficients and the estimate of the BCS coherence length (6.46), determining the size of Cooper pairs. We can see that the coherence length $\xi(T)$ (pair size) also diverges for $T \to T_c$ (pairs grow and become destroyed at $T = T_c$).

Dimensionless ratio of these characteristic lengths:

$$\kappa = \frac{\delta(T)}{\xi(T)} = \frac{mcb^{1/2}}{(2\pi)^{1/2}|e|\hbar} \tag{6.102}$$

defines the so called Ginzburg–Landau parameter. Depending on its value, all superconductors are divided into two classes with significantly different properties in an external magnetic field: superconductors with $\kappa < \frac{1}{\sqrt{2}}$ are called type I superconductors, while those with $\kappa > \frac{1}{\sqrt{2}}$ are called type II superconductors. Most superconductors used for practical applications, as well as all high-temperature superconductors are in fact type II superconductors.

Let us derive one remarkable result following from Equation (6.95) and the Meissner effect. Consider a superconductor forming a torus and place it in an external magnetic field. We assume that both diameters of the torus are much larger than the penetration depth and coherence length (macroscopic torus). Now we can show that the value of the magnetic flux through the torus is quantized – it may only be integer units of the elementary "flux quantum", expressed via the fundamental constants (flux quantization). Deep inside the superconducting torus (outside the border region defined by the penetration depth) the current density is obviously zero $\mathbf{j} = 0$ (there is no field to induce the current), while the vector potential is non-zero (only its rotor is zero, so that $\mathbf{B} = 0$). Consider some closed contour C, going around the torus inside its body, far from its surface. Circulation of $\mathbf{A}$ along the contour C coincides with the magnetic flux through the contour, i.e. with the flux Φ through the torus:

$$\oint \mathbf{A}d\mathbf{l} = \int \operatorname{rot}\mathbf{A}d\mathbf{f} = \int \mathbf{B}d\mathbf{f} \equiv \Phi. \tag{6.103}$$

On the other hand, taking (6.95) equal to zero and integrating it around the contour we get:

$$\oint \mathbf{A}d\mathbf{l} = \frac{\hbar c}{2e}\oint \nabla\phi d\mathbf{l} = \frac{\hbar c}{2e}\delta\phi, \tag{6.104}$$

where $\delta\phi$ is the change of the phase of the wave function while we go around the contour. Demanding the single-valuedness of the wave function, after we perform a total circulation (one or several times), we conclude that this change of phase can only be 2π, multiplied by an integer. Thus we obtain:

$$\Phi = n\phi_0 \quad \text{where} \quad \phi_0 = \frac{\pi\hbar c}{|e|} = 2\ 10^{-7}\,\text{Gauss cm}^2, \tag{6.105}$$

where n is an integer. The value of ϕ_0 represents an elementary quantum of magnetic flux. This remarkable result is directly confirmed by the experiments, which is, by the way, a direct proof that (super)current carriers in superconductors are quasiparticles with an electric charge equal to $2e$ (Cooper pairs).

If we consider a massive cylinder in an external (longitudinal) magnetic field B made of a type I superconductor, it will a undergo first order phase transition to the normal state, if we reach the thermodynamic critical field B_{ct} discussed above. For a type II superconductor, even before we reach the thermodynamic critical field B_{ct} it becomes favorable thermodynamically to form some small regions of normal phase inside the cylinder and the unusual penetration of the magnetic field to the body of the superconductor, in the form of the so-called Abrikosov's vortices of normal phase, oriented along the field, and allowing the magnetic field to penetrate inside. It only becomes possible, once the external field reaches the value of the so-called first (or lower) critical field B_{c1}. For $B < B_{c1}$ superconductor is in the usual Meissner state (no field inside). If we start with a metal in the normal state in a high external field, the lowering of this field up to some second (or upper) critical magnetic field $B_{c2} > B_c$ makes it favorable for finite regions of the superconducting phase to form inside the normal metal. Thus, in the field region $B_{c1} < B < B_{c2}$ a type II superconductor is in the mixed (Shubnikov) phase. The phase diagram of such a superconductor in a magnetic field is shown schematically in Figure 6.3.

The value of B_{c2} can be determined from GL theory. It is clear that for $B < B_{c2}$, but close to it, nuclei of the superconducting phase possess small values of the order parameter Ψ ($\Psi \to 0$ for $B \to B_{c2}$). Then we can write down the linearized GL equation:

$$\frac{1}{4m}\left(-i\hbar\nabla - \frac{2e}{c}\mathbf{A}\right)^2 \Psi = |a|\Psi, \tag{6.106}$$

which has the form of Schroedinger equation for a particle with mass $2m$ and charge $2e$ in a magnetic field. The value of $|a|$ in the right-hand side of this equation plays the role of an energy level. The boundary condition at infinity is $\Psi = 0$. Now remember the quantum mechanical (Landau) problem of a charged particle in a constant homogeneous magnetic field [5]. The minimal value of the energy of such a particle is $E_0 = \hbar\omega_B/2$, where the cyclotron frequency $\omega_B = 2|e|B/2mc = |e|B/mc$. Starting from this value we have a continuous energy spectrum. Thus our superconducting nuclei can exist only for:

$$|a| > \frac{|e|\hbar}{2mc}B \tag{6.107}$$

so that

$$B_{c2} = \frac{2mc|a|}{|e|\hbar} = \sqrt{2}\kappa B_c = \phi_0 \frac{1}{2\pi\xi^2(T)}, \tag{6.108}$$

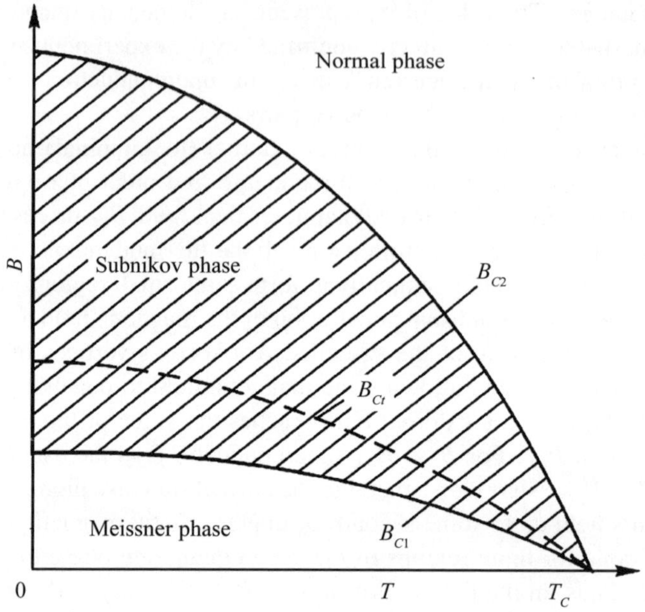

Figure 6.3. Phase diagram of a type II superconductor in a magnetic field. The dashed line shows the thermodynamic critical field B_{ct}.

where we have introduced $\phi_0 = \frac{\pi c \hbar}{|e|}$ – the elementary flux quantum of superconductivity theory introduced above, and also determining the magnetic flux through a single Abrikosov's vortex. During the derivation of the last equalities we have used Equations (6.88), (6.101) and (6.102).

The description of the vortex structure of the mixed state of type II superconductors by Abrikosov remains one of the most remarkable achievements of Ginzburg–Landau theory. Unfortunately, we have to limit ourselves to this qualitative discussion.

Finally, let us briefly discuss the limits of applicability of GL theory. First of all, it is necessary to satisfy the condition for $T_c - T \ll T_c$, which is equivalent to $\xi(T) \gg \xi_0$. Then we can use the Landau expansion. However, for $T \to T_c$ the validity of GL theory is limited also by the general condition for the applicability of Landau theory of phase transitions, connected with the growth of order parameter fluctuations in the immediate vicinity of T_c (in the so called critical region to be discussed later). In case of superconductivity this is a very weak limitation. Later, during the discussion of order parameter fluctuations in Landau theory we shall see that its region of validity (where we can neglect fluctuations) is expressed via GL coefficients by the following inequality:

$$T_c - T \gg \frac{b^2 T_c^2}{\alpha(\hbar^2/m)^3}. \tag{6.109}$$

Estimating the right-hand side here using microscopic values of these coefficients, derived in BCS theory, we get:

$$\frac{T_c - T}{T_c} \gg \left(\frac{T_c}{\varepsilon_F}\right)^4. \tag{6.110}$$

Due to the smallness of the ratio $T_c/\varepsilon_F \sim 10^{-3} - 10^{-4}$ in usual superconductors (usual metals), we can conclude that this limitation is practically irrelevant. The situation changes in high-temperature superconductors, where the critical region becomes observable.

Chapter 7

Fluctuations

7.1 Gaussian distribution

Physical observables, characterizing the macroscopic body, are, with high accuracy equal to their average values. However, small deviations from the average value always take place – there are fluctuations! Let us discuss the ways to find the probability distributions for fluctuations[1].

Consider an arbitrary closed system, and let x be some physical parameter, characterizing our system or its part. In the following it is convenient to assume that the average value $\langle x \rangle$ is already subtracted from x, so that below we always have $\langle x \rangle = 0$. In most cases $\langle x \rangle = x^*$, where x^* is the most probable value of x. During our general discussion of entropy, we have seen (cf. (1.180)), knowledge of the entropy as a function of some macroscopic parameters $x = (x_1, x_2, \ldots, x_n)$, we can find the probability of their specific values as:

$$w(x) = C \exp\{S(E, N, V, x)\}, \tag{7.1}$$

which is called Boltzmann's principle. Thus, the probability to find a value of some physical characteristic x in the interval $x, x + dx$ is proportional to $\exp S(x)$, where $S(x)$ is entropy as function of an exact value of x. This is a way to define the probability distribution of x in a most general way, allowing us to find the appropriate average values and fluctuations. Equation (7.1) is the starting point of the theory of fluctuations developed by Einstein.

If $\langle x \rangle$ is not subtracted from x, we should note that in equilibrium the entropy is equal to $S_0 = S(\langle x \rangle)$. Then the probability for the system to be in a state characterized by the value of x, belonging to the interval $\langle x \rangle, \langle x \rangle + dx$, takes the form:

$$dw = w(x)dx = \tilde{C} \exp[S(x) - S(\langle x \rangle)]dx = \tilde{C} e^{\Delta S} dx, \tag{7.2}$$

where ΔS is the entropy change due to the fluctuation dx. In Equation (7.1) the value of $e^{-S(\langle x \rangle)}$ is simply absorbed into the normalization constant C.

Consider the limits for the applicability of Equations (7.1) and (7.2). All previous arguments implicitly assumed the classical nature of x. Thus, it is necessary to find the condition for the quantum effects to be neglected. From quantum mechanics it is known [5] that the quantum indeterminacy of energy and some other physical variable

---

[1] Below we mainly follow [1].

x are related by the following relation:

$$\Delta E \Delta x \sim \hbar \dot{x}, \tag{7.3}$$

where $\dot{x}$ is the classical time derivative of x[2].

Let τ be a characteristic time of change of x, so that $\dot{x} \sim x/\tau$ and

$$\Delta E \Delta x \sim \frac{\hbar x}{\tau}. \tag{7.8}$$

It is clear that we can speak about a well defined value of x only if $\Delta x \ll x$, so that it is necessary to have

$$\Delta E \gg \frac{\hbar}{\tau} \tag{7.9}$$

i.e. the quantum indeterminacy of the energy must be large in comparison to $\hbar/\tau$. Then the entropy of the system has an indeterminacy

$$\Delta S \gg \frac{\hbar}{T\tau}. \tag{7.10}$$

For Equations (7.1) and (7.2) to be valid, it is necessary for the indeterminacy of entropy to be small in comparison to unity:

$$T \gg \frac{\hbar}{\tau} \quad \tau \gg \frac{\hbar}{T}. \tag{7.11}$$

This is the condition we seek. At low enough temperatures and in the case of very fast changes of x in time (small τ!), these fluctuations can not be considered as classical

---

[2] Consider two physical variables f and g, described by operators satisfying the commutation relations:

$$\hat{f}\hat{g} - \hat{g}\hat{f} = -i\hbar\hat{c}, \tag{7.4}$$

where $\hat{c}$ is also some operator. In the quasiclassical limit $\hbar \to 0$, in first approximation $\hat{c}$ can be replaced by c-number. Then:

$$\hat{f}\hat{g} - \hat{g}\hat{f} = -i\hbar c. \tag{7.5}$$

This commutation relation is similar to $p_x x - x p_x = -i\hbar$, but with $\hbar \to \hbar c$. Then, in analogy with the Heisenberg relation $\Delta x \Delta p_x \sim \hbar$, we can conclude that in the quasiclassical approximation f and g satisfy the following indeterminacy relation:

$$\Delta f \Delta g \sim \hbar c. \tag{7.6}$$

In particular, when one of the variables is the energy $f \equiv H$ and the second operator ($\hat{g}$) does not depend on time, using the $\dot{g} = \frac{i}{\hbar}(\hat{H}\hat{g} - \hat{g}\hat{H})$, we obtain $c = \dot{g}$ and quasiclassical indeterminacy relation takes the form:

$$\Delta E \Delta g \sim \hbar \dot{g}. \tag{7.7}$$

For $g = x$ it reduces to (7.3).

(thermodynamic), they become quantum fluctuations! Below we shall limit ourselves only to the case of classical fluctuations.

Let us return to Equation (7.1). The entropy S has a maximum at $x = \langle x \rangle = 0$. Then:

$$\left.\frac{\partial S}{\partial x}\right|_{x=0} = 0, \qquad \left.\frac{\partial^2 S}{\partial x^2}\right|_{x=0} < 0. \tag{7.12}$$

The value of x due to the fluctuation is small. Expanding $S(x)$ in powers of x up to the second order, we have:

$$S(x) = S(0) - \frac{\beta}{2}x^2; \quad \beta > 0. \tag{7.13}$$

Substituting this into (7.1), we obtain:

$$w(x)dx = Ae^{-\frac{\beta}{2}x^2}dx. \tag{7.14}$$

The normalization constant A is defined by $\int_{-\infty}^{\infty} dxw(x) = 1$, giving $A = \sqrt{\beta/2\pi}$.

We see that the probability distribution of fluctuations in x is given by *Gaussian law*:

$$w(x) = \sqrt{\frac{\beta}{2\pi}}e^{-\frac{\beta}{2}x^2}. \tag{7.15}$$

The average square of the fluctuation is equal to:

$$\langle x^2 \rangle = \int_{-\infty}^{\infty} dxx^2 w(x) = \frac{1}{\beta}. \tag{7.16}$$

Thus, the Gaussian distribution can also be written as:

$$w(x) = \frac{1}{\sqrt{2\pi\langle x^2 \rangle}} \exp\left(-\frac{x^2}{2\langle x^2 \rangle}\right). \tag{7.17}$$

Function $w(x)$ has a sharper maximum for smaller values of $\langle x^2 \rangle$.

Knowledge of $\langle x^2 \rangle$ allows us to find a similar characteristic for any function $\varphi(x)$. Due to the smallness of x we have:

$$\langle (\Delta\varphi)^2 \rangle = \left(\frac{d\varphi}{dx}\right)^2_{x=0} \langle x^2 \rangle. \tag{7.18}$$

Similarly, we can determine the probability of simultaneous fluctuations of several thermodynamic variables. Let us denote these deviations from equilibrium (average) values as $x_1, x_2, \ldots, x_n$. Introducing the entropy $S(x_1, x_2, \ldots, x_n)$ we write this probability distribution as $w dx_1 \cdots dx_n \sim \exp[S(x_1, \ldots, x_n)] dx_1 \cdots dx_n$. Expanding S in powers of x_i up to terms of second order, we get:

$$\Delta S = -\frac{1}{2} \sum_{i,k=1}^{n} \beta_{ik} x_i x_k = -\frac{1}{2} \beta_{ik} x_i x_k, \tag{7.19}$$

which is a negative quadratic form. Obviously $\beta_{ik} = \beta_{ki}$. In the last equality we assume the usual rule of summation over repeating indices. Then:

$$w = A \exp\left(-\frac{1}{2} \beta_{ik} x_i x_k\right), \tag{7.20}$$

where A is defined by the normalization $\int dx_1 \cdots dx_n w = 1$. Further calculations proceed as follows. Let us make a linear transformation of x_i:

$$x_i = a_{ik} x'_k \tag{7.21}$$

diagonalizing the quadratic form $\beta_{ik} x_i x_k$. To get:

$$\beta_{ik} x_i x_k = x_i'^2 \equiv x'_i x'_k \delta_{ik} \tag{7.22}$$

we require that the coefficients of our transformation (7.22) satisfy the condition:

$$\beta_{ik} a_{il} a_{km} = \delta_{lm}. \tag{7.23}$$

The determinant of the matrix in the left-hand side is equal to the product of the determinants:

$$\beta a^2 = 1 \quad \beta = \operatorname{Det} \beta_{ik} \quad a = \operatorname{Det} a_{ik}. \tag{7.24}$$

The Jacobian of the linear transformation $x_i \to x'_i$ is equal to a. Then, after the linear transformation in (7.21) the integral is factorized into the product of n identical integrals. Taking into account (7.24) we get:

$$Aa \left[\int_{-\infty}^{\infty} dx' \exp\left(-\frac{1}{2} x'^2\right)\right]^n = \frac{A}{\sqrt{\beta}} (2\pi)^{n/2} = 1. \tag{7.25}$$

Finally, the Gaussian distribution for several variables is written as:

$$w = \frac{\sqrt{\beta}}{(2\pi)^{n/2}} \exp\left(-\frac{1}{2} \beta_{ik} x_i x_k\right) \quad \beta = \operatorname{Det} |\beta_{ik}| \tag{7.26}$$

using it we can find:

$$\langle x_i x_k \rangle = \beta_{ik}^{-1}, \tag{7.27}$$

where β_{ik}^{-1} is the matrix element of the matrix inverse to β_{ik}. For statistically independent fluctuations in x_1 and x_2 the average of their product factorizes: $\langle x_1 x_2 \rangle = \langle x_1 \rangle \langle x_2 \rangle = 0$, so that $\beta_{12}^{-1} = 0$. In the case of a Gaussian distribution, the inverse theorem is also valid. If $\langle x_1 x_2 \rangle = 0$ (i.e. $\beta_{12}^{-1} = 0$), then fluctuations in x_1 and x_2 are statistically independent.

7.2 Fluctuations in basic physical properties

Let us calculate the mean square fluctuations in basic the thermodynamic variables of some separate small part of a macroscopic body. This small part is assumed to still contain a large enough number of particles.

For such variables as energy or volume, which also have a direct mechanical interpretation, the notion of fluctuation is obvious. However, it needs clarification for such variables as entropy and temperature, as a definition of these variables is necessarily connected with the system's evolution during finite time intervals.

The probability w can be written as:

$$w \sim \exp \Delta S, \tag{7.28}$$

where ΔS is the entropy change due to fluctuation. From thermodynamics we know [1] that

$$\Delta S = -\frac{R_{\min}}{T_0}, \tag{7.29}$$

where $R_{\min}$ is the minimal necessary work for a reversible change of thermodynamic variables in the given small part of the body (due to fluctuation), while the rest of the system plays the role of a bath with temperature T_0. Thus:

$$w \sim \exp\left(-\frac{R_{\min}}{T_0}\right). \tag{7.30}$$

Now we can substitute here (for fixed temperature and pressure of the bath):

$$R_{\min} = \Delta E - T_0 \Delta S + P_0 \Delta V, \tag{7.31}$$

where $\Delta E, \Delta S, \Delta V$ are changes of energy, entropy and volume of the small part of the system due to fluctuation, while T_0 and P_0 are the temperature and pressure of the bath, i.e. the temperature and pressure of our system in equilibrium. Below, we drop the index zero and understand that the coefficients are taken at equilibrium. The we obtain:

$$w \sim \exp\left(-\frac{\Delta E - T\Delta S + P\Delta V}{T}\right) \sim \exp\left(-\frac{\Delta \Phi}{T}\right), \tag{7.32}$$

where $\Delta\Phi$ is the change of thermodynamic potential due to fluctuation. For $\Delta V = 0$, i.e. in the absence of volume fluctuations, we have:

$$w \sim \exp\left(-\frac{\Delta F}{T}\right), \tag{7.33}$$

here ΔF is the free energy change due to fluctuation.

Note that Equations (7.32) and (7.33) are actually applicable to arbitrary fluctuations, both small and large. In case of small fluctuations we may proceed as follows. Expanding ΔE in a power series we get:

$$\Delta E - T\Delta S + P\Delta V = \frac{1}{2}\left[\frac{\partial^2 E}{\partial S^2}(\Delta S)^2 + 2\frac{\partial^2 E}{\partial S \partial V}\Delta S \Delta V + \frac{\partial^2 E}{\partial V^2}(\Delta V)^2\right], \quad (7.34)$$

where the first order terms in the expansion of ΔE canceled out as $\frac{\partial E}{\partial S} = T$ and $\frac{\partial E}{\partial V} = -P$. It is easily seen that (7.34) can be rewritten as:

$$\frac{1}{2}\left[\Delta S \Delta \left(\frac{\partial E}{\partial S}\right)_V + \Delta V \Delta \left(\frac{\partial E}{\partial V}\right)_S\right] = \frac{1}{2}(\Delta S \Delta T - \Delta P \Delta V). \quad (7.35)$$

Then we obtain the probability of fluctuation as:

$$w \sim \exp\left(\frac{\Delta P \Delta V - \Delta T \Delta S}{2T}\right). \quad (7.36)$$

From this general expression we can find the fluctuations in different thermodynamic variables.

First, let us choose V and T as independent variables. Then:

$$\Delta S = \left(\frac{\partial S}{\partial T}\right)_V \Delta T + \left(\frac{\partial S}{\partial V}\right)_T \Delta V = \frac{C_v}{T}\Delta T + \left(\frac{\partial P}{\partial T}\right)_V \Delta V, \quad (7.37)$$

$$\Delta P = \left(\frac{\partial P}{\partial T}\right)_V \Delta T + \left(\frac{\partial P}{\partial V}\right)_T \Delta V. \quad (7.38)$$

Substituting these expressions into (7.36) we can see, that terms with $\Delta V \Delta T$ cancel, and what remains is:

$$w \sim \exp\left\{-\frac{C_v}{2T^2}(\Delta T)^2 + \frac{1}{2T}\left(\frac{\partial P}{\partial V}\right)_T (\Delta V)^2\right\}. \quad (7.39)$$

This expression factorizes into two factors, depending only on ΔT or ΔV. Thus, the fluctuations in temperature and volume are statistically independent:

$$\langle \Delta T \Delta V \rangle = 0. \quad (7.40)$$

Comparing each of the two factors in Equation (7.39) with the general expression for the Gaussian distribution (7.17), we find the following expressions for the mean square fluctuations in temperature and volume:

$$\langle (\Delta T)^2 \rangle = \frac{T^2}{C_v}, \quad (7.41)$$

$$\langle (\Delta V)^2 \rangle = -T\left(\frac{\partial V}{\partial P}\right)_T. \quad (7.42)$$

Positivity of these expressions is guaranteed by the thermodynamic inequalities $C_v > 0$ and $(\partial P/\partial V)_T < 0$ [1].

Now, choose P and S as independent variables in Equation (7.36). Then:

$$\Delta V = \left(\frac{\partial V}{\partial P}\right)_S \Delta P + \left(\frac{\partial V}{\partial S}\right)_P \Delta S, \tag{7.43}$$

$$\Delta T = \left(\frac{\partial T}{\partial S}\right)_P \Delta S + \left(\frac{\partial T}{\partial P}\right)_S \Delta P = \frac{T}{C_p}\Delta S + \left(\frac{\partial T}{\partial P}\right)_S \Delta P. \tag{7.44}$$

But according to $dW = TdS + VdP$ we have $\left(\frac{\partial V}{\partial S}\right)_P = \frac{\partial^2 W}{\partial P \partial S} = \left(\frac{\partial T}{\partial P}\right)_S$, then:

$$\Delta V = \left(\frac{\partial V}{\partial P}\right)_S \Delta P + \left(\frac{\partial T}{\partial P}\right)_S \Delta S. \tag{7.45}$$

Substituting ΔV and ΔT into (7.36), we obtain:

$$w \sim \exp\left\{\frac{1}{2T}\left(\frac{\partial V}{\partial P}\right)_S (\Delta P)^2 - \frac{1}{2C_p}(\Delta S)^2\right\}. \tag{7.46}$$

As before, this expression factorizes in two factors, depending on ΔP and ΔS. Thus:

$$\langle(\Delta S)^2\rangle = C_p, \tag{7.47}$$

$$\langle(\Delta P)^2\rangle = -T\left(\frac{\partial P}{\partial V}\right)_S. \tag{7.48}$$

From relations obtained above, it is seen that mean square fluctuations in additive thermodynamic variables, such as volume and entropy, are proportional to the size (volume) of those part of the system, to which they are related. Accordingly, these fluctuations are $\sim\sqrt{V}$, while the relative fluctuations are $\sim 1/\sqrt{V}$. At the same time, for temperature and pressure the mean square fluctuations are already inversely proportional to the square root of the volume.

Expressions for the fluctuations in thermodynamic variables can also be obtained directly from the Gibbs distribution. As an example, let us consider the fluctuations of particle numbers. Using the grand canonical distribution we have:

$$\langle N\rangle = e^{\frac{\Omega}{T}} \sum_N N e^{\frac{\mu N}{T}} \sum_n e^{-\frac{E_{nN}}{T}}. \tag{7.49}$$

Differentiating this expression with respect to μ (at constant V and T), we get:

$$\begin{aligned}\frac{\partial\langle N\rangle}{\partial\mu} &= \frac{1}{T}e^{\frac{\Omega}{T}} \sum_N \left(N^2 + N\frac{\partial\Omega}{\partial\mu}\right) e^{\frac{\mu N}{T}} \sum_n e^{-\frac{E_{nN}}{T}} \\ &= \frac{1}{T}\left(\langle N^2\rangle + \langle N\rangle\frac{\partial\Omega}{\partial\mu}\right).\end{aligned} \tag{7.50}$$

But $\partial\Omega/\partial\mu = -\langle N\rangle$, so that:

$$\frac{\partial\langle N\rangle}{\partial\mu} = \frac{1}{T}(\langle N^2\rangle - \langle N\rangle^2) = \frac{1}{T}\langle(\Delta N)^2\rangle, \tag{7.51}$$

and accordingly:

$$\langle(\Delta N)^2\rangle = T\,(\partial\langle N\rangle/\partial\mu)_{T,V}\,. \tag{7.52}$$

From these expressions, it is clear that the mean square fluctuations in such variables as energy, volume and pressure tend to zero as $T \to 0$. This is a general property of all thermodynamic variables, which also have a direct mechanical meaning, but in general it is not so for such purely thermodynamic variables as entropy and temperature. According to Equation (7.41), for fixed energy, we can not attribute a well defined temperature to our system, since it fluctuates, and Equation (7.41) characterizes the limits for a precise determination of the temperature of an isolated system.

7.3 Fluctuations in ideal gases

Consider now the calculations of $\langle(\Delta N)^2\rangle$ from another point of view. According to Equation (7.42) fluctuations in volume are given by $\langle(\Delta V)^2\rangle = -T\left(\frac{\partial V}{\partial P}\right)_T$. Dividing both parts of this equality by N^2, we find the fluctuation in the volume per one particle:

$$\langle(\Delta\frac{V}{N})^2\rangle = -\frac{T}{N^2}\left(\frac{\partial V}{\partial P}\right)_T. \tag{7.53}$$

This allows us to find the fluctuation in the particle number in any separate volume inside the body. The volume V is fixed, so that $\Delta\frac{V}{N} = V\Delta\frac{1}{N} = -\frac{V}{N^2}\Delta N$, and substitution into Equation (7.53) gives:

$$\langle(\delta N)^2\rangle = -T\frac{N^2}{V^2}\left(\frac{\partial V}{\partial P}\right)_T. \tag{7.54}$$

Using now the equation of state of an ideal gas, giving $V = NT/P$, we obtain:

$$\langle(\Delta N)^2\rangle = N. \tag{7.55}$$

Then the relative fluctuation is:

$$\frac{\langle(\Delta N)^2\rangle^{1/2}}{N} = \frac{1}{\sqrt{N}}. \tag{7.56}$$

Consider now the fluctuations in the particle distribution over different quantum states. Let n_k be the number of particles in k-th quantum state. Due to the total independence of this (sub)system of particles from the rest of the system (gas), we may apply Equation (7.52):

$$\langle(\Delta n_k)^2\rangle = T\frac{\partial\langle n_k\rangle}{\partial\mu}. \tag{7.57}$$

For a Fermi-gas, after the substitution of $\langle n_k\rangle = [e^{(\varepsilon_k-\mu)/T}+1]^{-1}$ we obtain:

$$\langle(\Delta n_k)^2\rangle = \langle n_k\rangle(1-\langle n_k\rangle). \tag{7.58}$$

Similarly, for a Bose-gas:

$$\langle(\Delta n_k)^2\rangle = \langle n_k\rangle(1+\langle n_k\rangle). \tag{7.59}$$

For a Boltzmann gas, after substitution of $\langle n_k\rangle = e^{(\mu-\varepsilon_k)/T}$, we obtain:

$$\langle(\Delta n_k)^2\rangle = \langle n_k\rangle. \tag{7.60}$$

Previous expressions of Equations (7.58) and (7.59) reduce to (7.60) for $n_k \ll 1$. Let us sum (7.58) and (7.59) over the group of G_j close levels, containing $N_j = \sum n_k$ particles. Due to the statistical independence of fluctuations in different n_k we have:

$$\langle(\Delta N_j)^2\rangle = G_j\langle n_j\rangle(1\mp\langle n_j\rangle) = N_j\left(1\mp\frac{\langle N_j\rangle}{G_j}\right), \tag{7.61}$$

where $\langle n_j\rangle$ is the average value of $\langle n_k\rangle$ close to each other, and $\langle N_j\rangle = \langle n_j\rangle G_j$.

These expressions can be applied e.g. to a photon gas, putting in (7.59) $\mu = 0$. Consider the set of quantum states of photons (in volume V) with close frequencies, belonging to a small interval $\Delta\omega_j$. The number of relevant states is $G_j = V\omega_j^2\Delta\omega_j/\pi^2c^3$. The total energy of the quanta in this frequency interval is given by $E_{\Delta\omega_j} = N_j\hbar\omega_j$. Multiplying (7.61) by $(\hbar\omega_j)^2$ and dropping the index j, we obtain the following Einstein expression for the fluctuation in the energy $E_{\Delta\omega}$ of a photon gas, in the given frequency interval $\Delta\omega$:

$$\langle(\Delta E_{\Delta\omega})^2\rangle = \hbar\omega E_{\Delta\omega} + \frac{\pi^2c^3(E_{\Delta\omega})^2}{V\omega^2\Delta\omega}. \tag{7.62}$$

Let us consider also fluctuations in particle number within the given volume of an ideal gas V. In principle we can analyze large enough fluctuations with $N-\langle N\rangle$ of the order of $\langle N\rangle$. This is relevant only for a Boltzmann gas, as in Fermi and Bose gases the probability of such fluctuations becomes noticeable only in such small volumes, that quantum fluctuations become important. According to the grand canonical ensemble, the distribution of N particles of the gas over different quantum states is proportional to

$$\exp\left\{\frac{\Omega+\mu N-\sum\varepsilon_k}{T}\right\}, \tag{7.63}$$

where $\sum \varepsilon_k$ is the sum of the energies of the particles. To obtain the probability distribution w_N we have to sum this expression over all states of the particles in the given volume V. Performing the summation independently over the states of each particle, we have to divide the result by $N!$, so that:

$$w_N = \frac{e^{\Omega/T}}{N!}\left(\sum_k e^{\frac{\mu-\varepsilon_k}{T}}\right)^N. \tag{7.64}$$

The sum written in this expression is simply the average number of particles in the given volume:

$$\sum_k e^{\frac{\mu-\varepsilon_k}{T}} = \langle N\rangle. \tag{7.65}$$

Then:

$$w_N = \text{const}\frac{\langle N\rangle^N}{N!}, \tag{7.66}$$

and finding const $= e^{-\langle N\rangle}$ from normalization[3] we obtain:

$$w_N = \frac{\langle N\rangle^N e^{-\langle N\rangle}}{N!}, \tag{7.67}$$

which is the so-called Poisson distribution. Using it we can directly show [1], that the mean square fluctuation in particle number is again:

$$\langle(\Delta N)^2\rangle = \langle N\rangle \tag{7.68}$$

and this expression is valid not only for large, but also for arbitrary values of $\langle N\rangle$.

[3] This reduces to $\Omega = -PV = -\langle N\rangle T$, in accordance with the equation of state of an ideal gas.

Chapter 8

Phase transitions and critical phenomena

8.1 Mean-field theory of magnetism

This Chapter is devoted to an elementary introduction into theory of second order phase transitions and critical phenomena. The simplest microscopic model of such a phase transition is the mean (or molecular) field theory of Curie and Weiss, which gives a qualitative description of a phase transition in ferromagnets. This model allows us to study the main aspects of the general problem, which are also characteristic for all other types of second order phase transitions.

Consider first the statistical mechanics of free spins in an external magnetic field (e.g. paramagnet with localized magnetic moments). The Hamiltonian of the system of N non interacting spins $\mathbf{S}_i$ in an external magnetic field $\mathbf{H}$ is written as:

$$H = -g\mu_B \sum_{i=1}^{N} \mathbf{S}_i \mathbf{H}, \tag{8.1}$$

where g is gyromagnetic ratio and $\mu_B = \frac{e\hbar}{2mc}$ is the Bohr magneton. To shorten notations in the following we introduce $\tilde{\mu} = g\mu_B$. Quantum states of spin are defined by its projections on an external magnetic field, which are given $2S + 1$ possible values ($m_i = -S, -S + 1, \ldots, S - 1, S$).

The partition functions of this system of spins takes the form:

$$Z = \sum_S \exp\left(-\frac{\tilde{\mu}}{T} \sum_{i=1}^{N} \mathbf{S}_i \mathbf{H}\right) = \sum_{m_1=-S}^{m_1=S} \cdots \sum_{m_N=-S}^{m_N=S} \exp\left(x \sum_{i=1}^{N} m_i\right), \tag{8.2}$$

where

$$x \equiv \frac{\tilde{\mu} H}{T}. \tag{8.3}$$

The summation in Equation (8.2) is especially simple in the case of $S = 1/2$:

$$\begin{aligned} Z &= \prod_{i=1}^{N} \left\{ \sum_{m_i=-1/2}^{m_i=1/2} \exp(x m_i) \right\} \\ &= \prod_{i=1}^{N} 2ch\left(\frac{1}{2}x\right) = 2^N ch^N\left(\frac{1}{2}x\right). \end{aligned} \tag{8.4}$$

For arbitrary S we have:

$$Z = \left\{ \frac{\exp(-xS)[1 - \exp\{(2S+1)x\}]}{1 - \exp(x)} \right\}^N = \left[\frac{sh\{(S+1/2)x\}}{sh(x/2)} \right]^N. \tag{8.5}$$

The free energy is now given by:

$$F(T,H) = -T \ln Z = -NT \ln \left[\frac{sh\{(S+1/2)x\}}{sh(x/2)} \right]. \tag{8.6}$$

Then the magnetization is obtained as:

$$M(T,H) = -\left(\frac{\partial F}{\partial H} \right)_T = T \frac{\partial}{\partial H} \ln Z = M_0 B_S(Sx), \tag{8.7}$$

where $M_0 \equiv M(T=0, H=0) = NS\tilde{\mu} = NSg\mu_B$ is the maximal possible value of the magnetization, while

$$B_S(x) \equiv \frac{2S+1}{2S} cth\left(\frac{2S+1}{2S} x \right) - \frac{1}{2S} cth\left(\frac{1}{2S} x \right) \tag{8.8}$$

is the so called Brillouin function. This function relates the magnetization of the paramagnet to the value of an external magnetic field, which is shown graphically in Figure 8.1. For the case where $S = 1/2$ the Brillouin function is given by:

$$B_{1/2}\left(\frac{1}{2} x \right) = 2cth(x) - cth(x/2) = th(x/2). \tag{8.9}$$

From Figure 8.1 we see that $M = 0$ for $H = 0$, which is in fact obvious for a paramagnet state. In ferromagnets the situation is different, spins interact with each other and at low temperatures the system acquires a spontaneous magnetization, which exists also in the absence of an external magnetic field, i.e. for $H = 0$. The basic assumption of the mean field theory approach to magnetic ordering is that spin–spin interaction produces within the system some mean (or "molecular") magnetic field H_m, which is to be added to the external field H. It is also assumed that this field is just proportional to the internal magnetization of the system

$$H_m = \lambda M(T,H) \tag{8.10}$$

so that an effective field acting upon each spin is given by:

$$H_{\text{eff}} = H + \lambda M(T,H). \tag{8.11}$$

The parameter $\lambda > 0$ is called the molecular field parameter. All relations derived above remain valid and we only have to substitute $H \to H_{\text{eff}}$. In particular, after such a substitution, Equation (8.7) reduces to:

$$M = M_0 B_S \left[\frac{\tilde{\mu} S}{T} (H + \lambda M) \right]. \tag{8.12}$$

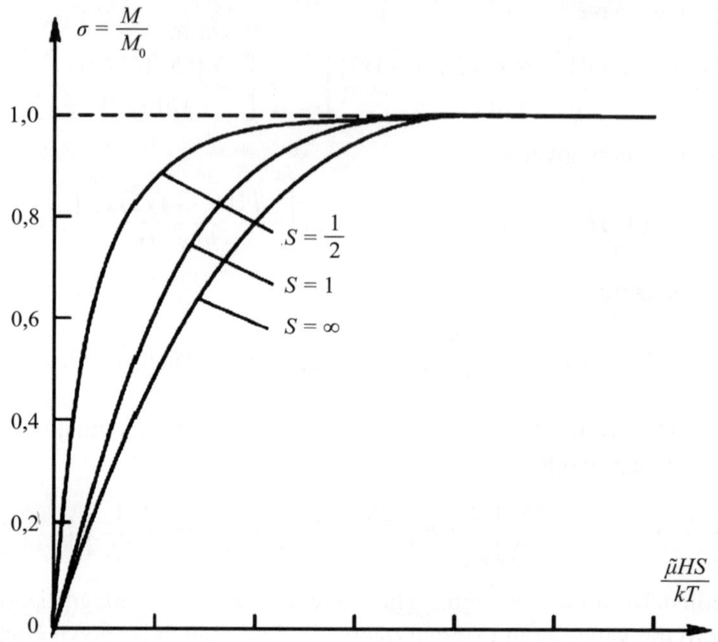

Figure 8.1. Dependence of relative magnetization of a paramagnet $\sigma = M/M_0$ on the parameter $\tilde{\mu}HS/T$, described by the Brillouin function for different values of spin S.

Setting now $H = 0$ we get the equation determining the magnetization M:

$$M = M_0 B_S\left(\frac{\tilde{\mu}\lambda M}{T} S\right). \tag{8.13}$$

A graphic solution of this equation is shown in Figure 8.2. Equation (8.13) possesses the trivial solution $M = 0$ for arbitrary values of the temperature T. However, there is also the possibility of a second (non trivial) solution for $M \neq 0$, when the initial slope of the curve, representing the right-hand side of Equation (8.13), is steeper than the left-hand side. To analyze this situation analytically, we perform a Taylor expansion of Brillouin function:

$$B_S(x) = \frac{S+1}{3S}x - \frac{S+1}{3S}\frac{2S^2+2S+1}{30S^2}x^3 + \cdots. \tag{8.14}$$

Then the initial slope of the curve, for the right-hand side of (8.13) is defined by:

$$M_0\left(\frac{S+1}{3S}\right)\frac{\tilde{\mu}S\lambda}{T} = C\frac{\lambda}{T}, \tag{8.15}$$

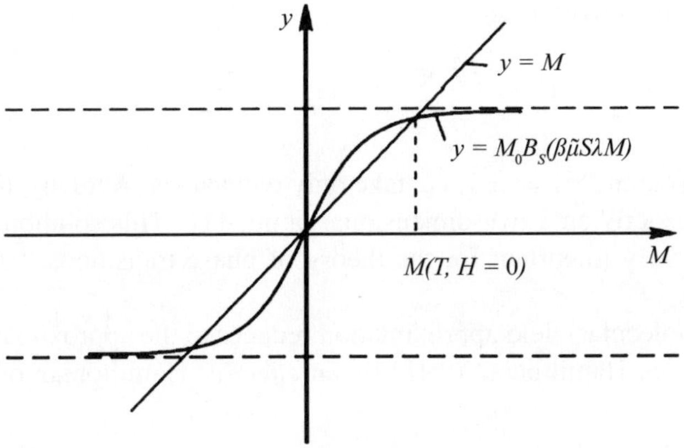

Figure 8.2. Graphic solution of the equation for the magnetization in molecular (mean) field theory ($\beta = 1/T$).

where we have introduced the so called Curie constant:

$$C \equiv \frac{N\tilde{\mu}^2 S(S+1)}{3} \tag{8.16}$$

expressing M_0 via microscopic parameters, in accordance with an expression after Equation (8.7). Now from Equation (8.15) we can see that the non trivial solution exists for $T < \lambda C$, giving the value of the critical temperature of the ferromagnetic phase transition in mean field theory:

$$T_c = \lambda C. \tag{8.17}$$

For lower temperatures $M \neq 0$ even in the absence of an external magnetic field. The transition temperature T_c obviously tends to zero as $\lambda \to 0$, when we return to the case of a paramagnet.

Let us consider the origin of the molecular field from the microscopic point of view. The majority of models for magnetic ordering are based upon the concept of the exchange interaction between spins, which in the simplest case can be described by the Heisenberg model, with the interaction Hamiltonian written as:

$$H = -\frac{1}{2}\sum_{i\neq j} J_{ij}\mathbf{S}_i\mathbf{S}_j - \tilde{\mu}\sum_i \mathbf{S}_i\mathbf{H}, \tag{8.18}$$

where J_{ij} is the so-called exchange integral, which is taken to be positive (the case of ferromagnetic ordering).

Very popular is also the simplified version of this model called the Ising model, described by the Hamiltonian (8.18) with only S_z spin components left. Usually, the

Ising Hamiltonian is written as:

$$H = -\frac{1}{2}\sum_{i \neq j} J_{ij} s_i s_j - \tilde{\mu}\sum_i s_i H, \tag{8.19}$$

where the Ising "spins" $s_i = \pm 1$, i.e. take only two values. Actually, the Ising model can be solved exactly on a two-dimensional lattice [1]. This solution, first obtained by Onsager, is very important for the theory of phase transitions, but we shall not describe it here.

The mean (molecular) field approximation reduces to the approximate replacement of the microscopic Hamiltonian (8.18) by an *effective* Hamiltonian of the following form:

$$H = -\sum_{i \neq j} J_{ij} \langle S_z \rangle S_{iz} - \tilde{\mu}\sum_i S_{iz} H, \tag{8.20}$$

where the external magnetic field is assumed to be oriented along the z-axis, while $\langle S_z \rangle$ denotes the average value of the z-component of the spin on an arbitrary lattice site. It is clear that (8.20) describes the system of free (non-interacting) spins in an effective (mean or molecular) field, oriented along the z-axis and given by:

$$H_{\text{eff}} = H + \frac{J_0}{\tilde{\mu}}\langle S_z \rangle = H + \frac{J_0}{N\tilde{\mu}^2}M, \tag{8.21}$$

where

$$J_0 = \sum_j J_{ij} \quad J_{ii} = 0. \tag{8.22}$$

It can be said that the molecular field on the given lattice site is actually the mean magnetic field, which is self-consistently created on this site by all other spins of the system. Comparing Equations (8.21) and (8.11), we can see that the molecular field constant λ in this model is determined by the following expression:

$$\lambda = \frac{J_0}{N\tilde{\mu}^2}. \tag{8.23}$$

From Equation (8.17) it now follows that the critical temperature of the ferromagnetic phase transition (Curie temperature) is given by:

$$T_c = \frac{1}{3} J_0 S(S+1). \tag{8.24}$$

In case of spins interacting with nearest neighbors only, i.e. for $J_{ij} = J$ when the site j is one of the z nearest neighbors of site i, while for other cases $J_{ij} = 0$, we have:

$$T_c = \frac{1}{3} zJS(S+1). \tag{8.25}$$

Let us return to the simplest case where $S = 1/2$. According to Equations (8.9) and (8.12) we can write:

$$M = M_0 th\left[\frac{1}{2T}\tilde{\mu}(H + \lambda M)\right]. \tag{8.26}$$

Introducing the dimensionless (relative) variables $\sigma = M/M_0$ and $t = T/T_c$, we can rewrite Equation (8.26) as:

$$\sigma = th\left(\frac{1}{2}\frac{\tilde{\mu}H}{T} + \frac{\sigma}{t}\right). \tag{8.27}$$

Using $th(x + y) = \frac{thx+thy}{1+(thx)(thy)}$ we rewrite (8.27) as:

$$h \equiv th\left(\frac{\tilde{\mu}H}{2T}\right) = \frac{\sigma - th(\sigma/t)}{1 - \sigma th(\sigma/t)}. \tag{8.28}$$

Near the critical point ($H = 0$, $M = 0$, $T = T_c$) all arguments of the hyperbolic functions in (8.28) are small and we can perform the Taylor expansions: $thx = x - \frac{1}{3}x^3 + \frac{2}{15}x^5 + \cdots$. Then:

$$h = \sigma\left(1 - \frac{1}{t}\right) + \sigma^3\left[\frac{1}{3t^3} + \frac{1 - 1/t}{t}\right]. \tag{8.29}$$

This is the so-called magnetic equation of state, which determines the behavior of all the relevant physical characteristics of the magnet close to the critical point.

For example, we can consider the magnetization and the magnetic susceptibility. From Equation (8.28) it is seen, that in a zero external field $h = 0$ and for $T < T_c$ Equation (8.29) takes the form:

$$\sigma^2 = \frac{T_c/T - 1}{\frac{T_c^3}{3T^3} + \frac{T_c}{T}\left(1 - \frac{T_c}{T}\right)} + \cdots \approx 3\left(\frac{T}{T_c}\right)^2 \frac{T_c - T}{T_c}. \tag{8.30}$$

Thus, we obtain the following behavior of the magnetization in a zero field close to T_c (for $T < T_c$):

$$\sigma \sim |\tau|^\beta \quad \tau = \frac{T - T_c}{T_c}, \tag{8.31}$$

where the *critical exponent* of the magnetization (order parameter) $\beta = 1/2$.

The isothermal susceptibility in a zero field $\chi_T = \left(\frac{\partial M}{\partial H}\right)_T$ satisfies the following relation:

$$\chi_T = \left(\frac{\partial M}{\partial \sigma}\right)_T \left(\frac{\partial \sigma}{\partial h}\right)_T \left(\frac{\partial h}{\partial H}\right)_T = \left(\frac{1}{2}N\tilde{\mu}\right)\left(\frac{\tilde{\mu}}{2T}\right)\left(\frac{\partial \sigma}{\partial h}\right)_T = \frac{C}{T}\left(\frac{\partial \sigma}{\partial h}\right)_T, \tag{8.32}$$

where the Curie constant was taken from (8.16) for the case of $S = 1/2$. Differentiating both sides of (8.29) with respect to h for $T \approx T_c$, we get:

$$1 = \frac{\partial \sigma}{\partial h}\left[\left(1 - \frac{1}{t}\right) + 3\sigma^2\left(\frac{1}{3t^3}\right)\right] \tag{8.33}$$

or, using (8.32),

$$\chi_T = \frac{C}{T}\left[\frac{\tau}{t} + \frac{\sigma^3}{t^3}\right]^{-1}. \tag{8.34}$$

Then, for $T > T_c$ we have $\sigma = 0$ for $H = 0$ and (8.34) reduces to:

$$\chi_T = \frac{C}{T}\left(\frac{T_c}{T}\frac{T - T_c}{T_c}\right)^{-1} = \frac{C}{T - T_c} \sim \tau^{-\gamma}, \tag{8.35}$$

where the critical exponent of susceptibility $\gamma = 1$. For $T < T_c$, according to (8.30) we have $\sigma^2 \approx -3\tau$, so that from (8.34) we get:

$$\chi_T \approx \frac{1}{2}\frac{C}{T}\frac{1}{(-\tau)} \sim |\tau|^{-1} \tag{8.36}$$

and the critical exponent of susceptibility for $T < T_c$ is also $\gamma' = 1$.

Direct calculations within the mean (molecular) field model show that the specific heat of the system at $T = T_c$ has a discontinuity $\Delta C_H = 3/2N$. Within this model we can also study the critical behavior of a number of other physical characteristics of the system, described by the appropriate critical exponents.

In general, the molecular field model (approximation) gives a rather satisfactory qualitative description of the ferromagnetic phase transition. It is easily generalized to the case of an antiferromagnetic transition. In fact, this model is the origin of a number of similar mean field models for the microscopic description of different phase transitions in many physical systems. For example, the BCS model of superconductivity, described above, is the typical mean field model, where the relevant "mean field" is described by the anomalous averages (6.17), (6.18), while the Hamiltonians (6.15) or (6.16) are direct analogs of (8.20) [1]. In superconductivity theory, this approach actually gives a very accurate description of the system's behavior close to T_c. For a majority of other phase transitions, e.g. in real magnetics, this description is only qualitative, the experimental values of the critical exponents are significantly different from mean field theory predictions. The physical reason for these discrepancies is the increasing role of the fluctuations in the critical region close to T_c. We shall return to this problem below.

[1] The BCS Hamiltonian can even be rewritten via some "pseudospin" operators (introduced by Anderson), when it is reduced to practically the same form as (8.20).

8.2 Quasi-averages*

The microscopic theory of phase transitions addresses the very important question of the degeneracy of the system's ground state and the closely related problem of the proper definition of statistical averages. Consider as an example a Heisenberg ferromagnet, described by the Hamiltonian (8.18). In the absence of an external magnetic field (for $\mathbf{H} = 0$) this Hamiltonian is obviously invariant with respect to rotations in three-dimensional space. It is clear, as in this case (8.18) depends only on scalar products of spins on different lattice sites. However, the ferromagnetic ground state is not invariant with respect to three-dimensional rotations – spontaneous magnetization has a definite direction in space and the system is invariant only with respect to rotations around this direction. At the same time, it is obvious that the other ground state of the same system, characterized by the other direction of the magnetization vector, corresponds to the same energy. Accordingly, there is an infinite set of ground states, differing only by the directions of the magnetization. The introduction of an external magnetic field (even infinitesimal) breaks this degeneracy and allows well defined calculations of all statistical averages. This leads to the concept of *quasi-averages* [23] – one of the central concepts in the theory of phase transitions.

Let us return to the Heisenberg model in the absence of an external magnetic field:

$$H = -\frac{1}{2}\sum_{i \neq j} J_{ij}\mathbf{S}_i\mathbf{S}_j. \tag{8.37}$$

The total spin of this system:

$$\mathbf{S} = \sum_j \mathbf{S}_j \tag{8.38}$$

is an integral of motion (this is valid for each of its components, in quantum mechanics each one commutes with the Hamiltonian of the system). Consider now the commutation relations:

$$\begin{aligned} S_x S_y - S_y S_x &= i S_z \\ S_y S_z - S_z S_y &= i S_x \\ S_z S_x - S_x S_z &= i S_y \end{aligned} \tag{8.39}$$

Using these relations we can write:

$$i\,\mathrm{Sp}\left(S_z e^{-\frac{H}{T}}\right) = \mathrm{Sp}\left[(S_x S_y - S_y S_x)e^{-\frac{H}{T}}\right]. \tag{8.40}$$

As S_x commutes with H we get:

$$\mathrm{Sp}\left(S_y S_x e^{-\frac{H}{T}}\right) = \mathrm{Sp}\left(S_y e^{-\frac{H}{T}} S_x\right) = \mathrm{Sp}\left(S_x S_y e^{-\frac{H}{T}}\right), \tag{8.41}$$

so that

$$\mathrm{Sp}\left(S_z e^{-\frac{H}{T}}\right) = 0. \tag{8.42}$$

Similarly we find that:

$$\mathrm{Sp}\left(S_x e^{-\frac{H}{T}}\right) = 0 \quad \mathrm{Sp}\left(S_y e^{-\frac{H}{T}}\right) = 0. \tag{8.43}$$

Let us introduce the magnetization of the unit volume as:

$$\mathbf{M} = \frac{\tilde{\mu}}{V} \sum_j \mathbf{S}_j = \frac{\tilde{\mu}}{V} \mathbf{S}. \tag{8.44}$$

Then:

$$\mathrm{Sp}\left(\mathbf{M} e^{-\frac{H}{T}}\right) = 0 \tag{8.45}$$

so that the average magnetization:

$$\langle \mathbf{M} \rangle = \lim_{V \to \infty} \frac{\mathrm{Sp}\left(\mathbf{M} e^{-\frac{H}{T}}\right)}{\mathrm{Sp}\left(e^{-\frac{H}{T}}\right)} = 0. \tag{8.46}$$

Thus, the standard definition of the statistical (Gibbs) average leads to zero average magnetization, which correspond to the invariance of the system with respect to three-dimensional rotations.

Let us stress that this result is valid for arbitrary temperatures, e.g. for temperatures below the Curie temperature. It may seem paradoxical, as for $T < T_c$ the system acquires a spontaneous magnetization. However, the direction of the magnetization vector in the absence of an external field is arbitrary, so that the (statistical) equilibrium state is actually infinitely degenerate.

Let us introduce the external magnetic field $\nu\mathbf{e}(\nu > 0, \mathbf{e}^2 = 1)$, replacing the Hamiltonian (8.37) by

$$H_{\nu e} = H + \nu V \mathbf{e}\mathbf{M}. \tag{8.47}$$

Then, for temperatures below the Curie temperature we have

$$\langle \mathbf{M} \rangle = \mathbf{e} M_\nu, \tag{8.48}$$

where M_ν will have a finite (nonzero) limit as the intensity ν of the external field tends to zero. Formally, we can say that here we observe a kind of "instability" of the usual definition of averages due to the addition to the Hamiltonian of a term with an infinitesimal external field[2] and the average value of $\langle \mathbf{M} \rangle$ acquires the finite value:

$$\mathbf{e}m \quad \text{where} \quad m = \lim_{\nu \to 0} M_\nu. \tag{8.49}$$

[2] It is assumed that we first perform the thermodynamic limit of statistical mechanics $V \to \infty$, and only after that we tend ν to zero.

Now it is convenient to introduce the concept of the *quasi-average*. Consider some dynamic variable A, built on spin operators. Then its quasi-average is defined as:

$$\prec A \succ = \lim_{\nu \to 0} \langle A \rangle_{\nu e}, \tag{8.50}$$

where $\langle A \rangle_{\nu e}$ is the usual statistical average of A with Hamiltonian $H_{\nu e}$.

Thus, the degeneracy is actually reflected in the quasi-averages via their dependence on the arbitrary direction of the unit vector **e**. The usual average is given by:

$$\langle A \rangle = \int \prec A \succ d\mathbf{e} \tag{8.51}$$

i.e. is obtained by integration over all directions of **e**. Obviously quasi-averages are more convenient and "physical", in comparison with usual averages, if we are dealing with degenerate equilibrium states. In fact, in practical calculations in phase transition theory we are always using quasi-averages (explicitly or implicitly).

As another example we can mention the BCS theory of superconductivity. As we noted above, the BCS state breaks the gauge symmetry related to particle number conservation, which is reflected in the appearance of anomalous averages like (6.17), (6.18). Here we do not have the real physical field, breaking this symmetry, as in the case of an external magnetic field breaking the rotational symmetry of a Heisenberg ferromagnet. However, we can instead introduce the fictitious infinitesimal "source" of Cooper pairs in the BCS Hamiltonian (6.12), writing it as:

$$H_\nu = H - \nu \sum_{\mathbf{p}} \left[a_{-\mathbf{p}\downarrow} a_{\mathbf{p}\uparrow} + a^+_{\mathbf{p}\uparrow} a^+_{-\mathbf{p}\downarrow} \right], \tag{8.52}$$

which explicitly breaks particle number conservation (gauge symmetry). Accordingly, all the averages in the superconducting state are to be understood as quasi-averages obtained with the Hamiltonian (8.52), with $\nu \to 0$ at the end of the calculations. Naturally, all these averages depend on the arbitrary phase angle ϕ. While discussing the superconducting state above we just assumed $\phi = 0$, which is quite similar to fixing the direction of magnetization of the Heisenberg ferromagnet in the mean field theory approach, which we oriented along the arbitrary direction of the z-axis, defined by the direction of an external magnetic field. Quite similarly we can analyze the Bose condensation [23].

In fact, discussing any kind of phase transition we always assume the introduction of an infinitesimal Bogolyubov's field or "source", lifting (breaking) the appropriate symmetry. Then, during all calculations we have to take into account appropriate anomalous averages, breaking the symmetry of the initial Hamiltonian. The "condensed" state after the phase transition (appearing for $T < T_c$) is characterized by finite values of the anomalous averages, which remain nonzero even after the external field (or "source") is put to zero, i.e. for $\nu \to 0$. In the "normal" phase (for $T > T_c$), anomalous averages tend to zero as $\nu \to 0$, and the appropriate symmetry remains unbroken. In this sense, all phase transitions of second order are associated with "spontaneous" breaking of some (usually continuous) symmetry.

8.3 Fluctuations in the order parameter

Let us discuss now fluctuations in the order parameter. We have already noted above that these fluctuations become important near the critical transition temperature, significantly modifying the results of mean field theories. Our analysis will be essentially based on Landau theory, as a typical mean field theory of second order phase transitions.

In most cases, the order parameter in Landau theory can be represented by an n-component vector, either in the usual coordinate space, or in some associated space, according to the nature of the symmetry breaking during the phase transition. In the Heisenberg model this is the usual three-component vector (magnetization), in Ginzburg–Landau superconductivity theory this is the complex (i.e. two-component) wave function of the Cooper pairs condensate etc. Below we shall analyze the simplest possible variant of the phase transition, described by a single-component order parameter η, which corresponds e.g. to the Ising model[3].

In thermodynamics, the minimal work necessary to create some fluctuation out of the equilibrium state of the system (at fixed pressure and temperature) is equal to the appropriate change of the thermodynamic potential $\Delta\Phi$. Thus, according to Equation (7.32) the probability of a fluctuation at fixed P and T is estimated as:

$$w \sim \exp\left(-\frac{\Delta\Phi}{T}\right). \tag{8.53}$$

Let us denote the equilibrium value of η as $\bar{\eta}$. For a small deviation from equilibrium write:

$$\Delta\Phi = \frac{1}{2}(\eta - \bar{\eta})^2 \left(\frac{\partial^2\Phi}{\cdot}\partial\eta^2\right)_{P,T} \tag{8.54}$$

The equilibrium value of the order parameter is determined by the Landau expansion:

$$\Phi(T, P, \eta) = \Phi_0(P, T) + at\eta^2 + B\eta^4 - \eta h V, \tag{8.55}$$

where $t = T - T_c(P)$, and h is an external field interacting with the order parameter (e.g. a magnetic field in the Ising model). Using Equation (8.55) we define the equilibrium value of the order parameter $\bar{\eta}$ from:

$$\left(\frac{\partial\Phi}{\partial\eta}\right)_{T,h} = 0, \tag{8.56}$$

which reduces to:

$$2at\bar{\eta} + 4B\bar{\eta}^3 = hV, \tag{8.57}$$

[3] We drop the discussion of the very important symmetry aspects of Landau theory, related to the specific type of crystal lattice [1] and assume our system to be homogeneous and isotropic.

which is equivalent to the result (8.29) derived from mean (molecular) field theory. The solution of Equation (8.57) for $h \to 0$ has the form:

$$\bar{\eta}^2 = 0 \qquad \text{for } t > 0$$
$$\bar{\eta}^2 = -\frac{at}{2B} \qquad \text{for } t < 0 \tag{8.58}$$

so that the critical exponent of the order parameter is equal to $1/2$, the same value as in Equation (8.31).

The susceptibility is defined as:

$$\chi = \left(\frac{\partial \bar{\eta}}{\partial h}\right)_{T;h\to 0}. \tag{8.59}$$

Differentiating (8.57), we obtain for $h \to 0$:

$$\frac{\partial \bar{\eta}}{\partial h} = \frac{V}{2at + 12B\bar{\eta}^2}. \tag{8.60}$$

Substituting now (8.58) we get:

$$\chi = \frac{V}{2at} \qquad \text{for } t > 0$$
$$\chi = \frac{V}{-4at} \qquad \text{for } t < 0, \tag{8.61}$$

which is similar to Equations (8.35), (8.36) and demonstrate the divergence of $\chi \sim |T - T_c|^{-1}$, so that the critical exponent of susceptibility $\gamma = \gamma' = 1$, as obtained from the molecular field approximation. In fact, Landau theory is a typical mean field theory and all critical exponents are obtained in the same way as in similar microscopic models.

Using (8.60) we can write:

$$\chi = V\left[\left(\frac{\partial^2 \Phi}{\partial \eta^2}\right)_{h=0}\right]^{-1}. \tag{8.62}$$

Thus, the probability of a fluctuation is determined from (8.53) and (8.54) by the following expression:

$$\Delta\Phi = \frac{1}{2}(\eta - \bar{\eta})^2 \frac{V}{\chi}, \tag{8.63}$$

$$w \sim \exp\left[-\frac{(\eta - \bar{\eta})^2 V}{2\chi T_c}\right]. \tag{8.64}$$

Now, in accordance with the general form of the Gaussian distribution (7.17), we obtain the mean square of the order parameter fluctuation as:

$$\langle(\Delta\eta)^2\rangle = \frac{T_c\chi}{V} \sim \frac{1}{|t|} \quad \text{for } T \to T_c. \tag{8.65}$$

We see that the fluctuations grow close to T_c and diverge as $\sim|T - T_c|^{-1}$.

For a deeper understanding of the physical nature of this phenomenon, it is useful to find the spatial correlation function of order parameter fluctuations. For an inhomogeneous system (fluctuations actually create inhomogeneities!) the thermodynamic potential is conveniently written as $\Phi = \int dV \Phi(\mathbf{r})$, where $\Phi(\mathbf{r})$ is its density (which is a function of the coordinate). We shall actually use the thermodynamic potential $\Omega(T, \mu)$ and consider some volume V within the body, containing a variable number of particles N. The potential $\Omega(T, \mu, \eta)$, for the unit volume, can be expanded in the usual Landau form, similar to (8.55):

$$\Omega(T, \mu, \eta) = \Omega_0(T, \mu) + \alpha t \eta^2 + b\eta^4 - \eta h, \tag{8.66}$$

where $\alpha = a/V$, $b = B/V$, $t = T - T_c(\mu)$. This form of expansion is valid for the homogeneous case. In inhomogeneous systems it must contain spatial derivatives of the order parameter η. For long-wavelength fluctuations we can limit ourselves to the lowest order derivatives and their lowest powers. Terms linear in the derivatives like $f(\eta)\frac{\partial\eta}{\partial x_i}$ reduce to surface integrals after volume integration, thus corresponding to irrelevant surface effects. We shall limit ourselves to the simplest case (valid for crystals with cubic symmetry), when the density of thermodynamic potential can be written as:

$$\Omega = \Omega_0 + \alpha t \eta^2 + b\eta^4 + g(\nabla\eta)^2 - \eta h. \tag{8.67}$$

For the homogeneous state to be stable, we have to require that $g > 0$. In the opposite case Ω does not have a minimum for $\eta = \text{const}$.

Considering fluctuations as fixed μ and T, we write the fluctuation probability as:

$$w \sim \exp\left(-\frac{\Delta\Omega}{T}\right) \tag{8.68}$$

because the minimal work required to bring the system out of equilibrium under these conditions is given by $R_{\min} = \Delta\Omega$.

Let us consider fluctuations in a symmetric (e.g. paramagnetic) phase (at $h = 0$), when $\bar{\eta} = 0$, so that $\Delta\eta = \eta$. Limiting ourselves to second order terms in the fluctuations, we can write the change of Ω as[4]:

$$\Delta\Omega = \int dV\{\alpha t(\eta)^2 + g(\nabla\eta)^2\}. \tag{8.70}$$

Let us introduce the Fourier expansion of $\eta(\mathbf{r})$:

$$\eta(\mathbf{r}) = \sum_{\mathbf{k}} \eta_{\mathbf{k}} e^{i\mathbf{kr}} \quad \eta_{-\mathbf{k}} = \eta_{\mathbf{k}}^*. \tag{8.71}$$

Then its gradient can be written as:

$$\nabla\eta(\mathbf{r}) = \sum_{\mathbf{k}} i\mathbf{k}\eta_{\mathbf{k}} e^{i\mathbf{kr}}. \tag{8.72}$$

Substitution of these expressions into Equation (8.70) and volume integration leaves only non-zero terms, containing the products, like $\eta_{\mathbf{k}}\eta_{-\mathbf{k}} = |\eta_{\mathbf{k}}|^2$. Then we obtain:

$$\Delta\Omega = V\sum_{\mathbf{k}}(gk^2 + \alpha t)|\eta_{\mathbf{k}}|^2 \tag{8.73}$$

so that:

$$\langle|\eta_{\mathbf{k}}|^2\rangle = \frac{T}{2V(gk^2 + \alpha t)}. \tag{8.74}$$

This expression is usually called the Ornstein–Zernike correlator. From this expression it is clear that only the long-wavelength fluctuations with $k \sim \sqrt{\alpha t/g}$ grow as $t \to 0$. Actually, the expression (8.74) is valid only for long enough wavelengths k^{-1}, which are large in comparison to the average interatomic distance a.

Let us define the correlation function in coordinate space as:

$$G(\mathbf{r}_1 - \mathbf{r}_2) = \langle\eta(\mathbf{r}_1)\eta(\mathbf{r}_2)\rangle. \tag{8.75}$$

This can be calculated as:

$$G(\mathbf{r}) = \sum_{\mathbf{k}}\langle|\eta_{\mathbf{k}}|^2\rangle e^{i\mathbf{kr}} = V\int\frac{d^3k}{(2\pi)^3}e^{i\mathbf{kr}}\langle|\eta_{\mathbf{k}}|^2\rangle. \tag{8.76}$$

---

4 Note that quite similar results can be obtained at the other side of the transition, in the broken symmetry phase. Here we have non-zero $\bar{\eta} = (-\alpha t/2b)^{1/2}$ and for the change of Ω, up to terms of the order of $\sim(\Delta\eta)^2$, we get:

$$\Delta\Omega = \int dV\{-2\alpha t(\Delta\eta)^2 + g(\nabla\eta)^2\}. \tag{8.69}$$

Thus, for any characteristics of the system, we obtain expressions, which differ from those for the symmetric phase by substitution of αt by $2\alpha|t|$.

Then from (8.74) we obtain[5]:

$$G(r) = \frac{T_c}{8\pi g r} \exp\left(-\frac{r}{\xi}\right), \tag{8.78}$$

where

$$\xi = \sqrt{\frac{g}{\alpha t}} \sim (T - T_c)^{-1/2}. \tag{8.79}$$

The parameter ξ is called the correlation length of the fluctuations and defines the characteristic distance for the decay of their correlations. We have already encountered this length in the Ginzburg–Landau theory, where it was called the coherence length. The divergence of ξ for $T \to T_c (T > T_c)$ corresponds to the appearance (at $T = T_c$) of *long range order*. The correlation length critical exponent $\nu = 1/2$, which is again the standard result of the mean field theory.

For $r = 0$ the integral in (8.76) determines the average square of the order parameter fluctuation $\eta(\mathbf{r})$ at the given point of space. Its divergence is directly related to the inapplicability of Equation (8.74) for large $k \sim a^{-1}$. This is easily avoided by the introduction of the cutoff:

$$G(0) = \frac{T}{4\pi^2} \int_0^{k_0} dk k^2 \frac{1}{gk^2 + \alpha t}, \tag{8.80}$$

where $k_0 \sim 1/a$. Here we observe a significant dependence on the spatial dimensions. For d-dimensional space instead of (8.80) we have to write:

$$G(0) \sim \int_0^{k_0} dk k^{d-1} \frac{1}{k^2 + \xi^{-2}}. \tag{8.81}$$

This integral is easily estimated as:

$$G(0) \sim \int_{\xi^{-1}}^{k_0} dk k^{d-3} \sim \begin{cases} k_0 - \xi^{-1} & d = 3 \\ \ln(k_0 \xi) & d = 2 \\ \xi - \frac{1}{k_0} & d = 1. \end{cases} \tag{8.82}$$

[5] Here we use the following expressions for the Fourier transformation:

$$\int dV \frac{e^{-\kappa r}}{r} e^{i\mathbf{kr}} = \frac{4\pi}{k^2 + \kappa^2}$$
$$\int \frac{d^3k}{(2\pi)^3} \frac{e^{i\mathbf{kr}}}{k^2 + \kappa^2} = \frac{e^{-\kappa r}}{4\pi r}. \tag{8.77}$$

These are most easily obtained if we note that $\varphi(r) = \frac{e^{-\kappa r}}{4\pi r}$ satisfies the differential equation: $\nabla^2 \varphi - \kappa^2 \varphi = -4\pi\delta(r)$. Multiplying both sides of this equation by $e^{-i\mathbf{kr}}$ and integrating over all space (performing partial integration of $e^{-i\mathbf{kr}} \nabla^2 \varphi$ twice) we obtain the required result.

From this estimate we see that for $T \to T_c$, when $\xi \to \infty$, the average square of the order parameter fluctuation at the given point is finite for $d = 3$ and diverges for $d = 1, 2$. This reflects the impossibility of the existence of long range order in one-dimensional and two-dimensional systems [1]. Let us stress that here the relevant divergence of the integral in (8.82) is at the lower integration limit ("infrared" divergence), not at the upper limit, where it is regularized by a cutoff. In the theory of critical phenomena, a spatial dimensionality $d = 2$ is called the *lower critical dimensionality*. The reasoning presented here is rather crude, but qualitatively valid. More accurate proof of the impossibility of long range order in low-dimensional systems also requires an analysis of the situation for $T < T_c$ [26]. In fact, the lower critical dimensionality $d = 2$ is valid only for phase transitions breaking the continuous symmetry, while for Ising-like single-component order parameter the lower critical dimension $d = 1$. This is clear, for example, from an exact Onsager solution for the two-dimensional Ising model, which demonstrates the existence of the phase transition for $d = 2$ [1].

To avoid confusion we note that Equation (8.65) determines fluctuations in the order parameter η, averaged over the volume V with linear dimensions $L \gg \xi$. Let us denote it by $\langle \eta^2 \rangle_V$. The average of $\eta(\mathbf{r})$ over the volume V is given by $\eta_{\mathbf{k}=0}$. Thus, it is natural that for $k = 0$ (8.74) coincides with (8.65), so that:

$$\chi = \frac{V}{T_c} \int d\mathbf{r} G(\mathbf{r}). \tag{8.83}$$

The value of $\langle \eta^2 \rangle_V$ can also be directly obtained from the correlation function:

$$\langle \eta^2 \rangle_V = \frac{1}{V^2} \int d\mathbf{r}_1 d\mathbf{r}_2 \langle \eta(\mathbf{r}_1)\eta(\mathbf{r}_2) \rangle = \frac{1}{V} \int dV G(r). \tag{8.84}$$

Now we can formulate the criterion for the applicability of the Landau theory of phase transitions (or mean field theory), based on the expansion (8.67). For the validity of this theory we have to demand that the mean square fluctuations in the order parameter η, averaged over the correlation volume $\sim \xi^3$, be small compared with the equilibrium value of the order parameter $\bar{\eta}^2 \sim \alpha|t|/b$. Using (8.65) with $V \sim \xi^3$ we arrive at the condition:

$$\frac{T_c \chi}{\xi^3} \ll \frac{\alpha|t|}{b} \tag{8.85}$$

or, taking χ and ξ from (8.61) and (8.79):

$$\alpha|t| \gg \frac{T_c^2 b^2}{g^3}. \tag{8.86}$$

This condition is usually called the Ginzburg criterion for the applicability of the Landau theory of phase transitions[6]. This inequality defines the size of the so-called

[6] Expansion in powers of $t = T - T_c$ in Landau coefficients also requires the validity of condition $t \ll T_c$. For this to be in agreement with (8.86) it is also necessary to satisfy: $\frac{T_c b^2}{\alpha g^3} \ll 1$.

critical region around T_c, where fluctuations are large and significantly change the mean field picture of the phase transition, e.g. the critical exponents[7]. The description of the system within the critical region belongs to the field of the theory of critical phenomena [26]. Some aspects of this theory will be discussed in the next section.

8.4 Scaling

The theory of critical phenomena introduces the following standard set of characteristics of the system and appropriate critical exponents, determining the singular behavior of these characteristics at the critical point, as a function of the parameter $\tau = \frac{T-T_c}{T_c} \to 0$.

Order parameter:

$$\bar{\eta} \sim |\tau|^{\beta} \quad T \to T_c - 0, \tag{8.87}$$

$$\bar{\eta} \sim h^{\frac{1}{\delta}} \quad T = T_c. \tag{8.88}$$

Susceptibility:

$$\chi \sim \begin{cases} \tau^{-\gamma} & T \to T_c + 0 \\ |\tau|^{-\gamma'} & T \to T_c - 0. \end{cases} \tag{8.89}$$

Correlation function of the order parameter (d is spatial the dimensionality):

$$G(r) \sim \frac{\exp(-r/\xi)}{r^{d-(2-\eta)}}, \tag{8.90}$$

where the correlation length:

$$\xi \sim \begin{cases} \tau^{-\nu} & T \to T_c + 0 \\ |\tau|^{-\nu'} & T \to T_c - 0. \end{cases} \tag{8.91}$$

At the critical point itself:

$$G(r) \sim \frac{1}{r^{d-(2-\eta)}}, \tag{8.92}$$

$$G(k) \sim \frac{1}{k^{2-\eta}}. \tag{8.93}$$

The critical exponent α of the specific heat is introduced in a similar way:

$$C(\tau, h = 0) = \frac{A^+}{\alpha}[\tau^{-\alpha} - 1] + B^+ \quad T \to T_c + 0, \tag{8.94}$$

$$C(\tau, h = 0) = \frac{A^-}{\alpha'}[|\tau|^{-\alpha'} - 1] + B^- \quad T \to T_c - 0 \tag{8.95}$$

with $\alpha = 0$ corresponding to a logarithmic singularity.

[7] Above we already mentioned the Ginzburg criterion while discussing the limits of the Ginzburg–Landau theory of superconductivity. We have seen that in superconductors the size of the critical region is negligible.

The theoretical problem of the description of critical phenomena reduces to the derivation of these expressions and the calculation of the critical exponents $\alpha, \alpha', \beta, \gamma, \gamma', \delta, \eta, \nu, \nu'$.

Significant progress in the studies of critical phenomena was achieved after the introduction of the concept of *scaling* or scale invariance. This is essentially based on the idea that the growth of the correlation length close to T_c leads to significant interaction of the fluctuations, which defines the singular behavior of the physical characteristics at the critical point. At the same time, as the correlation length becomes much larger than the interatomic spacing $\xi \gg a$, the microscopic details of the interactions are probably not so important. The hypothesis of scale invariance (scaling) assumes that the singular dependence of the physical characteristics on $T - T_c$ is controlled by the divergence of the correlation length ξ, and it becomes the only relevant parameter of length in the problem.

Let us discuss scaling, using the simple qualitative arguments due to Kadanoff. For simplicity we consider the system of N Ising spins (cf. (8.19)) in a d-dimensional lattice, with interaction parameter J, different from zero only between nearest neighbors. The external magnetic field is H. Then the Hamiltonian (8.19) can be rewritten in units of T as:

$$\frac{H}{T} = -K \sum_{\langle ij \rangle} s_i s_j - h \sum_{i=1}^{N} s_i, \tag{8.96}$$

where we have introduced the dimensionless parameters $K = J/2T$ and $h = \tilde{\mu} H/T$.

Let us break the lattice into cells with linear dimensions La, where a is the lattice parameter and L is an arbitrary integer ($L \gg 1$). (cf. Figure 8.3). Then we obtain a total of $\mathcal{N} = N/L^d$ cells, each containing L^d spins. Below, we consider only temperatures close enough to T_c, so that the correlation length ξ is much larger than the size of a cell, i.e. $\xi \gg La$. It guarantees that each cell, containing L^d spins, with $1 \ll L \ll \xi/a$, contains only spins oriented "up" or "down". Then the total magnetic moment of each cell s_α ($\alpha = 1, 2, \ldots, \mathcal{N}$) can, in some sense, be considered as similar to the single site moment s_i. This assumption is qualitatively valid if the given cell is inside the group of correlated spins. The resulting moment of this cell is given by L^d, with $\pm$ sign. It is convenient to introduce $\tilde{s}_\alpha = s_\alpha / L^d$, i.e. normalize the spin of the cell to unity. Then, if we try to rewrite the Hamiltonian as a function of cell moments s_α (not site moments s_i), we can expect it to be of the same form as (8.96) for the standard Ising model, but with different values of the parameters, i.e. with K and h replaced by some K_L and h_L:

$$\frac{H}{T} = -K_L \sum_{\langle \alpha, \alpha' \rangle} \tilde{s}_\alpha \tilde{s}_{\alpha'} - h_L \sum_{\alpha} \tilde{s}_\alpha, \tag{8.97}$$

where the summation is performed over the Kadanoff cells numbered by α.

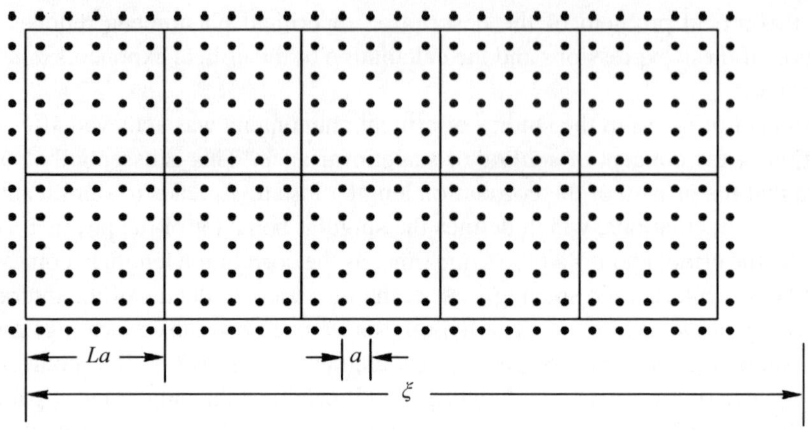

Figure 8.3. Kadanoff construction for an Ising lattice.

If the external magnetic field $h \to 0$, the effective field h_L in the cell formulation obviously also tends to zero. Similarly, as $T \to T_c$ and $K \to K_c$, with $K_c = \frac{J}{2T_c}$ given by the initial Ising model, we should get $K_L \to K_c$. Thus, we can assume the following scaling relations:

$$\tau_L = \tau L^y \quad \text{for } K_L = K_c - \tau L^y, \tag{8.98}$$

$$h_L = hL^x, \tag{8.99}$$

where $\tau = K_c - K$, $\tau_L = K_c - K_L$. Critical values of the interaction parameters are the same in both formulations, as we assumed their equivalence[8]. The critical exponents x and y remain undetermined, but we shall see that all other (physical) critical exponents can be expressed via these, so that only two critical exponents are independent.

Consider the change of free energy of the system under a small change of h. Let us assume that the magnetic field is different at different sites of the lattice, but these changes are smooth enough, so that it is effectively constant within each Kadanoff cell. Then, the change of the free energy is given by:

$$\delta\left(\frac{F}{T}\right) = -\sum_i \langle s_i\rangle \delta h_i = -\sum_\alpha \langle s_\alpha\rangle \delta h_{L\alpha}, \tag{8.100}$$

where $\langle s_i\rangle$ is an average spin at the lattice site, and $\langle s_\alpha\rangle$ is an average spin of a cell. Both expressions should be equivalent. Due to the assumption of a smooth change of

[8] Parameter τ, defined here, has the same meaning as above in the case where $J = \text{const}$. In principle, we can also consider the phase transition with the change of J at fixed temperature.

magnetic field in space, we can write within each cell:

$$L^d \langle s_i \rangle \delta h_i = \langle s_\alpha \rangle \delta h_{L\alpha}. \quad (8.101)$$

Using (8.99), we obtain:

$$\langle s_i \rangle = L^{x-d} \langle s_\alpha \rangle. \quad (8.102)$$

Consider now the homogeneous field, independent of site number i. Then the magnetization at the site (which is equivalent to the order parameter $\bar{\eta}$) is a function of τ and h only:

$$\langle s_i \rangle = F(\tau, h). \quad (8.103)$$

According to our basic assumption, in terms of s_α we are describing the same system, but with new values of τ_L and h_L, so that the value of $\langle s_\alpha \rangle$ is represented by the *same* function, depending on *new* variables:

$$\langle s_\alpha \rangle = F(\tau_L, h_L). \quad (8.104)$$

Then from Equations (8.100), (8.102), (8.103), (8.104) we can see that the order parameter can be written as:

$$\bar{\eta} = \langle s \rangle = F(\tau, h) = L^{x-d} F(L^y \tau, L^x h). \quad (8.105)$$

Now, the length L introduced above is a purely mathematical invention and should cancel from all physical characteristics of the system! This is possible only if the function $F(\tau, h)$ has the following form:

$$\bar{\eta} = \left(\frac{h}{|h|}\right) |\tau|^{\frac{d-x}{y}} f\left(\frac{\tau}{|h|^{\frac{y}{x}}}\right). \quad (8.106)$$

The factor $h/|h|$ here is added just to guarantee the change in sign of the magnetization with the sign of an external magnetic field.

The explicit form of function $f(z)$, entering (8.106), is unknown. However, these arguments allowed us to transform an unknown function of two variables τ and h into a function of a single variable $z = \tau/|h|^{\frac{y}{x}}$. Remarkably, this is sufficient to express all physical critical exponents of our system via the exponents x and y, or, in other words, express all physical critical exponents via any two of them (which can be determined from experiments).

For example, remembering (8.87), i.e. $\bar{\eta} \sim |\tau|^\beta$, which is valid for small negative values of τ and $h \to 0$, we note that $f(-\infty) = \text{const}$ and

$$\beta = \frac{d - x}{y}. \quad (8.107)$$

Differentiating (8.106) with respect to h for $h \to 0$, we get the susceptibility: $\chi \sim |\tau|^{\frac{d-x}{y}} \frac{\partial}{\partial h} f(\tau/|h|^{\frac{y}{x}}) \sim |\tau|^{\frac{d-x}{y}+1} |h|^{-\frac{y}{x}-1} f'(z)$. However, the dependence on h in χ should cancel for $h \to 0$. Then it is clear that $f'(z) \sim z^{-\frac{x}{y}-1}$ and $\chi \sim |\tau|^{-\gamma} \sim |\tau|^{\frac{d-2x}{y}}$. Thus we obtain:

$$\gamma = \gamma' = \frac{2x-d}{y}. \tag{8.108}$$

Similarly, for $\tau = 0$ according to (8.88) we should have $\bar{\eta} \sim h^{\frac{1}{\delta}}$. Equation (8.106) for $\tau = 0$, should become independent of τ, which is only possible if $f(z \to 0) \sim z^{\frac{x-d}{y}}$. Then from (8.106) we immediately obtain $\bar{\eta} \sim |h|^{\frac{d-x}{x}}$, so that

$$\delta = \frac{x}{d-x}. \tag{8.109}$$

From these relations we get:

$$d/y = \gamma + 2\beta = \beta(\delta + 1), \tag{8.110}$$

which gives the scaling relation between experimentally measurable exponents β, γ, δ.

Integrating $\bar{\eta} \sim \frac{\partial F}{\partial h} \sim |\tau|^{\frac{d-x}{y}} f(\tau/|h|^{y/x})$ it is easy to get

$$F \sim |\tau|^{\frac{d-x}{y}} \int dh f(\tau/|h|^{y/x}) \sim |\tau|^{\frac{d}{y}} \int dz \tilde{f}(z).$$

Then the specific heat is:

$$C \sim -T \frac{\partial^2 F}{\partial T^2} \sim |\tau|^{\frac{d}{y}-2}. \tag{8.111}$$

Comparing with (8.95), we obtain:

$$\alpha = \alpha' = 2 - \frac{d}{y} \quad \text{or} \quad \frac{d}{y} = 2 - \alpha \tag{8.112}$$

so that comparison with (8.110) gives:

$$\gamma + 2\beta = \beta(\delta + 1) = 2 - \alpha. \tag{8.113}$$

Consider now the correlation function, which is in general defined as:

$$G(\mathbf{r}_i - \mathbf{r}_j) = G(R, \tau, h) = \langle [s_i - \langle s \rangle][s_j - \langle s \rangle] \rangle, \tag{8.114}$$

where R is the distance between two lattice sites: $R = |\mathbf{r}_i - \mathbf{r}_j|/a$. In a similar way we can write down the correlation function in terms of the cell variables s_α, defined in (8.102). This expression is to be identical to $G(R,\tau,h)$, but with different scales of length, τ and h:

$$\begin{aligned} R &\to R/L \\ \tau &\to \tau_L = \tau L^y \\ h &\to h_L = hL^x. \end{aligned} \tag{8.115}$$

From here we get:

$$G(R,\tau,h) = L^{2(x-d)}\langle [s_\alpha - \langle s_\alpha\rangle][s'_\alpha - \langle s_\alpha\rangle]\rangle = L^{2(x-d)}G(R/L,\tau L^y,hL^x) \tag{8.116}$$

and $G(R,\tau,h)$ is independent of an arbitrary parameter L if we take:

$$G(R,\tau,h) = |\tau|^{2(d-x)/y}\tilde{G}(R|\tau|^{\frac{1}{y}},\tau/|h|^{y/x}) \tag{8.117}$$

for $R \gg 1$, $|\tau| \ll 1$ and $h \ll 1$.

Equation (8.117) determines the critical exponents ν, ν', η. We immediately observe (cf. (8.90), (8.91)) that for $h = 0$ the correlation length $\xi \sim |\tau|^{-1/y}$. Accordingly, its critical exponent is given by:

$$\frac{1}{y} = \nu = \nu' = \frac{2-\alpha}{d}. \tag{8.118}$$

Finally, the last of the critical exponents η is determined from (cf. (8.93)):

$$G(R,\tau=0,h=0) \sim \frac{1}{R^{d-2+\eta}}. \tag{8.119}$$

Then, demanding cancellation of the τ-dependence in (8.117) for $\tau \to 0$, we obtain $G(R) \sim R^{2(x-d)} \sim R^{2-d-\eta}$, so that:

$$-(d-2+\eta) = 2(x-d). \tag{8.120}$$

From Equation (8.109) we have $x = \frac{d\delta}{\delta+1}$, then from (8.120), using (8.113), we get:

$$d-2+\eta = \frac{2d}{\delta+1} = \frac{2d\beta}{2-\alpha} = \frac{2\beta}{\nu} \tag{8.121}$$

or

$$\beta = \frac{1}{2}(d-2+\eta)\nu. \tag{8.122}$$

From (8.110) and (8.118) we have $\gamma = \frac{d}{y} - 2\beta = d\nu - 2\beta$, and using (8.122) we obtain one more scaling relation:

$$(2-\eta)\nu = \gamma. \tag{8.123}$$

It is rather easy to also derive the following relations:

$$\frac{d\gamma}{2-\eta} = 2-\alpha,$$
$$\delta = \frac{d+2-\eta}{d-2+\eta}. \tag{8.124}$$

In the conclusion of our discussion we give the summary of the most widely used scaling relations between physical critical exponents:

$$\nu = \nu' = \frac{\gamma}{2-\eta}, \tag{8.125}$$
$$\alpha = \alpha' = 2 - \nu d, \tag{8.126}$$
$$\beta = \frac{1}{2}\nu(d-2+\eta). \tag{8.127}$$

Remarkably, all experiments in the critical region of widely different physical systems, undergoing phase transitions of the second order, confirm the scaling relations for critical exponents, derived here.

The theoretical problem of the *calculation* of the values of the critical exponents remained, for a rather long time, one of the most difficult problems of statistical physics. The physical reason for these difficulties was the strong interaction between fluctuations in the critical region and the absence of a natural small parameter for the development of some kind of perturbation theory. This problem was successfully solved by Wilson using a *renormalization group* approach, originating from quantum field theory. Renormalization group transformations are actually the modern and rigorous realization of scaling transformations, extending the elementary discussion given above. We shall not discuss this formalism here, referring the reader to special literature on the modern theory of critical phenomena [26] and limiting ourselves only to some qualitative results of this theory.

First of all, note that the values of the critical exponents obtained in Landau theory (mean field approximation):

$$\nu = \frac{1}{2} \quad \gamma = 1 \quad \eta = 0$$
$$\alpha = 0 \quad \beta = \frac{1}{2} \quad \delta = 3 \tag{8.128}$$

do not satisfy the scaling relations (8.127) and most experiments in real three-dimensional systems. At the same time, it is easy to see that the Landau theory exponents (8.128) *satisfy* the scaling relations if we formally take the space dimensionality $d = 4$. In this sense we can say that Landau theory gives the correct description of critical phenomena for spatial dimensionality $d = 4$ and, as is actually shown in modern theory [26], for all $d > 4$. The spatial dimensionality $d = 4$ is usually called the

Table 8.1. Critical exponents for the model with $n = 1$ (Ising).

Exponent	Wilson	Numerical	Landau
ν	0.626	0.642	0.5
η	0.037	0.055	0
γ	1.244	1.250	1
α	0.077	0.125	0
β	0.340	0.312	0.5
δ	4.460	5.15	3

upper critical dimension. A remarkable result of the modern theory of critical phenomena is the *universality* of critical behavior – the values of the critical exponents in different physical systems actually are determined only by the spatial dimensionality of the system and the number of components n of the order parameter (i.e. by the type of the symmetry broken at the phase transition).

Wilson proposed an original method to calculate critical exponents, based on perturbation theory with respect to an artificial small parameter $\varepsilon = 4 - d$ – a small deviation from the upper critical dimension $d = 4$, for which the critical exponents coincide with predictions of Landau (mean field) theory (ε-expansion). Below we present the theoretical values of the critical exponents up to terms of the order of $\sim\varepsilon^2$ with an n-component order parameter [26]:

$$\gamma = 1 + \frac{n+2}{n+8}\frac{\varepsilon}{2} + \frac{n+2}{n+8}\frac{n^2+22n+52}{(n+8)^2}\frac{\varepsilon^2}{4} + \cdots, \tag{8.129}$$

$$2\nu = 1 + \frac{n+2}{n+8}\frac{\varepsilon}{2} + \frac{n+2}{n+8}\frac{n^2+23n+60}{(n+8)^2}\frac{\varepsilon^2}{4} + \cdots, \tag{8.130}$$

$$\eta = \frac{n+2}{2(n+8)^2}\varepsilon^2 + \frac{n+2}{2(n+8)^2}\left[\frac{6(3n+14)}{(n+8)^2} - \frac{1}{4}\right]\varepsilon^3 + \cdots, \tag{8.131}$$

$$\delta = 3 + \varepsilon + \left[\frac{1}{2} - \frac{n+2}{(n+8)^2}\right]\varepsilon^2 + \cdots, \tag{8.132}$$

$$\beta = \frac{1}{2} - \frac{3}{n+8}\frac{\varepsilon}{2} + \frac{(n+2)(2n+1)}{2(n+8)}\varepsilon^2 + \cdots, \tag{8.133}$$

$$\alpha = \frac{4-n}{n+8}\frac{\varepsilon}{2} + \cdots. \tag{8.134}$$

In Table 8.1 we compare the values of the critical exponents, obtained from these expressions for the case of $d = 3$ ($\varepsilon = 1$) and $n = 1$ (Ising case), with the results of numerical calculations (high temperature expansions) for the three-dimensional Ising model. In the Table, we also give the values of the critical exponents from Landau theory. We can see that the ε-expansion gives a rather satisfactory agreement with the results of numerical analysis[9].

Modern methods of calculation significantly improve the results of the simplest form of the ε-expansion, taking into account the higher orders and asymptotic behavior of the appropriate perturbation series, produce the values of critical exponents in full agreement with the results of numerical calculations and experiments.

[9] Another effective method for the calculation of critical exponents is based on a perturbation expansion in powers of the inverse number of order parameter components $1/n$ [26], as for $n \to \infty$ it can be shown that the critical exponents are also given by the mean field approximation (Landau theory).

Chapter 9

Linear response

9.1 Linear response to mechanical perturbation

Up to now we mainly discussed the problems of equilibrium statistical mechanics. Actually, there is a wide class of problems related to non-equilibrium processes, which can be rigorously formulated and solved within the general formalism of equilibrium theory. We are speaking about the rather common situation where the system is initially in an equilibrium state, but later it is perturbed by some weak external perturbation. This class of problems is analyzed within linear response theory, which gives a well-developed and general approach to the to solution of such non-equilibrium problems[1].

There are two major types of external perturbations, which can be applied to an arbitrary physical system at equilibrium. First of all, we may consider mechanical perturbations, corresponding to the action of some external physical fields, which can be introduced by additional terms in the Hamiltonian describing the physical interactions with these fields. Perturbations, which can not be described in this way, are called, in non-equilibrium statistical mechanics, thermal perturbations. Typical examples are temperature or concentration gradients. For simplicity, below we are dealing only with mechanical perturbations, though the general formalism of linear response theory is also well developed for thermal perturbations.

Consider the response of a quantum Gibbs ensemble, corresponding to the time independent Hamiltonian H, toward an external perturbation H_t^1, explicitly dependent on time. The total Hamiltonian of the system is given by:

$$\mathcal{H} = H + H_t^1. \tag{9.1}$$

Let us assume that at $t = -\infty$ the external perturbation was absent, so that:

$$H_t^1|_{t=-\infty} = 0. \tag{9.2}$$

In the majority of practical cases the perturbation H_t^1 can be written as:

$$H_t^1 = -\sum_j B_j F_j(t), \tag{9.3}$$

where $F_j(t)$ are some functions of time (c-numbers, external fields), while B_j are operators with no explicit time dependence, which are "conjugated" to the fields $F_j(t)$. Explicit examples will be given below.

[1] Below we follow mainly [3].

For definiteness we shall consider an adiabatic "switching on" of a periodic (in time) external perturbation written as:

$$H_t^1 = -\sum_\omega e^{\varepsilon t - i\omega t} B_\omega \quad (\varepsilon \to +0), \tag{9.4}$$

where $B_\omega^+ = B_{-\omega}$ due to Hermiticity of the Hamiltonian.

In the general case, the statistical operator (density matrix) of the system ρ satisfies the quantum Liouville equation:

$$i\hbar \frac{\partial \rho}{\partial t} = [H + H_t^1, \rho] \tag{9.5}$$

and the initial condition is, in our case, written as:

$$\rho|_{t=-\infty} = \rho_0 = \frac{1}{Z} e^{-\frac{H}{T}} \tag{9.6}$$

which simply means that at initial moment $t = -\infty$ our system is at the state of thermodynamic (statistical) equilibrium and described by the canonical Gibbs ensemble. Of course, the grand canonical ensemble can also be used to describe the initial state of the system.

Let us perform now a canonical transformation of the following form:

$$\rho_1 = e^{\frac{iHt}{\hbar}} \rho e^{-\frac{iHt}{\hbar}}. \tag{9.7}$$

Then the Liouville equation is reduced to the following form:

$$i\hbar \frac{\partial \rho_1}{\partial t} = [H_t^1(t), \rho_1] \tag{9.8}$$

with the initial condition:

$$\rho_1|_{t=-\infty} = \rho_0. \tag{9.9}$$

Here we introduced

$$H_t^1(t) = e^{\frac{iHt}{\hbar}} H_t^1 e^{-\frac{iHt}{\hbar}} \tag{9.10}$$

the perturbation operator in the Heisenberg representation with Hamiltonian H, so that with respect to the total Hamiltonian (9.1) this defines the so called interaction representation.

Equation (9.8) with the initial condition given by (9.9) can be integrated and written as a single integral equation:

$$\rho_1(t) = \rho_0 + \int_{-\infty}^{t} dt' \frac{1}{i\hbar} [H_{t'}^1(t'), \rho_1(t')] \tag{9.11}$$

or, making the transformation to the initial form of the density matrix $\rho(t)$ using (9.7):

$$\rho(t) = \rho_0 + \int_{-\infty}^{t} dt' e^{-\frac{iH(t-t')}{\hbar}} \frac{1}{i\hbar}[H^1_{t'}, \rho] e^{\frac{iH(t-t')}{\hbar}}, \tag{9.12}$$

where we have also used (9.10).

If the perturbation H^1_t is small, the solution of Equation (9.12) can be obtained by iterations, taking ρ_0 as initial value. In the first order (linear) approximation we get:

$$\rho = \rho_0 + \int_{-\infty}^{t} dt' \frac{1}{i\hbar}[H^1_{t'}(t'-t), \rho_0]. \tag{9.13}$$

The second term in the right-hand side represents a *non-equilibrium* correction to the density matrix, calculated in linear approximation over the external perturbation. Up to now we have not used the explicit form of ρ_0. Now we can do it, taking into account the explicit form of the canonical distribution (9.6).

Let us use the so called Kubo identity, which is valid for any quantum operator A:

$$[A, e^{-\beta H}] = -e^{-\beta H} \int_0^{\beta} d\lambda e^{\lambda H} [A, H] e^{-\lambda H}. \tag{9.14}$$

The proof of this identity will be given shortly, but now we can use it to rewrite (9.13) as:

$$\rho = \rho_0 \left\{ 1 - \int_0^{\beta} d\lambda \int_{-\infty}^{t} dt' e^{\lambda H} \dot{H}^1_{t'}(t'-t) e^{-\lambda H} \right\}, \tag{9.15}$$

where

$$\dot{H}^1_{t'}(t'-t) = \frac{1}{i\hbar}[H^1_{t'}(t'-t), H]. \tag{9.16}$$

If we take ρ_0 in the form of the grand canonical distribution, Equation (9.15) remains valid, and we only have to make the replacement $H \to H - \mu N$.

Let us derive now the Kubo identity. We write:

$$[A, e^{-\beta H}] = e^{-\beta H} S(\beta), \tag{9.17}$$

where $S(\beta)$ is an operator to be determined. Differentiating (9.17) with respect to β, we obtain a differential equation for $S(\beta)$:

$$\frac{\partial S}{\partial \beta} = -e^{\beta H}[A, H] e^{-\beta H} \tag{9.18}$$

with the initial condition $S|_{\beta=0} = 0$. Integrating with this initial condition we get (9.14).

Equations (9.13) and (9.15) allow us to calculate (in linear approximation over ïî H_t^1) the average value of an arbitrary physical variable, represented by some operator A:

$$\langle A\rangle = \operatorname{Sp}\rho A$$

$$\langle A\rangle = \langle A\rangle_0 + \int_{-\infty}^{t} dt'\frac{1}{i\hbar}\langle[A(t), H_{t'}^1(t')]\rangle_0, \tag{9.19}$$

where we used (9.13) and have taken into account the invariance of Sp with respect to cyclic permutation of operators[2] and

$$A(t) = e^{\frac{iHt}{\hbar}} A e^{-\frac{iHt}{\hbar}} \tag{9.20}$$

is the A operator in Heisenberg representation, and $\langle\cdots\rangle_0 = \operatorname{Sp}\rho_0\cdots$ is the averaging with the *equilibrium* density matrix. This means that the non-equilibrium problem of a linear response is reduced to equilibrium problem, as all the averages to be calculated now are, in fact, calculated for the equilibrium state. This remarkable result (Kubo) allows the application of the powerful apparatus of equilibrium statistical mechanics to the solution of this kind of (weakly) non-equilibrium problems.

Equation (9.19) describes the response of the average value of an operator A to an external perturbation $H_{t'}^1$. Note that here we are dealing with the retarded response – it appears at the moments in time after the perturbation is switched on. This reflects the causality principle, which is basic to all physical processes. Formally extending the integration over time in (9.20) to $+\infty$, which may be done by the introduction of a step – like $\theta(t-t')$-function, it is convenient to rewrite (9.19) as:

$$\langle A\rangle = \langle A\rangle_0 + \int_{-\infty}^{\infty} dt'\langle\langle A(t) H_{t'}^1(t')\rangle\rangle, \tag{9.21}$$

where we have introduced the *retarded* double-time (commutator) Green's function (Bogolyubov, Tyablikov), defined for the pair of arbitrary operators A and B as [27]:

$$\langle\langle A(t), B(t')\rangle\rangle = \theta(t-t')\frac{1}{i\hbar}\langle[A(t), B(t')]\rangle_0, \tag{9.22}$$

where

$$\theta(t-t') = \begin{cases} 1 & \text{for } t \ge t' \\ 0 & \text{for } t < t'. \end{cases} \tag{9.23}$$

As a result the problem is reduced to the calculation of appropriate double-time Green's functions, using the well-developed mathematical formalism [27].

[2] We have $\operatorname{Sp}[H_{t'}^1(t'-t), \rho_0]A = \operatorname{Sp}\rho_0[A, H_{t'}^1(t'-t)]$ etc. The expression for $A(t)$ appears here with on account of (9.10) and further permutations of operators under Sp.

The response to external perturbations can be expressed also in another form, using the time correlation functions. Let us use the Kubo identity (9.14). Then:

$$\begin{aligned}\langle A\rangle &= \langle A\rangle_0 - \int_0^{\beta} d\lambda \int_{-\infty}^{t} dt' \langle e^{\lambda H} \dot{H}^1_{t'}(t') e^{-\lambda H} A(t)\rangle_0 \\ &= \langle A_0\rangle + \int_0^{\beta} d\lambda \int_{-\infty}^{t} dt' \langle e^{\lambda H} H^1_{t'}(t') e^{-\lambda H} \dot{A}(t)\rangle_0,\end{aligned} \tag{9.24}$$

where we have used the so-called stationarity condition:

$$\langle A\dot{H}^1_{t'}(t'-t)\rangle_0 = -\langle \dot{A}(t-t')H^1_{t'}\rangle_0. \tag{9.25}$$

The last equality follows from the fact that the equilibrium average of the product of dynamic variables depends only on the time difference:

$$\langle AH^1_{t'}(t'-t)\rangle_0 = \langle A(t-t')H^1_{t'}\rangle_0, \tag{9.26}$$

which is obtained by cyclic permutations of operators like $e^{\frac{iHt}{\hbar}}$ in the averaging. Differentiating (9.26) with respect to t we obtain (9.25).

Equation (9.24) can also be rewritten as:

$$\begin{aligned}\langle A\rangle &= \langle A\rangle_0 - \int_0^{\beta} d\lambda \int_{-\infty}^{t} dt' \langle \dot{H}^1_{t'}(t'-i\hbar\lambda)A(t)\rangle_0 \\ &= \langle A\rangle_0 + \int_0^{\beta} d\lambda \int_{-\infty}^{t} dt' \langle H^1_{t'}(t'-i\hbar\lambda)\dot{A}(t)\rangle.\end{aligned} \tag{9.27}$$

Equations (9.21) and (9.27) give the general expressions for a linear response of the system to a mechanical perturbation. For an external perturbation (9.3) these can be written as:

$$\langle A\rangle = \langle A\rangle_0 - \sum_j \int_{-\infty}^{\infty} dt' \langle\langle A(t)B_j(t')\rangle\rangle F_j(t'), \tag{9.28}$$

$$\langle A\rangle = \langle A\rangle_0 + \sum_j \int_{-\infty}^{t} dt' \int_0^{\beta} d\lambda \langle e^{\lambda H} B_j(t') e^{-\lambda H} A(t)\rangle_0 F_j(t'). \tag{9.29}$$

These are the so-called Kubo formulas for the linear response of a quantum mechanical system, which reduce the non-equilibrium problem to calculations of equilibrium correlators. This last task is, in general, quite nontrivial and requires the development of a special formalism, such as e.g. the theory of double-time commutator Green's functions.

The physical meaning of the retarded double-time Green function can be easily understood considering the reaction of the system toward instantaneous δ-like perturbation:

$$H^1_t = B\delta(t-t_1) \tag{9.30}$$

substituting this into (9.21) gives:

$$\langle A \rangle = \langle A \rangle_0 + \langle\langle A(t)B(t_1)\rangle\rangle. \tag{9.31}$$

There is a number of well developed methods to calculate such Green's functions. Here we briefly describe the approach based on the method of equations of motion (chain equations) [27]. The equation of motion for Green's function (9.22):

$$G_{AB}(t,t') \equiv \langle\langle A(t), B(t')\rangle\rangle = \theta(t-t')\frac{1}{i\hbar}\langle [A(t), B(t')]\rangle_0 \tag{9.32}$$

can be easily obtained from the general equation of motion for an arbitrary quantum operator in Heisenberg representation:

$$i\hbar\frac{dA}{dt} = [A, H] = AH - HA. \tag{9.33}$$

The right-hand side of this equation can be calculated for each concrete problem, using the explicit form of the Hamiltonian and the commutation relations for the operators. Differentiating (9.32) with respect to t we obtain the equation:

$$i\hbar\frac{dG_{AB}}{dt} = \frac{d\theta(t-t')}{dt}\langle [A(t), B(t')]\rangle_0 + \left\langle\!\!\left\langle i\hbar\frac{dA(t)}{dt}, B(t')\right\rangle\!\!\right\rangle. \tag{9.34}$$

Taking into account the obvious relation of the $\theta(t)$ step-like function to the δ-function of t:

$$\theta(t) = \int_{-\infty}^{t'} dt\delta(t') \tag{9.35}$$

as well as the equations of motion for the operator A (9.33), we can write the equation of motion for the Green's function in the following form:

$$i\hbar\frac{dG_{AB}}{dt} = \delta(t-t')\langle [A(t), B(t')]\rangle_0 + \langle\langle \{A(t)H(t) - H(t)A(t)\}, B(t')\rangle\rangle. \tag{9.36}$$

The right-hand side of Equation (9.36), in general, contains double-time Green's functions of higher order than the initial one, which is connected with the nontrivial interaction in any many-particle system. For these Green's functions we can again write equations of motion similar to (9.36) and obtain the *chain* of interconnected equations of motion for the set of Green's functions of higher and higher orders. This chain of equations is, in the general case, infinite, so that we are dealing with an infinite system of integro–differential equations, which can not be solved. However, in most practical cases this chain of equations can be approximately "decoupled", expressing in some way the higher-order Green's functions via the lower-order ones. Then we obtain a finite system of equations (or sometime even a single equation), which is much easier to solve. Unfortunately, there is no general theoretical recipe for decoupling this chain of equations, everything depends on the skills and abilities of a theorist, trying to solve the problem. Examples of successful decouplings and solutions of a number of physical models by using this method can be found in the literature [27].

9.2 Electrical conductivity and magnetic susceptibility

Consider the reaction of the system to an external electric field. The perturbation (9.3) can in this case be written as:

$$H_t^1 = -\sum_j e_j(\mathbf{E}\mathbf{x}_j)\cos\omega t e^{\varepsilon t} = -(\mathbf{E}\mathbf{P})\cos\omega t e^{\varepsilon t}, \tag{9.37}$$

where e_j is the charge of the j-th particle, $\mathbf{x}_j$ is its coordinate, $\mathbf{E}$ is the electric field, playing the role of an external (c-number) "force",

$$\mathbf{P} = \sum_j e_j\mathbf{x}_j \tag{9.38}$$

is the polarization vector, considered here as a quantum mechanical operator. This perturbation induces the electric current, which according to (9.21) it can be written as:

$$\langle J_\alpha\rangle = \int_{-\infty}^{\infty} dt'\langle\langle J_\alpha(t), H_{t'}^1(t')\rangle\rangle. \tag{9.39}$$

Here, we do not have the constant term, as in equilibrium the electric current is just zero, $\langle J_\alpha\rangle = 0$. Also in Equation (9.39) we have:

$$H_t^1(t) = -(\mathbf{E}\mathbf{P}(t))\cos\omega t e^{\varepsilon t} \quad J_\alpha(t) = \sum_j e_j\dot{x}_{j\alpha}(t) = \dot{P}_\alpha(t), \tag{9.40}$$

where J_α is an electric current operator, $\dot{x}_{j\alpha}$ is the appropriate velocity component of the j-th particle.

Taking into account (9.40), expression (9.39) can be written as:

$$\langle J_\alpha\rangle = -\sum_\beta \int_{-\infty}^{\infty} dt'\langle\langle J_\alpha(t)P_\beta(t')\rangle\rangle E_\beta\cos\omega t' e^{\varepsilon t'}. \tag{9.41}$$

Accordingly:

$$\langle J_\alpha\rangle = \sum_\beta \mathrm{Re}\,\{\sigma_{\alpha\beta}(\omega)e^{-i\omega t+\varepsilon t}\}E_\beta, \tag{9.42}$$

where

$$\sigma_{\alpha\beta}(\omega) = -\int_{-\infty}^{\infty} dt e^{-i\omega t+\varepsilon t}\langle\langle J_\alpha P_\beta(t)\rangle\rangle \tag{9.43}$$

is the conductivity tensor in a periodic external field. The limit of $\varepsilon \to 0$ is to be taken here after the thermodynamic limit $V \to \infty$, $N \to \infty$ ($V/N \to$ const).

Thus, an adiabatic switching on of the electric field leads to the appearance of an electric current in a system with finite conductivity (irreversible process). Static conductivity can be obtained from (9.43) taking the limit of $\omega \to 0$:

$$\sigma_{\alpha\beta} = \lim_{\varepsilon\to 0} \int_{-\infty}^{\infty} dt e^{\varepsilon t} \langle\langle J_\alpha P_\beta(t)\rangle\rangle. \tag{9.44}$$

Let us rewrite (9.43) as (making permutations of operators in Sp):

$$\sigma_{\alpha\beta}(\omega) = -\frac{1}{i\hbar} \int_{-\infty}^{0} dt e^{-i\omega t+\varepsilon t} \mathrm{Sp}\,\{[P_\beta(t), \rho_0] J_\alpha\} \tag{9.45}$$

and apply the Kubo identity:

$$[P_\beta(t), \rho_0] = -i\hbar\rho_0 \int_0^\beta d\lambda e^{\lambda H} \dot{P}_\beta(t) e^{-\lambda H}. \tag{9.46}$$

Then we obtain:

$$\begin{aligned} \sigma_{\alpha\beta} &= \int_0^\beta d\lambda \int_0^\infty dt e^{i\omega t-\varepsilon t} \langle e^{\lambda H} J_\beta e^{-\lambda H} J_\alpha(t)\rangle_0 \\ &= \int_0^\beta d\lambda \int_0^\infty dt e^{i\omega t-\varepsilon t} \langle J_\beta J_\alpha(t+i\hbar\lambda)\rangle_0, \end{aligned} \tag{9.47}$$

which is the notorious Kubo formula for conductivity.

In the static limit we have:

$$\sigma_{\alpha\beta} = \lim_{\varepsilon\to 0} \int_0^\beta \int_0^\infty dt e^{-\varepsilon t} \langle J_\beta J_\alpha(t+i\hbar\lambda)\rangle_0. \tag{9.48}$$

Thus, the problem of the calculation of conductivity is reduced to the calculation of the time correlation functions of the currents in thermodynamic equilibrium. In concrete systems this is obviously a rather complicated task, which can be analyzed and solved by different methods, which we shall not discuss here.

Consider now the response of the system to the switching on of a homogeneous (in space) time-dependent (periodic) magnetic field (periodic) $\mathbf{H}(t)$ with frequency ω:

$$\mathbf{H}(t) = \mathbf{H} \cos\omega t e^{\varepsilon t} = \mathrm{Re}\, e^{-i\omega t+\varepsilon t}\mathbf{H}. \tag{9.49}$$

This perturbation is described by the operator (9.3) of the following form:

$$H_t^1 = -\mathbf{M}\mathbf{H}(t) = -\mathbf{M}\mathbf{H}\cos\omega t e^{\varepsilon t}, \tag{9.50}$$

where $\mathbf{M}$ is the operator for the (total) magnetic moment the system. Under the influence of this perturbation, the magnetic moment of the system changes, according to (9.21), as:

$$\langle M_\alpha\rangle = \langle M_\alpha\rangle_0 + \int_{-\infty}^{\infty} dt' \langle\langle M_\alpha(t) H_{t'}^1(t')\rangle\rangle, \tag{9.51}$$

where $\langle M_\alpha \rangle_0$ is the average projection of the magnetic moment on α-axis at equilibrium. If there is a magnetic field present at equilibrium we have $\langle M_\alpha \rangle_0 \neq 0$. Expression (9.51) can be written as:

$$\langle M_\alpha \rangle = \langle M_\alpha \rangle_0 + \sum_\beta \mathrm{Re}\,\{\chi_{\alpha\beta}(\omega) e^{-i\omega t+\varepsilon t}\} H_\beta, \tag{9.52}$$

where

$$\chi_{\alpha\beta}(\omega) = -\int_{-\infty}^{\infty} dt e^{-i\omega t+\varepsilon t} \langle\langle M_\alpha M_\beta(t)\rangle\rangle \tag{9.53}$$

is the tensor of the magnetic susceptibility in the periodic magnetic field. With the help of the Kubo identity Equation (9.53) can be rewritten also as:

$$\chi_{\alpha\beta} = \int_0^\beta d\lambda \int_0^\infty dt e^{i\omega t-\varepsilon t} \langle \dot{M}_\beta M_\alpha(t + i\hbar\lambda)\rangle. \tag{9.54}$$

These expressions are widely used e.g. in the theory of magnetic resonance.

As an elementary example of the use of the Kubo formulas we consider electric conductivity, making the simplest assumptions for the time behavior of the correlation functions. Using Equations (9.22), (9.44) we get:

$$\sigma_{\alpha\beta} = -\lim_{\varepsilon\to 0} \frac{1}{i\hbar} \int_{-\infty}^{0} dt e^{\varepsilon t} \langle [J_\alpha, P_\beta(t)]\rangle_0. \tag{9.55}$$

Let us assume that

$$\langle [J_\alpha, P_\beta(t)]\rangle_0 \approx \langle [J_\alpha, P_\beta]\rangle_0 e^{-\frac{|t|}{\tau}}, \tag{9.56}$$

where τ is some relaxation time. The correlation function at coinciding times can be found in an elementary way as:

$$\begin{aligned} \langle [J_\alpha, P_\beta]\rangle_0 &= \left\langle \left[\sum_i \frac{e}{m} p_i^\alpha, \sum_j e x_j^\beta \right] \right\rangle_0 \\ &= \frac{e^2}{m} \sum_i [p_i^\alpha, x_i^\beta] = -i\hbar\delta_{\alpha\beta} \frac{e^2}{m} N, \end{aligned} \tag{9.57}$$

where N is the total number of particles, and we used the standard commutation relation $[x_i^\beta, p_i^\alpha] = i\hbar\delta_{\alpha\beta}$. Then we find:

$$\sigma_{\alpha\beta} = \frac{Ne^2}{m}\delta_{\alpha\beta} \lim_{\varepsilon\to 0} \int_{-\infty}^{0} dt e^{(\varepsilon+1/\tau)t} = \frac{Ne^2}{m}\tau\delta_{\alpha\beta} \tag{9.58}$$

or, per unit volume:

$$\sigma_{\alpha\beta} = \frac{ne^2}{m}\tau\delta_{\alpha\beta}, \tag{9.59}$$

which is the usual Drude expression for conductivity. Let us stress that the real problem for the theory is, of course, the derivation of behavior like that given by Equation (9.56) from some microscopic model, which also allows the calculation of the dependencies of τ on temperature (or the concentration of impurities) for different mechanisms of scattering. These problems can be solved by modern theoretical methods, such as the Green's functions formalism.

9.3 Dispersion relations

Now we shall discuss some general properties of response functions. Consider again a time-depending mechanical perturbation, which is switched on adiabatically is and described by the following term in the Hamiltonian:

$$H_t^1 = -\sum_{j=1}^{n} F_j(t)B_j, \tag{9.60}$$

where $F_j(t) \sim e^{\varepsilon t}$ for $t \to -\infty$, $\varepsilon \to +0$, B_j are some dynamical variables (operators), while $F_j(t)$ are c-number "forces", representing external fields acting upon the variables B_j. For simplicity we assume below, that in the equilibrium state (for $F_j = 0$) we have $\langle A_j\rangle_0 = 0$, so that the response of the system to an external perturbation (9.60) is written, according to (9.21), in the following form:

$$\langle A_i\rangle = \int_{-\infty}^{t} dt'\kappa_{ij}(t-t')F_j(t'), \tag{9.61}$$

where

$$\kappa_{ij}(t-t') = -\langle\langle A_i(t)B_j(t')\rangle\rangle \tag{9.62}$$

is the generalized response matrix. The retarded Green's function is different from zero only for the positive values of time difference, so that:

$$\kappa_{ij}(t-t') = 0 \quad \text{for } t < t', \tag{9.63}$$

which reflects the *causality*: the response of the system can not be earlier in time than the perturbation due to which it appears.

Let us make a Fourier expansion of $F_j(t)$ and $\langle A_i\rangle$:

$$\langle A_i\rangle = \frac{1}{2\pi}\int_{-\infty}^{\infty} d\omega e^{-i\omega t} A_i,(\omega) \tag{9.64}$$

$$F_j(t) = \frac{1}{2\pi}\int_{-\infty}^{\infty} d\omega e^{-i\omega t} F_j,(\omega) \tag{9.65}$$

where the Fourier components:

$$A_i(\omega) = \int_{-\infty}^{\infty} e^{i\omega t}\langle A_i(t)\rangle, \tag{9.66}$$

$$F_j(\omega) = \int_{-\infty}^{\infty} dt e^{i\omega t} F_j(t). \tag{9.67}$$

Making the Fourier transformation in (9.61) we reduce the integral relation to an algebraic one:

$$A_i(\omega) = \kappa_{ij}(\omega) F_j(\omega), \tag{9.68}$$

where

$$\begin{aligned}\kappa_{ij}(\omega) &= \int_{-\infty}^{\infty} dt e^{i\omega t}\kappa_{ij}(t) = -\langle\langle A_i|B_j\rangle\rangle_\omega \\ &= \int_0^{\infty} dt e^{-i\omega t-\varepsilon t}\int_0^{\beta} d\lambda \langle \dot{B}_j A_i(t+i\hbar\lambda)\rangle\end{aligned} \tag{9.69}$$

is the Fourier-transformed generalized susceptibility matrix. The last expression is sometimes called Kubo's fluctuation–dissipation theorem[3].

As both A_i and F_j are real, we have:

$$A_i(\omega) = A_i^\star(-\omega) \quad F_j(\omega) = F_j^\star(-\omega) \tag{9.70}$$

so that

$$\kappa_{ij} = \kappa_{ij}^\star(-\omega) \tag{9.71}$$

and we obtain

$$\begin{aligned}\operatorname{Re}\kappa_{ij}(\omega) &= \operatorname{Re}\kappa_{ij}(-\omega) \\ \operatorname{Im}\kappa_{ij}(\omega) &= -\operatorname{Im}\kappa_{ij}(-\omega).\end{aligned} \tag{9.72}$$

We see that the real part of the generalized susceptibility $\kappa_{ij}(\omega)$ is even, while the imaginary part is odd over frequency ω[4].

Due to causality (cf. (9.63)) the first integral in (9.69) is in fact reduced to (for brevity, we drop indices i, j in the following):

$$\kappa(\omega) = \int_0^{\infty} dt \kappa(t) e^{i\omega t}. \tag{9.73}$$

[3] The fluctuation–dissipation theorem can be written in different forms and gives the relation between the susceptibilities (or transport coefficients) and the appropriate equilibrium correlators (fluctuations).

[4] It can be shown that $\operatorname{Im}\kappa_{ij}$ determines the dissipation of energy of an external field, so that $\operatorname{Im}\kappa_{ij}(\omega > 0) > 0$.

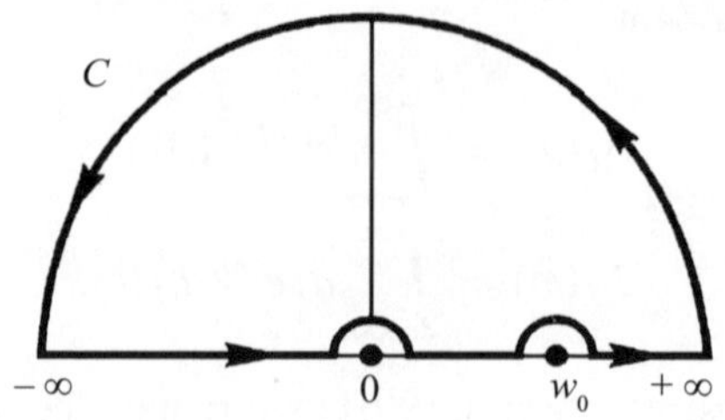

Figure 9.1. Contour of integration used in the derivation of the Kramers–Kronig relations.

From this fact alone we can obtain some quite general relations for $\kappa(\omega)$, considering it as a function of the complex frequency $\omega = \omega' + i\omega''$. Consider the properties of $\kappa(\omega)$ in the upper half-plane of ω. From (9.73) and the fact that $\kappa(t)$ is finite for all positive values of t it follows, that $\kappa(\omega)$ is a finite single-valued function in the whole upper half-plane of ω, where it never becomes infinite, i.e. it does not have any singularities there. The proof is simple: for $\omega'' > 0$ there is an exponential dumping factor of $\exp(-t\omega'')$ in the integrand of (9.73), the function $\kappa(t)$ is finite over the whole range of integration, so that the integral in (9.73) converges. Let us stress that the conclusion about the absence of singularities of $\kappa(\omega)$ in the upper half-plane, from a physical point of view is a direct consequence of causality. Causality alone transforms the integration in (9.73) to the limits from 0 to ∞ (instead of $-\infty$ to ∞). The function $\kappa(\omega)$ is non-singular also along the real axis of the frequency ($\omega'' = 0$), except probably at the origin ($\omega = 0$).

Let us derive now the general formulas connecting the real and imaginary parts of $\kappa(\omega)$. Let us choose some real and positive value of $\omega = \omega_0$ and integrate $\frac{\kappa(\omega)}{\omega-\omega_0}$ over the contour C, shown in Figure 9.1. At infinity $\kappa \to 0$ so that $\frac{\kappa(\omega)}{\omega-\omega_0}$ tends to zero faster than $1/\omega$. Thus the integral $\int_C d\omega \frac{\kappa(\omega)}{\omega-\omega_0}$ converges. The function $\kappa(\omega)$ does not have singularities in the upper half-plane and point $\omega = \omega_0$ is excluded from integration, so that $\frac{\kappa(\omega)}{\omega-\omega_0}$ is analytic inside contour C, so that our integral is just zero (Cauchy theorem).

The integral over the semicircle becomes zero at infinity, due to fast dumping of the integrand. The point ω_0 is surpassed by a small semicircle (with radius $\rho \to 0$). This encirclement is performed clockwise and leads to a contribution $-i\pi\kappa(\omega_0)$ (the integral over the complete circle gives $-2i\pi\kappa(\omega_0)$). If $\kappa(0)$ is finite, surpassing the origin is excessive and integration along the real axis leads to:

$$\lim_{\rho\to 0}\left\{\int_{-\infty}^{\omega_0-\rho} d\omega \frac{\kappa(\omega)}{\omega-\omega_0} + \int_{\omega_0+\rho}^{\infty} d\omega \frac{\kappa(\omega)}{\omega-\omega_0}\right\} - i\pi\kappa(\omega_0) = 0. \tag{9.74}$$

The first term here is the integral from $-\infty$ to ∞, understood as a principal value, thus we obtain:

$$i\pi\kappa(\omega_0) = P\int_{-\infty}^{\infty} d\omega \frac{\kappa(\omega)}{\omega - \omega_0}. \tag{9.75}$$

This relation is obtained immediately if we consider the integral $\frac{\kappa(\omega)}{\omega-\omega_0+i\delta}$ along the real axis and use the famous relation for generalized functions:

$$\frac{1}{x+i\delta} = P\frac{1}{x} - i\pi\delta(x) \quad \delta \to +0. \tag{9.76}$$

The previous discussion in fact just gives the derivation of this useful relation.

The integration variable ω in (9.75) takes on only real values. Let us denote it ξ, and from now on use ω to denote the fixed real value of the frequency ω_0. Then, separating the real and imaginary parts in (9.75), we obtain:

$$\mathrm{Re}\,\kappa(\omega) = \frac{1}{\pi}P\int_{-\infty}^{\infty} d\xi \frac{\mathrm{Im}\,\kappa(\xi)}{\xi - \omega}, \tag{9.77}$$

$$\mathrm{Im}\,\kappa(\omega) = -\frac{1}{\pi}P\int_{-\infty}^{\infty} d\xi \frac{\mathrm{Re}\,\kappa(\xi)}{\xi - \omega}. \tag{9.78}$$

These are the notorious Kramers–Kronig relations. The only property of $\kappa(\omega)$ used in our derivation was the absence of singularities of this function in the upper half-plane[5]. Thus, we may say, that the Kramers–Kronig relations directly follow from the causality principle.

Using the oddness of $\mathrm{Im}\,\kappa(\xi)$, we can rewrite the first of these relations as:

$$\mathrm{Re}\,\kappa(\omega) = \frac{1}{\pi}P\int_{0}^{\infty} d\xi \frac{\mathrm{Im}\,\kappa(\xi)}{\xi - \omega} + P\int_{0}^{\infty} d\xi \frac{\mathrm{Im}\,\kappa(\xi)}{\xi + \omega} \tag{9.79}$$

or

$$\mathrm{Re}\,\kappa(\omega) = \frac{2}{\pi}\int_{0}^{\infty} d\xi \frac{\xi \mathrm{Im}\,\kappa(\omega)}{\xi^2 - \omega^2}. \tag{9.80}$$

If $\kappa(\omega)$ has a pole at $\omega = 0$, so that close to it $\kappa = iA/\omega$, surpassing this pole over the semicircle produces an additional $-A/\omega$ contribution to the integral, which is to be added to the left-hand side of (9.75). Accordingly, a similar term will appear in (9.78):

$$\mathrm{Im}\,\kappa(\omega) = -\frac{1}{\pi}P\int_{-\infty}^{\infty} d\xi \frac{\mathrm{Re}\,\kappa(\xi)}{\xi - \omega} + \frac{A}{\omega}. \tag{9.81}$$

The Kramers–Kronig relations are the most important exact expressions allowing to control theoretical models and calculations, with important experimental applications: measurements of $\mathrm{Re}\,\kappa(\omega)$ in a wide frequency interval allow to restore the values of $\mathrm{Im}\,\kappa(\omega)$ (and vice versa), performing numerical integration of experimental data.

[5] As to the property of $\kappa \to 0$ for $\omega \to \infty$, it is not so important: if the limit of κ_∞ is finite, we can simply consider the difference $\kappa - \kappa_\infty$ instead of κ, with appropriate changes in all expressions.

Chapter 10

Kinetic equations

10.1 Boltzmann equation

The theory of linear response is appropriate to describe the reaction of a system to weak external perturbations, moving it slightly outside of thermodynamic equilibrium. In principle, it can be applied to systems of a quite general nature. Another problem is the description of arbitrary nonequilibrium states. Up to now there is no general theory of this kind applicable to arbitrary systems of particles. However, much progress was made in the studies of general nonequilibrium behavior of gases of weakly interacting (or rarefied) particles (or quasiparticles). Historically, this was the first branch of nonequilibrium statistical mechanics, started in works of Boltzmann. This is often called *physical kinetics* or the theory of kinetic equations.

Here we shall rather briefly discuss the derivation of the basic equation of the kinetic theory of gases, determining the distribution function $f(\mathbf{p}, \mathbf{r}, t)$ of separate particles in the general nonequilibrium case[1]. This equation is essential to the solution of plenty of problems of the physical kinetics of gases [4, 15]. Similar quantum kinetic equations describe nonequilibrium processes in gases of quasiparticles in quantum liquids and solids at low temperatures.

If we neglect atomic collisions each atom represents a closed subsystem and its distribution function satisfies the Liouville equation:

$$\frac{df}{dt} = 0. \tag{10.1}$$

The total derivative here denotes differentiation along the phase trajectory of an atom, determined by the equations of motion. In the absence of an external field, the value of a freely moving atom remains constant, only its coordinates $\mathbf{r}$ change. Then:

$$\frac{df}{dt} = \frac{\partial f}{\partial t} + \mathbf{v}\nabla f, \tag{10.2}$$

where $\mathbf{v}$ is the velocity. If our gas is in an external field, defined by the potential $U(\mathbf{r})$, we have:

$$\frac{df}{dt} = \frac{\partial f}{\partial t} + \mathbf{v}\nabla f + \mathbf{F}\frac{\partial f}{\partial \mathbf{p}}, \tag{10.3}$$

[1] Previously, during our discussion of Boltzmann's statistics (cf. e.g Equations (3.8), (3.28) etc.), we denoted this function as $n(p, q)$. For simplicity we limit ourselves to one-atom gases.

where $\mathbf{F} = -\nabla U$ is the force, acting upon an atom due to this field. In the following, for brevity, we assume that an external field is absent, so that $\mathbf{F} = 0$.

Atomic collisions break the equality in Equation (10.1) and the distribution function is not conserving along the phase trajectories, so that instead of (10.1) we have to write:

$$\frac{df}{dt} = St\ f, \tag{10.4}$$

where $St\ f$ denotes the rate of the change of the distribution function due to these collisions. Using Equation (10.2) we can write:

$$\frac{\partial f}{\partial t} = -\mathbf{v}\nabla f + St\ f, \tag{10.5}$$

which defines the total change of the distribution function at a given point in phase space, where the first term in the right-hand side determines the number of atoms leaving the given phase space element due to free motion. The most important term $St\ f$ is called the *collision integral*, while Equation (10.4) itself is called the *kinetic equation*[2].

Obviously, the kinetic equation becomes well defined only after we establish the explicit form of the collision integral. For qualitative estimates of the kinetics in gases, a very common (and crude) form of the collision term can be introduced using the concept of *mean free time* τ, i.e. the average time between two successive atomic collisions (the so called τ-approximation):

$$St\ f \approx -\frac{f - f_0}{\tau}, \tag{10.6}$$

where f_0 denotes the equilibrium distribution function. The numerator of this expression guarantees the vanishing of the collision integral in equilibrium, while the minus sign reflects the fact that collisions, in general, lead to the equilibrium state of the system, i.e. diminish the deviation of the distribution function from its equilibrium value. In this sense the value of τ plays the role of relaxation time for the establishment of equilibrium in each elementary volume of the gas.

The consistent derivation of the collision integral for a classical gas can be performed using Bogolyubov's method, which gives the regular procedure for the derivation of not only the simplest Boltzmann's equation (which can also be obtained from a purely heuristic approach [15]), but also corrections to it. However, below we limit ourselves to the derivation of Boltzmann's collision integral, which is sufficient for us to illustrate the general method.

The starting point of Bogolyubov's approach is the use of the chain of equations for partial distribution functions (1.93):

$$\frac{\partial F_s}{\partial t} = \{H^{(s)}, F_s\} + \frac{N}{V}\sum_{i=1}^{s}\int \frac{\partial U(|\mathbf{r}_i - \mathbf{r}_{s+1}|)}{\partial \mathbf{r}_i}\frac{\partial F_{s+1}}{\partial \mathbf{p}_i} d\mathbf{r}_{s+1} d\mathbf{p}_{s+1}. \tag{10.7}$$

---

2 Sometimes it is also called the transport equation.

Our aim is to construct the *closed* equation for the one-particle distribution function $f(\mathbf{p},\mathbf{r},t) = \frac{N}{V} F_1(\mathbf{r},\mathbf{p},t)$ [3].

Using the definition of the Poisson brackets and Equation (10.7) we immediately obtain the first equation of the chain for $F_1(\mathbf{r},\mathbf{p},t)$ as:

$$\frac{\partial F_1(t,\tau_1)}{\partial t} + \mathbf{v}_1 \frac{\partial F_1(t,\tau_1)}{\partial \mathbf{r}_1} = \frac{N}{V} \int \frac{\partial U_{12}}{\partial \mathbf{r}_1} \frac{\partial F_2(t,\tau_1,\tau_2)}{\partial \mathbf{p}_1} d\tau_2, \tag{10.8}$$

where for brevity we introduced the variables $\tau = \mathbf{r},\mathbf{p}$.

Similarly, the second equation of the chain takes the form:

$$\begin{aligned} &\frac{\partial F_2}{\partial t} + \mathbf{v}_1 \frac{\partial F_2}{\partial \mathbf{r}_1} + \mathbf{v}_2 \frac{\partial F_2}{\partial \mathbf{r}_2} - \frac{\partial U_{12}}{\partial \mathbf{r}_1} \frac{\partial F_2}{\partial \mathbf{p}_1} - \frac{\partial U_{12}}{\partial \mathbf{r}_2} \frac{\partial F_2}{\partial \mathbf{p}_2} \\ &= \frac{N}{V} \int d\tau_3 \left[\frac{\partial F_3}{\partial \mathbf{p}_1} \frac{\partial U_{13}}{\partial \mathbf{r}_1} + \frac{\partial F_3}{\partial \mathbf{p}_2} \frac{\partial U_{23}}{\partial \mathbf{r}_2} \right]. \end{aligned} \tag{10.9}$$

It is not difficult to see that the integral in the r.h.s. of the last equation is small. In fact, the interaction potential $U(r)$ is effectively nonzero only within the limits defined by the radius of the forces it creates, which we denote by d, i.e. for $r < d$. Thus, integration over the coordinates in $d\tau_3$ is performed over the region defined by $|\mathbf{r}_1 - \mathbf{r}_3| < d$ or $|\mathbf{r}_2 - \mathbf{r}_3| < d$, i.e. the volume of the order of $\sim d^3$. Using (1.81) we have $\frac{1}{V} \int F_3 d\tau_3 = F_2$, where integration is over the whole phase space. Then we get the following estimate:

$$\frac{N}{V} \int \left[\frac{\partial F_3}{\partial \mathbf{p}_1} \frac{\partial U_{13}}{\partial \mathbf{r}_1} \right] d\tau_3 \sim \frac{\partial U(r)}{\partial r} \frac{\partial F_2}{\partial p_1} \frac{d^3}{a^3}, \tag{10.10}$$

where a is the average distance between the particles in our gas. Then it is clear that the r.h.s. of Equation (10.9) is small with respect to the parameter $(d/a)^3$ (we assume the gas to be rarefied!), as compared with terms containing $\partial U/\partial r$ in the l.h.s. Thus the r.h.s. can be neglected. The sum of all terms in the l.h.s. of the equation in fact represents the total derivative dF_2/dt, where $\mathbf{r}_1, \mathbf{r}_2, \mathbf{p}_1, \mathbf{p}_2$ are considered as functions of time, satisfying the equations of motion for the two particle problem, defined by the Hamiltonian:

$$H = \frac{\mathbf{p}_1^2}{2m} + \frac{\mathbf{p}_2^2}{2m} + U(|\mathbf{r}_1 - \mathbf{r}_2|). \tag{10.11}$$

Thus, we have:

$$\frac{d}{dt} F_2(t,\tau_1,\tau_2) = 0. \tag{10.12}$$

[3] The distribution function $f(\mathbf{p},\mathbf{r})$ is normalized to the total number of particles (3.28), while $F_1(\mathbf{r},\mathbf{p},t)$ is normalized to unity, according to (1.80).

Up to now our analysis was purely mechanical. To derive the kinetic equation we have to make some assumptions of a *statistical* nature. Let us assume that all colliding particles of the gas are statistically independent. This assumption will be used as a kind of initial condition for the differential equation (10.12). This assumption introduces time asymmetry, which leads to an *irreversible* kinetic equation, despite our use of the time-reversible equations of motion of classical mechanics. The essence of the problem here is that any correlation between coordinates and momenta of the particles in the gas appears only during the pretty short collision time of the order of $\sim d/v$ (v is the average velocity of gas particles), and affects particles up to distances of the order of d only.

Let t_0 be some moment in time before the collision, when the two particles are rather far from each other, so that ($|\mathbf{r}_{10} - \mathbf{r}_{20}| \gg d$, where the subscript zero denotes the values at this given moment). The statistical independence of the colliding particles means, that at this moment t_0 the two particle distribution function F_2 is factorized into the product of one particle functions F_1. Then, integration of Equation (10.12) from t_0 to t gives:

$$F_2(t, \tau_1, \tau_2) = F_1(t_0, \tau_{10}) F_1(t_0, \tau_{20}). \tag{10.13}$$

Here $\tau_{10} = (\mathbf{r}_{10}, \mathbf{p}_{10})$ and $\tau_{20} = (\mathbf{r}_{20}, \mathbf{p}_{20})$ are to be understood as the values of the coordinates and momenta, which the particles should have had at the moment t_0 to achieve the given values of $\tau_1 = (\mathbf{r}_1, \mathbf{p}_1)$ and $\tau_2 = (\mathbf{r}_2, \mathbf{p}_2)$ at time t. In this sense τ_{10} è τ_{20} are functions of τ_1, τ_2 and $t - t_0$. Furthermore, only $\mathbf{r}_{10}$ and $\mathbf{r}_{20}$ depend on $t - t_0$, while the values of $\mathbf{p}_{10}$ and $\mathbf{p}_{20}$, related to the free-moving particles before the collision do not depend on $t - t_0$.

Let us return to Equation (10.8) – the future kinetic equation. The left-hand side here is of the required form, but we are interested in the right hand side, which should become the collision integral. Let us there substitute F_2 from (10.13) and introduce $f(\mathbf{p}, \mathbf{r}, t) = \frac{N}{V} F_1(\mathbf{r}, \mathbf{p}, t)$ instead of F_1. Then we obtain:

$$\frac{\partial f(t, \tau_1)}{\partial t} + \mathbf{v}_1 \frac{\partial f(t, \tau_1)}{\partial \mathbf{r}_1} = St\ f, \tag{10.14}$$

where

$$St\ f = \int d\tau_2 \frac{\partial U_{12}}{\partial \mathbf{r}_1} \frac{\partial}{\partial \mathbf{p}_1} [f(t_0, \tau_{10}) f(t_0, \tau_{20})]. \tag{10.15}$$

In (10.15) the relevant region for integration is determined by $|\mathbf{r}_2 - \mathbf{r}_1| \sim d$, i.e. by the region where the real collision takes place. In this region, in first approximation, we can simply neglect the coordinate dependence of f, as it significantly changes only on the scale of the order of the mean free path l, which is much greater than d. The final form of the collision integral does not change at all, if we consider from the very beginning only the spatially homogeneous case, assuming that f does not depend on coordinates. In accordance with previous remarks this means that in the functions $f(t_0, \mathbf{p}_{10})$ and $f(t_0, \mathbf{p}_{10})$ we can just neglect an explicit time dependence via $\mathbf{r}_{10}(t)$ and $\mathbf{r}_{20}(t)$.

Let us transform the integrand in (10.15) using (10.12) and taking into account the absence of an explicit dependence on time:

$$\frac{d}{dt} f(t_0, \mathbf{p}_{10}) f(t_0, \mathbf{p}_{20}) \\ = \left(\mathbf{v}_1 \frac{\partial}{\partial \mathbf{r}_1} + \mathbf{v}_2 \frac{\partial}{\partial \mathbf{r}_2} - \frac{\partial U_{12}}{\partial \mathbf{r}_1} \frac{\partial}{\partial \mathbf{p}_1} - \frac{\partial U_{12}}{\partial \mathbf{r}_2} \frac{\partial}{\partial \mathbf{p}_2} \right) f(t_0, \mathbf{p}_{10}) f(t_0, \mathbf{p}_{20}) = 0. \tag{10.16}$$

Now, we can express the derivative with respect to $\mathbf{p}_1$ via the derivatives with respect to $\mathbf{r}_1, \mathbf{r}_2$ è $\mathbf{p}_2$ and substitute the expression obtained in this way into (10.15). The term with $\partial/\partial \mathbf{p}_2$ disappears after transformation into a surface integral in momentum space (using Gauss' theorem). After that we get:

$$St\ f(t, \mathbf{p}_1) = \int \mathbf{v}_{12} \frac{\partial}{\partial \mathbf{r}} [f(t_0, \mathbf{p}_{10}) f(t_0, \mathbf{p}_{20})] d^3 r d^3 p_2, \tag{10.17}$$

where we have introduced the relative velocity of the particles $\mathbf{v}_{12} = \mathbf{v}_1 - \mathbf{v}_2$, and taken into account that both $\mathbf{p}_{10}$ and $\mathbf{p}_{20}$ and, correspondingly, the whole expression in square brackets depends not on $\mathbf{r}_1$ and $\mathbf{r}_2$ separately, but only on the difference $\mathbf{r} = \mathbf{r}_1 - \mathbf{r}_2$. Let us introduce, instead of $\mathbf{r} = (x, y, z)$, cylindrical coordinates z, ρ, φ, with the z-axis along $\mathbf{v}_{12}$. Noting that $\mathbf{v}_{12}\partial/\partial \mathbf{r} = v_{12}\partial/\partial z$, and performing integration over dz, we rewrite (10.17) as:

$$St\ f(t, \mathbf{p}_1) = \int [f(t_0, \mathbf{p}_{10}) f(t_0, \mathbf{p}_{20})] |_{-\infty}^{\infty}\ v_{12} \rho d\rho d\varphi d^3 p_2, \tag{10.18}$$

where the limits $z = \pm\infty$ should be understood as distances large in comparison with d, but small in comparison with the mean free path l. This is due to our use of Equation (10.16) during transformation from (10.15) to (10.18), which is valid until the particles under consideration do not collide anymore.

Remember now that $\mathbf{p}_{10}$ and $\mathbf{p}_{20}$ are the initial momenta (at the moment t_0) of particles, which at the final moment t possess $\mathbf{p}_1$ and $\mathbf{p}_2$. If in the final moment $z = z_1 - z_2 = -\infty$, the particles 1 and 2 are at a distance, which is obviously greater than d and do not interact with each other, and there were no collisions between them, so that initial and final momenta just coincide: $\mathbf{p}_{10} = \mathbf{p}_1$, $\mathbf{p}_{20} = \mathbf{p}_2$ for $z = -\infty$. If at the final moment $z = +\infty$, there was a collision and the particles acquired momenta $\mathbf{p}_1$ and $\mathbf{p}_2$ as a result of it. In this case we denote $\mathbf{p}_{10} = \mathbf{p}'_1(\rho)$ and $\mathbf{p}_{20} = \mathbf{p}'_2(\rho)$ for $z = \infty$. These values for the momenta are functions of the coordinate ρ, which is actually the impact parameter, while the product

$$\rho d\rho d\varphi = d\sigma \tag{10.19}$$

represents the classical differential scattering cross section [11].

Note finally, that the explicit dependence of the functions $f(t_0, \mathbf{p}_{10})$ and $f(t_0, \mathbf{p}_{20})$ on t_0 can be replaced, on the same level of approximation, by a similar dependence on t. In fact, the validity of (10.13) requires only that $t - t_0 \gg d/v$: at the moment t_0 the distance between the particles must be great in comparison with the effective radius of the forces d. At the same time, the time difference $t - t_0$ can be chosen to satisfy $t - t_0 \ll l/v$, where l is the mean free path. The ratio of l/v gives the mean free time, which is just a characteristic time for a significant change of distribution function. Then. the change of distribution function during the time interval $t - t_0$ will be relatively small and can be neglected.

Taking into account these remarks we can reduce (10.18) to the final form:

$$St\ f(t, \mathbf{p}_1) = \int [f(t, \mathbf{p'}_1) f(t, \mathbf{p'}_2) - f(t, \mathbf{p}_1) f(t, \mathbf{p}_2)] v_{12} d\sigma d^3 p_2, \qquad (10.20)$$

which is called Boltzmann's collision integral. The kinetic equation (10.5) with such a collision integral is called Boltzmann's kinetic equation.

Boltzmann obtained his collision integral from simple heuristic considerations, based on the so called *Stosszahlansatz*. It is clear that the collision integral can be written as $St\ f = \bar{R} - R$, where $\bar{R}$ represents the growth rate of the distribution function $f(\mathbf{r}, \mathbf{p}_1, t)$ due to atomic collisions in the gas, while R is its drop rate due to similar collisions. Let us determine first R. Consider some atom within the volume element $d^3 r$ surrounding point $\mathbf{r}$, with its momentum belonging to some element $d^3 p_1$ of momentum space around $\mathbf{p}_1$. Within the same spatial volume we have atoms with arbitrary momenta $\mathbf{p}_2$, which can be considered as a beam of particles, scattered by an atom with momentum $\mathbf{p}_1$. The flow of these scattered atoms is given by:

$$I = f(\mathbf{r}, \mathbf{p}_2, t) d^3 p_2 |\mathbf{v}_1 - \mathbf{v}_2|. \qquad (10.21)$$

According to the *Stosszahlansatz*, the distribution function f in (10.21) *coincides* with our distribution function, to be determined from the kinetic equation. This seems an almost obvious assumption, but actually it is the central point of our derivation, with no rigorous justification. The number of collisions like $\mathbf{p}_2, \mathbf{p}_2 \to \mathbf{p'}_1, \mathbf{p'}_2$, taking place in volume element $d^3 r$ during one unit of time, is given by:

$$I d\sigma = f(\mathbf{r}, \mathbf{p}_2, t) |\mathbf{v}_1 - \mathbf{v}_2| d\sigma. \qquad (10.22)$$

The drop rate R of the distribution function is obtained by summation of (10.22) over all values of $\mathbf{p}_2$ and multiplication of the result by the density of the atoms in volume element $d^3 p_1$ in velocity space:

$$R = f(\mathbf{r}, \mathbf{p}_1, t) \int d^3 p_2 d\sigma |\mathbf{v}_1 - \mathbf{v}_2| f(\mathbf{r}, \mathbf{p}_2, t). \qquad (10.23)$$

In a similar way we can determine also the value of $\bar{R}$. Consider the collisions $\mathbf{p'}_1, \mathbf{p'}_2 \to \mathbf{p}_1, \mathbf{p}_2$, where momentum $\mathbf{p}_1$ is considered as fixed. Consider the beam of atoms with momenta $\mathbf{p'}_2$, colliding with an atom possessing the momentum $\mathbf{p'}_1$. The flow density of this beam is given by:

$$f(\mathbf{r}, \mathbf{p'}_2, t) d^3 p'_2 |\mathbf{v'}_2 - \mathbf{v'}_1|. \tag{10.24}$$

The number of collision of this type per unit of time is:

$$f(\mathbf{r}, \mathbf{p'}_2, t) d^3 p'_2 |\mathbf{v'}_2 - \mathbf{v'}_1| d\sigma'. \tag{10.25}$$

The growth rate of the distribution function $\bar{R}$ is determined by the integral:

$$\bar{R} d^3 p_1 = \int d^3 p'_2 d\sigma' |\mathbf{v'}_2 - \mathbf{v'}_1| [f(\mathbf{r}, \mathbf{p'}_1, t) d^3 p'_1] f(\mathbf{r}, \mathbf{p'}_2, t). \tag{10.26}$$

Due to the time invariance of the equations of motion, the differential cross sections of the direct and inverse scatterings are the same: $d\sigma = d\sigma'$. Besides that, the conservation laws (we consider only elastic scatterings!) give:

$$\begin{gathered} |\mathbf{v}_1 - \mathbf{v}_2| = |\mathbf{v'}_1 - \mathbf{v'}_2| \\ d^3 p_1 d^3 p_2 = d^3 p'_1 d^3 p'_2. \end{gathered} \tag{10.27}$$

Then:

$$\bar{R} = \int d^3 p_2 d\sigma |\mathbf{v}_1 - \mathbf{v}_2| f(\mathbf{r}, \mathbf{p'}_1, t) f(\mathbf{r}, \mathbf{p'}_2, t). \tag{10.28}$$

It is necessary to note that the momentum $\mathbf{p}_1$ here is fixed, while $\mathbf{p'}_1$ and $\mathbf{p'}_2$ are functions of $\mathbf{p}_1, \mathbf{p}_2$.

Using the derived expressions for R and $\bar{R}$, and introducing the obvious shortened notations, we obtain:

$$St\, f = \bar{R} - R = \int d^3 p_2 d\sigma |\mathbf{v}_1 - \mathbf{v}_2| (f'_1 f'_2 - f_1 f_2), \tag{10.29}$$

which coincides with Equation (10.20).

10.2 *H*-theorem

A nonequilibrium gas freely evolving with no external perturbations tends to equilibrium. Similar behavior is characteristic for any closed macroscopic system. This should be accompanied by a corresponding entropy growth. This is experimentally observed behavior, and the evolution of the nonequilibrium distribution function, following from the kinetic equation should satisfy this observation. In fact, we can derive this (irreversible!) behavior directly from the Boltzmann's equation.

We have shown above (cf. Equations (3.25), (3.30)), that the entropy of an ideal gas in the nonequilibrium (macroscopic) state, described by the distribution function f, is equal to:

$$S = \int f \ln \frac{e}{f} dV d^3 p. \tag{10.30}$$

Differentiating this expression with respect to time, we can write:

$$\frac{dS}{dt} = \int \frac{\partial}{\partial t}\left(f \ln \frac{e}{f}\right) dV d^3 p = -\int \ln f \frac{\partial f}{\partial t} dV d^3 p. \tag{10.31}$$

An equilibrium state in a gas is achieved via atomic (molecular) collisions, and the corresponding entropy growth should be related precisely to the change in the distribution functions due to these collisions. The change in the distribution function due to the free motion of atoms cannot change the entropy of the gas. This change is determined (for a gas in an external field) by the first two terms in the right-hand side of:

$$\frac{\partial f}{\partial t} = -\mathbf{v}\nabla f - \mathbf{F}\frac{\partial f}{\partial \mathbf{p}} + St\ f. \tag{10.32}$$

The corresponding contribution to dS/dt is given by:

$$-\int \ln f \left[-\mathbf{v}\frac{\partial f}{\partial \mathbf{r}} - \mathbf{F}\frac{\partial f}{\partial \mathbf{p}}\right] dV d^3 p = \int \left[\mathbf{v}\frac{\partial}{\partial \mathbf{r}} + \mathbf{F}\frac{\partial}{\partial \mathbf{p}}\right]\left(f \ln \frac{f}{e}\right) dV d^3 p. \tag{10.33}$$

The integral over dV of the term with derivative $\partial/\partial\mathbf{r}$ is transformed via Gauss' theorem to the surface integral at infinity, which is actually zero, as outside the volume occupied by gas, we have $f = 0$. Similarly, the term with derivative $\partial/\partial\mathbf{p}$ integrated over $d^3 p$ is transformed to the surface integral at infinity in momentum space and is also just equal to zero.

Thus, we obtain the rate of change of entropy as:

$$\frac{dS}{dt} = -\int \ln f St\ f d^3 p dV. \tag{10.34}$$

Substituting here Boltzmann's collision integral (10.29), we get:

$$\frac{dS}{dt} = -\int d^3 p_1 \int dp_2^3 d\sigma |\mathbf{v}_1 - \mathbf{v}_2| (f_2' f_1' - f_2 f_1) \ln f_1. \tag{10.35}$$

The integral here does not change after permutation of the variables $\mathbf{p}_1 and \mathbf{p}_2$, as the scattering cross section is invariant with respect to this permutation. Performing this change of integration variables and taking the half-sums of the new and previous expression (10.35), we obtain:

$$\frac{dS}{dt} = -\frac{1}{2}\int d^3p_1 \int d^3p_2 d\sigma |\mathbf{v}_2 - \mathbf{v}_1|(f_2'f_1' - f_2 f_1)\ln(f_1 f_2). \qquad (10.36)$$

This integral is also invariant with respect to the mutual permutation of $\mathbf{p}_1, \mathbf{p}_2$ and $\mathbf{p'}_1, \mathbf{p'}_2$, as with each direct collision process correspond the inverse collision with the same cross section. Accordingly we can write:

$$\frac{dS}{dt} = -\frac{1}{2}\int d^3p_1' \int d^3p_2' d\sigma' |\mathbf{v'}_2 - \mathbf{v'}_1|(f_2 f_1 - f_2'f_1')\ln(f_1'f_2'). \qquad (10.37)$$

Noting that $d^3p_1'd^3p_2' = d^3p_1 d^3p_2$ and $|\mathbf{v'}_2 - \mathbf{v'}_1| = |\mathbf{v}_2 - \mathbf{v}_1|$ and $d\sigma' = d\sigma$, we take the half-sum of Equations (10.36), (10.37), and obtain:

$$\frac{dS}{dt} = -\frac{1}{4}\int d^3p_1 \int d^3p_2 d\sigma |\mathbf{v}_2 - \mathbf{v}_1|(f_2'f_1' - f_2 f_1)[\ln(f_1 f_2) - \ln(f_1'f_2')]. \qquad (10.38)$$

The integrand in (10.38) is never positive, which is clear from the previously used inequality $x \ln x > x - 1$ (valid for $x > 0$). Thus, we have proved the notorious Boltzmann's H-theorem: $\frac{dS}{dt} \geq 0$, which is equivalent to the law of entropy growth[4].

It is easy to see, that $\frac{dS}{dt} = 0$ only in case of the integrand in (10.38) being identically zero. This is only so when all distribution functions, entering the collision integral, are equal to their corresponding equilibrium (Boltzmann distribution) values. It is now also clear, that the arbitrary initial (nonequilibrium) distribution function $f(\mathbf{p}, t)$ tends to the equilibrium value as $t \to \infty$.

10.3 Quantum kinetic equations*

Let us now consider the derivation of the quantum kinetic equations. Our task now is to derive the closed equation for the one-particle density matrix from Bogolyubov's chain of equations for partial density matrices (1.163). The version of Bogolyubov's approach discussed below was proposed by Zyrianov [28].

Let us start from the quantum Liouville equation (1.128) for a general (N-particle) density matrix:

$$i\hbar\frac{\partial \rho}{\partial t} = [H, \rho] \equiv H\rho - \rho H. \qquad (10.39)$$

[4] The name H-theorem is historical, as Boltzmann used the notation $H = -S$.

We shall work in the secondary quantization representation, built upon eigenfunctions of the Hamiltonian of "free" particles (quasiparticles):

$$H_0|\nu\rangle = E_\nu|\nu\rangle, \tag{10.40}$$

$$H_0 = \sum_\nu E_\nu a_\nu^+ a_\nu, \tag{10.41}$$

where a_ν^+, a_ν are the creation and annihilation operators of Fermions or Bosons in the quantum state $|\nu\rangle$. Here ν denotes the quantum numbers, characterizing elementary excitations in our system. In most cases these correspond to free particles with definite momenta (or quasi-momenta) and spin: $|\nu\rangle = |\mathbf{p},\sigma\rangle = \chi_\sigma e^{i\mathbf{pr}/\hbar}$, where χ_σ is the spinor part of the wave function. In absence of external fields $E_\nu \equiv E_p = p^2/2m$. However, within this formalism, we can also discuss less trivial cases. For example, ν may correspond to the set of Landau quantum numbers of an electron in an external (homogeneous) magnetic field: $\nu = \{n, p_z, \sigma\}$, or these may be some quantum numbers of the energy levels for some other exactly solvable model, when the Hamiltonian can be written in diagonal form (10.41).

The operators of second quantization satisfy the usual commutation relations:

$$[a_\nu, a_{\nu'}^+]_\pm = \delta_{\nu\nu'}, \tag{10.42}$$

$$[a_\nu, a_{\nu'}]_\pm = 0 \quad [a_\nu^+, a_{\nu'}^+]_\pm = 0, \tag{10.43}$$

where $\pm$ refers to Fermions and Bosons respectively. It is supposed here that these operators are written in Schroedinger representation and are time-independent.

Our aim is to derive an equation for the one-particle density matrix, defined in Equation (1.163) as:

$$F_{1\nu\nu'} = \langle\nu|F_1|\nu'\rangle = \mathrm{Sp}\,\rho a_\nu^+ a_{\nu'} = \langle a_\nu^+ a_{\nu'}\rangle. \tag{10.44}$$

Naturally, we are going to discuss the case of interacting particles, when the total Hamiltonian is written as:

$$H = H_0 + V, \tag{10.45}$$

where V represents some interaction Hamiltonian, which is also written in secondary quantization representation.

Using the Liouville equation (10.39), we can write:

$$\begin{aligned} i\hbar\frac{\partial}{\partial t}\mathrm{Sp}\,\rho a_\nu^+ a_{\nu'} &= i\hbar\frac{\partial}{\partial t}\langle a_\nu^+ a_\nu\rangle = \mathrm{Sp}[H,\rho]a_\nu^+ a_{\nu'} \\ &= \mathrm{Sp}\,\rho[a_\nu^+ a_{\nu'}, H] = \langle[a_\nu^+ a_{\nu'}, H]\rangle, \end{aligned} \tag{10.46}$$

where we have performed an obvious cyclic permutation of the operators in Sp. Thus, our problem is reduced to calculating the average value of the commutator, standing in the r.h.s. of this equation. Now we have to introduce some specific model of interaction.

10.3.1 Electron-phonon interaction

Consider (not the simplest possible case!) the system of electrons (Fermions), occupying the states $|\nu>$, interacting with phonons (Bosons), with states characterized by a quasi-momentum $|\mathbf{k}>$. Then we write:

$$H_0 = H^0_{el} + H^0_{ph}, \tag{10.47}$$

$$H^0_{el} = \sum_\nu E_\nu a^+_\nu a_\nu \quad H^0_{ph} = \sum_{\mathbf{k}} \hbar\omega_{\mathbf{k}} b^+_{\mathbf{k}} b_{\mathbf{k}}, \tag{10.48}$$

$$V = H_{el-ph} = \sum_{\nu,\nu',\mathbf{k}} A(\nu',\nu,\mathbf{k}) a^+_\nu a_{\nu'}(b_{\mathbf{k}} + b^+_{-\mathbf{k}}), \tag{10.49}$$

where $A(\nu',\nu,\mathbf{k}) = g_{\mathbf{k}}\langle\nu|e^{i\mathbf{kr}}|\nu'\rangle$ is the matrix element of the electron–phonon interaction and $g_{\mathbf{k}}$ is the corresponding coupling constant.

In this problem we actually have to construct the system of interconnected kinetic equations for one-particle density matrices of electrons (10.44) and phonons:

$$N_{1\mathbf{kk}'} = \langle\mathbf{k}|N_1|\mathbf{k}'\rangle = \mathrm{Sp}\,\rho b^+_{\mathbf{k}} b_{\mathbf{k}'} = \langle b^+_{\mathbf{k}} b_{\mathbf{k}'}\rangle. \tag{10.50}$$

Consider first the exact equations of motion similar to Equation (10.46):

$$i\hbar\frac{\partial}{\partial t}\langle a^+_\kappa a_{\kappa'}\rangle = \langle[a^+_\kappa a_{\kappa'}, H^0_{el} + H^0_{ph} + H_{el-ph}]\rangle, \tag{10.51}$$

$$i\hbar\frac{\partial}{\partial t}\langle b^+_{\mathbf{k}} b_{\mathbf{k}'}\rangle = \langle[b^+_{\mathbf{k}} b_{\mathbf{k}'}, H^0_{el} + H^0_{ph} + H_{el-ph}]\rangle. \tag{10.52}$$

Now, it is necessary to perform explicit calculations of the different commutators here, using the basic commutation relations (10.42), (10.43). In particular, it is pretty easy to get the following relations:

$$[a^+_\nu a_{\nu'}, a^+_\kappa a_{\kappa'}] = a^+_\nu a_{\kappa'}\delta_{\nu'\kappa} - a^+_\kappa a_{\nu'}\delta_{\nu\kappa'} \tag{10.53}$$

$$[b^+_{\mathbf{q}} b_{\mathbf{q}'}, b^+_{\mathbf{k}} b_{\mathbf{k}'}] = b^+_{\mathbf{q}} b_{\mathbf{q}'}\delta_{\mathbf{q}'\mathbf{k}} - b^+_{\mathbf{k}} b_{\mathbf{q}'}\delta_{\mathbf{qk}'}. \tag{10.54}$$

Then, using (10.53) in (10.51), we obtain:

$$\left(i\hbar\frac{\partial}{\partial t} + E_\kappa - E_{\kappa'}\right) F_{1\kappa\kappa'} = \sum_{\nu\nu'\mathbf{q}} \Big\{A(\nu',\nu,\mathbf{q})[\delta_{\nu\kappa'} H^\star_{\kappa\nu'\mathbf{q}} - \delta_{\kappa\nu'} H^\star_{\nu\kappa'\mathbf{q}}]$$
$$+ A^\star(\nu',\nu,\mathbf{q})[\delta_{\nu\kappa'} H_{\kappa\nu'\mathbf{q}} - \delta_{\kappa\nu'} H_{\nu\kappa\mathbf{q}}]\Big\}, \tag{10.55}$$

where we have introduced:

$$H^\star_{\kappa\kappa'\mathbf{q}} = \langle a^+_\kappa a_{\kappa'} b_{\mathbf{q}}\rangle \quad H_{\kappa\kappa'\mathbf{q}} = \langle a^+_\kappa a_{\kappa'} b^+_{\mathbf{q}}\rangle. \tag{10.56}$$

Similarly, using (10.54) in (10.52) we obtain:

$$\left(i\hbar\frac{\partial}{\partial t}+\hbar\omega_{\mathbf{k}}-\hbar\omega_{\mathbf{k}'}\right)N_{1\mathbf{k}\mathbf{k}'}=\sum_{\nu\nu'}\{A^{\star}(\nu',\nu,\mathbf{k})H_{\nu\nu'\mathbf{k}}-A(\nu',\nu,\mathbf{k})H^{\star}_{\nu\nu'\mathbf{k}'}\}. \tag{10.57}$$

These are the first equations of Bogolyubov's chain. In the next step we have to write down the equations of motion for $H_{\kappa\kappa'\mathbf{q}}$ and $H^{\star}_{\kappa,\kappa'\mathbf{q}}$:

$$\begin{aligned} i\hbar\frac{\partial}{\partial t}\langle a^{+}_{\kappa}a_{\kappa}b_{\mathbf{q}}\rangle &= \langle[a^{+}_{\kappa}a_{\kappa'}b_{\mathbf{q}},H^{0}_{el}+H^{0}_{ph}+H_{el-ph}]\rangle \\ i\hbar\frac{\partial}{\partial t}\langle a^{+}_{\kappa}a_{\kappa'}b^{+}_{\mathbf{q}}\rangle &= \langle[a^{+}_{\kappa}a_{\kappa'}b^{+}_{\mathbf{q}},H^{0}_{el}+H^{0}_{ph}+H_{el-ph}]\rangle. \end{aligned} \tag{10.58}$$

Using again (10.53) and (10.54) and also

$$[b_{\mathbf{q}},b^{+}_{\mathbf{k}}b_{\mathbf{k}}]=b_{\mathbf{k}}\delta_{\mathbf{k}\mathbf{q}}\quad [b^{+}_{\mathbf{q}},b^{+}_{\mathbf{k}}b_{\mathbf{k}}]=-b^{+}_{\mathbf{k}}\delta_{\mathbf{k}\mathbf{q}} \tag{10.59}$$

we get:

$$\begin{aligned} \left(i\hbar\frac{\partial}{\partial t}+E_{\kappa}-E_{\kappa'}+\hbar\omega_{\mathbf{q}}\right)H_{\kappa',\kappa,\mathbf{q}} &= \sum_{\gamma\gamma'\mathbf{q}'}A(\gamma',\gamma,\mathbf{q}')\Big\{\langle a^{+}_{\kappa}a_{\gamma'}b_{\mathbf{q}'}b^{+}_{\mathbf{q}}\rangle\delta_{\gamma\kappa'} \\ &+\langle a^{+}_{\kappa}a_{\gamma'}b^{+}_{-\mathbf{q}'}b^{+}_{\mathbf{q}}\rangle\delta_{\gamma\kappa'}-\langle a^{+}_{\gamma}a_{\kappa'}b_{\mathbf{q}'}b^{+}_{\mathbf{q}}\rangle\delta_{\gamma'\kappa} \\ &-\langle a^{+}_{\gamma}a_{\kappa'}b^{+}_{-\mathbf{q}'}b^{+}_{\mathbf{q}}\rangle\delta_{\gamma'\kappa}-\langle a^{+}_{\kappa}a_{\kappa'}a^{+}_{\gamma}a_{\gamma'}\rangle\delta_{\mathbf{q}\mathbf{q}'}\Big\}, \end{aligned} \tag{10.60}$$

$$\begin{aligned} \left(i\hbar\frac{\partial}{\partial t}+E_{\kappa}-E_{\kappa'}-\hbar\omega_{\mathbf{q}}\right)H^{\star}_{\kappa',\kappa,\mathbf{q}} &= \sum_{\gamma\gamma'\mathbf{q}'}A^{\star}(\gamma',\gamma,\mathbf{q}')\Big\{\langle a^{+}_{\kappa}a_{\gamma'}b_{\mathbf{q}'}b_{\mathbf{q}}\rangle\delta_{\gamma\kappa'} \\ &+\langle a^{+}_{\kappa}a_{\gamma'}b^{+}_{-\mathbf{q}'}b_{\mathbf{q}}\rangle\delta_{\gamma\kappa'}-\langle a^{+}_{\gamma}a_{\kappa'}b_{\mathbf{q}'}b_{\mathbf{q}}\rangle\delta_{\gamma'\kappa} \\ &-\langle a^{+}_{\gamma}a_{\kappa'}b^{+}_{-\mathbf{q}'}b_{\mathbf{q}}\rangle\delta_{\gamma'\kappa}-\langle a^{+}_{\kappa}a_{\kappa'}a^{+}_{\gamma}a_{\gamma'}\rangle\delta_{\mathbf{q}\mathbf{q}'}\Big\}. \end{aligned} \tag{10.61}$$

In principle this procedure can be continued and we shall obtain the next equations in the chain, but for most practical problems it is sufficient to limit ourselves to the equations derived above (at least for weak enough interactions). The only way to "cut" Bogolyubov's chain is to use some approximate "decoupling" of the higher-order correlators (density matrices) via lower-order correlators (e.g. factorize higher-order density matrices into products of lower-order density matrices). Unfortunately, in the general case, this procedure is not completely unambiguous and there may be several ways to perform such a "decoupling". For the problem under consideration, the two-particle correlators may be expressed via the one-particle correlators in the following way[5]:

$$\begin{aligned}
&\langle a_\kappa^+ a_{\kappa'} b_{\mathbf{k}}^+ b_{\mathbf{k}'}\rangle \approx F_{1\kappa\kappa'} N_{1\mathbf{k}\mathbf{k}'}\\
&\langle a_\kappa^+ a_{\kappa'} b_{\mathbf{k}} b_{\mathbf{k}'}^+\rangle \approx < a_\kappa^+ a_{\kappa'} > \langle \delta_{\mathbf{k}\mathbf{k}'} + b_{\mathbf{k}'}^+ b_{\mathbf{k}}\rangle = F_{1\kappa\kappa'}(\delta_{\mathbf{k}\mathbf{k}'} + N_{1\mathbf{k}\mathbf{k}'})\\
&\langle a_\kappa^+ a_{\kappa'} a_\nu^+ a_{\nu'}\rangle \approx F_{1\kappa\kappa'} F_{1\nu\nu'} + F_{1\kappa\nu'}(\delta_{\kappa'\nu} - F_{1\nu\kappa'})\\
&\langle a_\kappa^+ a_{\kappa'} b_{\mathbf{k}'} b_{\mathbf{k}}\rangle = \langle a_\kappa a_\kappa b_{\mathbf{k}'}^+ b_{\mathbf{k}}^+\rangle = 0.
\end{aligned} \tag{10.62}$$

Using (10.62) in (10.60) and performing formal integration we get:

$$\begin{aligned}
H_{\kappa\kappa'\mathbf{q}}(t) = e^{\frac{i}{\hbar}(E_\kappa - E_{\kappa'} + \hbar\omega_{\mathbf{q}})(t-t_0)} \Big\{ & H_{\kappa\kappa'\mathbf{q}}(t_0)\\
&+ \frac{1}{i\hbar}\int_{t_0}^{t} dt' e^{\frac{i}{\hbar}(E_\kappa - E_{\kappa'} + \hbar\omega_{\mathbf{q}})(t-t')} I_{\kappa\kappa'\mathbf{q}}^{FN}(t')\Big\},
\end{aligned} \tag{10.63}$$

where t_0 is the initial moment in time, and we used the notation:

$$\begin{aligned}
I_{\kappa\kappa'\mathbf{q}}^{FN}(t') = \sum_{\gamma\gamma'\mathbf{q}} A(\gamma', \gamma, \mathbf{q}') \Big[& (F_{1\kappa\gamma'}\delta_{\gamma\kappa'} - F_{1\gamma\kappa'}\delta_{\gamma'\kappa})(\delta_{\mathbf{q}\mathbf{q}'} + N_{1\mathbf{q}\mathbf{q}'})\\
&- (F_{1\kappa\kappa'} F_{1\kappa\gamma'}(\delta_{\gamma\kappa'} - F_{1\gamma\kappa'}))\delta_{\mathbf{q}\mathbf{q}'}\Big]_{t'},
\end{aligned} \tag{10.64}$$

where the last index denotes that all the density matrices (correlators) in square brackets are taken at the time t'.

Let us introduce now Bogolyubov's condition for "weak correlations " far away in the past:

$$\lim_{t_0 \to -\infty} H_{\kappa\kappa'\mathbf{q}}(t_0) = 0. \tag{10.65}$$

The choice for this condition in the past is obviously connected with causality and Equation (10.65) explicitly introduces the "arrow of time".

Then Equation (10.63) can be rewritten as:

$$H_{\kappa\kappa'\mathbf{q}}(t) = \frac{1}{i\hbar}\int_{-\infty}^{t} dt' e^{-\frac{i}{\hbar}(E_\kappa - E_{\kappa'} + \hbar\omega_{\mathbf{q}})(t'-t)} I_{\kappa\kappa'\mathbf{q}}^{FN}(t') \tag{10.66}$$

so that after the change of the variable $t' - t = \tau$ we obtain:

$$H_{\kappa\kappa'\mathbf{q}}(t) = \frac{1}{i\hbar}\int_{-\infty}^{0} d\tau e^{-i(E_\kappa - E_{\kappa'} + \hbar\omega_{\mathbf{q}})\frac{\tau}{\hbar}} I_{\kappa\kappa'\mathbf{q}}^{FN}(t+\tau). \tag{10.67}$$

[5] It is easy to see that here we have taken all possible combinations of the pair products of the creation and annihilation operators (called "pairings"), the average values of those giving the one-particle density matrices. Such decoupling is equivalent to the use of the so called Wick theorem, which is strictly valid if we average over the equilibrium density matrix [29].

Thus, in principle, the values of the correlator $H_{\kappa\kappa'\mathbf{q}}$ at time t are determined by the values of the density matrices F_1 and N_1 in all previous moments in time (a solution with a "memory"). Following Bogolyubov we shall assume that the characteristic time scale of this "memory" is of the order of the typical (microscopic) time scale of the electron–phonon interaction τ_0, so that afterwards, the time evolution of all (kinetic) variables is determined only by the time dependence of the one-particle density matrices. Then, being interested only in the evolution of the system on the time scale $t \gg \tau_0$, we can totally neglect the "memory" effects in Equation (10.67). Accordingly, using

$$\begin{aligned}\frac{1}{i}\int_{-\infty}^{0} dt e^{\mp ixt} &= \lim_{\varepsilon\to 0+}\frac{1}{i}\int_{-\infty}^{0} dt e^{\mp i(x\pm i\varepsilon)t}\\ &= \lim_{\varepsilon\to 0+}\frac{1}{x\pm i\varepsilon} = P\frac{1}{x}\mp i\delta(x)\end{aligned} \tag{10.68}$$

we immediately obtain:

$$\begin{aligned}H_{\kappa\kappa'\mathbf{q}} = \lim_{\varepsilon\to 0+}&\frac{1}{E_\kappa - E_{\kappa'} + \hbar\omega_{\mathbf{q}} + i\varepsilon}\sum_{\gamma\gamma'\mathbf{q}'} A(\gamma',\gamma,\mathbf{q}')\Big\{(F_{1\kappa\gamma'}\delta_{\gamma\kappa'}\\ &- F_{1\gamma\kappa'}\delta_{\gamma'\kappa})(\delta_{\mathbf{q}\mathbf{q}'} + N_{1\mathbf{q}\mathbf{q}'}) - (F_{1\kappa\kappa'}F_{1\gamma\gamma'} + F_{1\kappa\gamma'}(\delta_{\gamma\kappa'} - F_{1\gamma\kappa'}))\delta_{\mathbf{q}\mathbf{q}'}\Big\}.\end{aligned} \tag{10.69}$$

Similarly, from Equation (10.61) we get:

$$\begin{aligned}H^{\star}_{\kappa\kappa'\mathbf{q}} = \lim_{\varepsilon\to 0+}&\frac{1}{E_\kappa - E_{\kappa'} - \hbar\omega_{\mathbf{q}} + i\varepsilon}\sum_{\gamma\gamma'\mathbf{q}'} A^{\star}(\gamma',\gamma,\mathbf{q}')\Big\{(F_{1\kappa\gamma'}\delta_{\gamma\kappa'}(N_{1\mathbf{q}'\mathbf{q}} + \delta_{\mathbf{q}\mathbf{q}'})\\ &- F_{1\gamma\kappa'}\delta_{\gamma'\kappa}N_{1\mathbf{q}\mathbf{q}'}) - (F_{1\gamma\kappa'}F_{1\kappa\gamma'} - F_{1\kappa\kappa'}F_{1\gamma\gamma'})\delta_{\mathbf{q}\mathbf{q}'}\Big\}.\end{aligned} \tag{10.70}$$

Note that the solutions (10.67) and (10.67) follow immediately from (10.60) and (10.61) (after the decoupling (10.62)), if we assume the absence of an *explicit* time dependence of H and $H^{\star}$, which allows us to perform, in Equations (10.60) and (10.61), the formal substitution $i\hbar\frac{\partial}{\partial t} \to i\varepsilon$[6].

We see that the substitution of (10.69) and (10.70) into (10.55) and (10.57) already produces the closed system of kinetic equations for F_1 and N_1. However, their general form becomes much simpler, if we assume the diagonal nature of the one-particle density matrices:

$$F_{1\kappa\kappa'} = F_{1\kappa}\delta_{\kappa\kappa'} \quad N_{1\mathbf{k}\mathbf{k}'} = N_{1\mathbf{k}}\delta_{\mathbf{k}\mathbf{k}'}. \tag{10.71}$$

---

[6] This corresponds to Bogolyubov's assumption, that on time scales typical for kinetic phenomena, higher order density matrices (or distribution functions) depend on time only through the appropriate time dependence of one-particle density matrices (distribution functions).

The validity of this assumption actually depends on the properties of the system under consideration and on the quantum number of the electrons ν, as well as on the possibility to neglect spatial inhomogeneities in the phonon gas. If these conditions are satisfied, the quantum kinetic equations for electrons and phonons acquire the final form:

$$\frac{\partial F_{1\kappa}}{\partial t} = \frac{2\pi}{\hbar} \sum_{\nu \mathbf{q}} \Big\{ |A(\kappa, \nu, \mathbf{q})|^2 \delta(E_\kappa - E_\nu + \hbar\omega_{\mathbf{q}})$$
$$\times [F_{1\nu}(1 - F_{1\kappa})(N_{1\mathbf{q}} + 1) - F_{1\kappa}(1 - F_{1\nu})N_{1\mathbf{q}}] + |A(\nu, \kappa, \mathbf{q})|^2 \delta(E_\kappa - E_\nu - \hbar\omega_{\mathbf{q}})$$
$$\times [F_{1\nu}(1 - F_{1\kappa})N_{1\mathbf{q}} - F_{1\kappa}(1 - F_{1\nu})(N_{1\mathbf{q}} + 1)] \Big\}, \qquad (10.72)$$

$$\frac{\partial}{\partial t} N_{1\mathbf{k}} = \frac{2\pi}{\hbar} \sum_{\nu\nu'} |A(\nu, \nu', \mathbf{k})|^2 \delta(E_{\nu'} - E_\nu - \hbar\omega_{\mathbf{k}})$$
$$\times \Big\{ [F_{1\nu'} - F_{1\nu}] N_{1\mathbf{k}} + F_{1\nu'}(1 - F_{1\nu}) \Big\}. \qquad (10.73)$$

These kinetic equations (collision integrals) for the electron–phonon system form the basis for the solution of numerous problems of physical kinetics in solids [28].

10.3.2 Electron–electron interaction

Let us discuss briefly the derivation of the quantum kinetic equation for the case of interacting electrons (Fermions), described by the Hamiltonian:

$$H = \sum_\nu E_\nu a_\nu^+ a_\nu + \sum_{\mu\mu'\nu\nu'} \langle \mu\nu | U | \mu'\nu' \rangle a_\nu^+ a_\mu^+ a_{\nu'} a_{\mu'}, \qquad (10.74)$$

where we assume that the interaction is of a short-range (screened) nature (the case of Coulomb interaction requires special treatment). The matrix element of this interaction can be written as:

$$\langle \mu\nu | U(\mathbf{r}) | \mu'\nu' \rangle = \int \frac{d^3k}{(2\pi)^3} U(\mathbf{k}) \langle \mu | e^{i\mathbf{kr}} | \mu' \rangle \langle \nu | e^{i\mathbf{kr}} | \nu' \rangle. \qquad (10.75)$$

Let us introduce again the partial density matrices:

$$F_{1\kappa\kappa'} = \mathrm{Sp}\, \rho a_\kappa^+ a_{\kappa'} = \langle a_\kappa^+ a_{\kappa'} \rangle \qquad (10.76)$$

$$\langle \kappa\kappa' | F_2 | \nu\nu' \rangle = \mathrm{Sp}\, \rho a_\kappa^+ a_{\kappa'}^+ a_\nu a_{\nu'} = \langle a_\kappa^+ a_{\kappa'}^+ a_\nu a_{\nu'} \rangle. \qquad (10.77)$$

Then the appropriate Bogolyubov's chain looks like:

$$\left(i\hbar\frac{\partial}{\partial t}+E_\kappa-E_{\kappa'}\right)F_{1\kappa\kappa'}=\sum_{\nu\nu'\mu\mu'}\langle\mu\nu|U|\mu'\nu'\rangle[\langle\nu\mu|F_2|\mu'\nu'\rangle\delta_{\kappa\mu'}$$
$$-\langle\nu\mu|F_2|\mu'\kappa'\rangle\delta_{\nu'\kappa}+\langle\kappa\mu|F_2|\mu'\nu'\rangle\delta_{\kappa'\nu}-\langle\kappa\nu|F_2|\mu'\nu'\rangle\delta_{\mu\kappa'}], \tag{10.78}$$
$$\left(i\hbar\frac{\partial}{\partial t}+E_{\kappa'}+E_\kappa-E_\gamma-E_{\gamma'}\right)\langle\kappa\kappa'|F_2|\gamma\gamma'\rangle$$
$$=\sum_{\mu\mu'\nu\nu'}\langle\mu\nu|U|\mu'\nu'\rangle\langle[a_\nu^+a_\mu^+a_{\nu'}a_{\mu'},a_\kappa^+a_{\kappa'}^+a_\gamma a_{\gamma'}]\rangle. \tag{10.79}$$

After calculations of the commutators in (10.79) we obtain the averages of the product of three creation and three annihilation operators, which can be decoupled in the following way:

$$\langle a_\nu^+a_\mu^+a_{\nu'}a_{\kappa'}^+a_\gamma a_{\gamma'}\rangle\approx F_{1\nu\gamma'}F_{1\mu\nu'}F_{1\kappa'\gamma}+F_{1\nu\nu'}F_{1\mu\gamma}F_{1\kappa'\gamma'}$$
$$+F_{1\nu\gamma}F_{1\mu\gamma'}(\delta_{\nu'\kappa'}-F_{1\kappa'\nu'})$$
$$\langle a_\kappa^+a_\nu^+a_\mu^+a_{\mu'}a_\gamma a_{\gamma'}\rangle\approx F_{1\mu\mu'}F_{1\nu\gamma}F_{1\kappa\gamma'}+F_{1\kappa\mu'}F_{1\nu\gamma'}F_{1\kappa'\gamma'}+F_{1\kappa\gamma}F_{1\nu\mu'}F_{1\mu\gamma'}. \tag{10.80}$$

As before, analyzing only the slow-enough kinetic processes, in Equation (10.79), we can replace $i\hbar\frac{\partial}{\partial t}\to i\varepsilon$, which allows (taking into account (10.80)) an immediate solution. After substitution of this solution into Equation (10.78), we obtain the kinetic equation for $F_{1\kappa\kappa'}$. Assuming the diagonal nature of $F_{1\kappa\kappa'}=F_{1\kappa}\delta_{\kappa\kappa'}$, we can reduce our kinetic equation to the following form:

$$\frac{\partial F_{1\kappa}}{\partial t}=\frac{2\pi}{\hbar}\sum_{\nu\nu'\kappa'}|\langle\nu\nu'|U|\kappa\kappa'\rangle|^2\delta(E_\nu+E_{\nu'}-E_\kappa-E_{\kappa'})$$
$$\times[F_{1\nu}(1-F_{1\kappa})F_{1\nu'}(1-F_{1\kappa'})-F_{1\kappa}(1-F_{1\nu})F_{1\kappa'}(1-F_{1\nu'})]. \tag{10.81}$$

In momentum representation:

$$|\kappa\rangle=\frac{1}{\sqrt{V}}e^{\frac{i}{\hbar}\mathbf{pr}} \tag{10.82}$$
$$F_{1\kappa}\to n_{\mathbf{p}} \tag{10.83}$$
$$E_\kappa\to\varepsilon(\mathbf{p})=p^2/2m \tag{10.84}$$
$$\langle\mu|e^{i\mathbf{kr}}|\nu\rangle=\frac{1}{V}\int d\mathbf{r}e^{\frac{i}{\hbar}(\mathbf{p}'-\mathbf{p}+\mathbf{k})\mathbf{r}}\quad\text{etc.} \tag{10.85}$$

so that the kinetic equation for electrons is written as:

$$\frac{\partial n_{\mathbf{p}}}{\partial t} = \frac{2\pi}{(2\pi\hbar)^6\hbar}\int d\mathbf{p}'_1 d\mathbf{p}'_2 d\mathbf{p}_2 |U(\mathbf{p}_1 - \mathbf{p}'_1)|^2 \delta(\mathbf{p}_1 + \mathbf{p}_2 - \mathbf{p}'_1 - \mathbf{p}'_2)$$
$$\times\, \delta(\varepsilon(\mathbf{p}_1) + \varepsilon(\mathbf{p}_2) - \varepsilon(\mathbf{p}'_1) - \varepsilon(\mathbf{p}'_2))\Big[n_{\mathbf{p}'_1} n_{\mathbf{p}'_2}(1 - n_{\mathbf{p}_1})(1 - n_{\mathbf{p}_2})$$
$$- n_{\mathbf{p}_1} n_{\mathbf{p}_2}(1 - n_{\mathbf{p}'_1})(1 - n_{\mathbf{p}'_2})\Big], \tag{10.86}$$

where $U(\mathbf{p})$ is the Fourier transform of the interaction potential.

Writing the entropy of the electron gas as in Equation (4.15):

$$S = 2\int \frac{d^3p}{(2\pi)^3}[(1 - n_{\mathbf{p}})\ln(1 - n_{\mathbf{p}}) - n_{\mathbf{p}} \ln n_{\mathbf{p}}] \tag{10.87}$$

and using (10.86) we can (after some tedious calculations) prove the quantum version of the H-theorem: $\frac{dS}{dt} \geq 0$.

The equilibrium Fermi distribution

$$n^0_{\mathbf{p}} = \frac{1}{e^{\frac{\varepsilon(p)-\mu}{T}} + 1} \tag{10.88}$$

leads to a zero value of the collision integral in Equation (10.86), which can be checked by direct calculations, taking into account the energy conservation law for scattering particles, expressed by the δ-function in the collision integral. It can be easily seen in this case, that the combination of (equilibrium) distribution functions in square brackets in (10.86) becomes identically zero.

The derived expression for the collision integral for the system of interacting electrons plays a major role in studies of low temperature kinetics in metals and other Fermi-liquids.

Chapter 11

Basics of the modern theory of many-particle systems

11.1 Quasiparticles and Green's functions

We have seen above the major role played by the concept of quasiparticles in the modern theory of condensed matter. A rigorous justification of this concept is achieved within the formalism of Green's functions, which is at present the standard apparatus of the theory of many particle systems. The Green's functions approach allows a clear formulation of the criteria for the existence of quasiparticles in concrete systems (models) and gives the universal method for calculating practically any physical characteristics of many particle systems, taking into account different interactions. This method first appeared in quantum field theory, where it was widely accepted after the formulation of a quite effective and convenient approach, based on the use of *Feynman diagrams*. Later it was applied to general many particle systems, which in fact lead to to the creation of modern condensed matter theory [2]. Obviously, here we are unable to give the complete and coherent presentation of the Green's functions formalism; our aim is only to introduce some of its major definitions and give a qualitative illustration of some simple applications[1].

Below, we mainly consider the case of temperature $T = 0$. A generalization of the Green's functions approach to finite temperatures is rather straightforward and we shall briefly discuss it at the end of this Chapter. Let us start from the case of a single quantum particle, described by the Schroedinger equation[2]:

$$i\frac{\partial\psi(\mathbf{r},t)}{\partial t} - H\psi(\mathbf{r},t) = 0. \tag{11.1}$$

Instead of this equation, we may introduce the equation of motion for the Green's function $G(\mathbf{r},t;\mathbf{r}',t')$:

$$i\frac{\partial G}{\partial t} - HG = i\delta(\mathbf{r}-\mathbf{r}')\delta(t-t') \tag{11.2}$$

with initial condition $G(\mathbf{r},t+0;\mathbf{r}',t) = \delta(\mathbf{r}-\mathbf{r}')$. Green's function represents the *probability amplitude* of particle transition from point $\mathbf{r}'$ at time t to point $\mathbf{r}$ at time t.

[1] The most clear presentation of the Green's functions method, as well as the Feynman diagram technique, with applications to the problems of statistical physics, was given in the classic book by Abrikosov, Gorkov and Dzyaloshinskii [29]. Rather detailed material can be found in [2]. A more elementary presentation on Green's functions is given in [30, 31, 32].

[2] Below we use the system of units with $\hbar = 1$, which is standard in most modern texts. If necessary, the value of $\hbar$ can be easily restored in the final expressions.

The squared modulus of this amplitude gives the probability of this transition. We can see this by expressing the ψ-function at time $t+\tau$ via ψ-function at time t:

$$\psi(\mathbf{r}, t+\tau) = \int d\mathbf{r}' G(\mathbf{r}, t+\tau; \mathbf{r}'t)\psi(\mathbf{r}', t). \tag{11.3}$$

It is easily seen, that $\psi(\mathbf{r}, t+\tau)$ written in this way, satisfies the Schroedinger equation (11.1), while for $\tau \to 0$ it transforms into $\psi(\mathbf{r}, t)$ due to the initial condition $G(\mathbf{r}, t+0; \mathbf{r}', t) = \delta(\mathbf{r}-\mathbf{r}')$. Besides that, we assume (by definition) that $G = 0$ for $\tau < 0$ (causality!).

Consider the eigenfunctions and eigenvalues of our Schroedinger equation:

$$H\varphi_\lambda(\mathbf{r}) = \varepsilon_\lambda \varphi_\lambda(\mathbf{r}). \tag{11.4}$$

The physical meaning of the quantum numbers λ may be different, depending on the nature of the problem under discussion. In a translationally invariant system $\lambda \to \mathbf{p}$ – the momentum, for an electron in an external magnetic field λ represents the Landau quantum numbers etc. Let us consider a particle in a potential field:

$$H = \frac{p^2}{2m} + V(\mathbf{r}). \tag{11.5}$$

In particular, this may correspond to the nontrivial problem of nucleons in potential well – an atomic nucleus [31], so that λ represents the quantum numbers of the shell model. Any solution of the Schroedinger equation can be expanded over this (complete) system of eigenfunctions:

$$\psi(\mathbf{r}, t) = \sum_\lambda c_\lambda(t)\varphi_\lambda(\mathbf{r}) \tag{11.6}$$

so that (11.3) can be written as:

$$c_\lambda(t+\tau) = \sum_{\lambda'} G_{\lambda\lambda'}(\tau) c_{\lambda'}(t), \tag{11.7}$$

$$G_{\lambda\lambda'}(\tau) = \int d^3r d^3r' G(\mathbf{r}, \mathbf{r}'\tau)\varphi^\star_\lambda(\mathbf{r})\varphi_{\lambda'}(\mathbf{r}'), \tag{11.8}$$

which gives the Green's function in λ-representation. As φ_λ is an eigenfunction of the Hamiltonian H, which is time-independent, there are no transitions to other states, so that $c_\lambda(t+\tau) = e^{-i\varepsilon_\lambda\tau} c_\lambda(t)$, i.e.

$$G_{\lambda\lambda'}(\tau) = G_\lambda(\tau)\delta_{\lambda\lambda'} = e^{-i\varepsilon_\lambda\tau}\theta(\tau), \tag{11.9}$$

where $\theta(\tau) = 1$ for $\tau \geq 0$ and $\theta(\tau) = 0$ for $\tau < 0$. Let us make the Fourier transformation:

$$G_\lambda(\varepsilon) = \frac{1}{i}\int_{-\infty}^{\infty} d\tau e^{i\varepsilon\tau} G_\lambda(\tau), \tag{11.10}$$

$$G_\lambda(\tau) = i\int_{-\infty}^{\infty} \frac{d\varepsilon}{2\pi} e^{-i\varepsilon\tau} G_\lambda(\varepsilon). \tag{11.11}$$

Then, after an elementary integration we obtain:

$$G_\lambda(\varepsilon) = \frac{1}{\varepsilon - \varepsilon_\lambda + i\delta} \quad \delta \to +0. \tag{11.12}$$

The sign of $\delta \to 0$ is chosen to guarantee $G_\lambda(\tau) = 0$ for $\tau < 0$. In fact, we have:

$$\begin{aligned} G_\lambda(\tau) &= i\int_{-\infty}^{\infty} \frac{d\varepsilon}{2\pi} \frac{e^{-i\varepsilon\tau}}{\varepsilon - \varepsilon_\lambda + i\delta} \\ &= \begin{cases} e^{-i\varepsilon_\lambda\tau} & \text{for } \tau > 0 \\ 0 & \text{for } \tau < 0. \end{cases} \end{aligned} \tag{11.13}$$

The integrand here possesses a pole at $\varepsilon = \varepsilon_\lambda - i\delta$. Then, for $\tau > 0$ we can perform integration over ε closing the integration contour in the lower half-plane of the complex variable ε (as the factor of $e^{-i\varepsilon\tau}$ guarantees exponential damping of the integrand on the semicircle at infinity), so that the pole is inside the integration contour and the integral is easily calculated using the Cauchy theorem, giving the result shown above. For $\tau < 0$, in a similar way, to make the contribution of semicircle zero at infinity, it is necessary to close the integration contour in the upper half-plane of ε. Then, there is no pole inside the integration contour and the integral is zero.

In the mixed $(\mathbf{r}, \varepsilon)$ representation we get:

$$\begin{aligned} G(\mathbf{r}, \mathbf{r}', \varepsilon) &= \sum_{\lambda,\lambda'} G_{\lambda\lambda'}(\varepsilon)\varphi_\lambda(\mathbf{r})\varphi^\star_{\lambda'}(\mathbf{r}') \\ &= \sum_\lambda \frac{\varphi_\lambda(\mathbf{r})\varphi^\star_\lambda(\mathbf{r}')}{\varepsilon - \varepsilon_\lambda + i\delta}. \end{aligned} \tag{11.14}$$

Here the sum over λ is performed over all bound states of a particle in a field, as well as over the continuous spectrum. We see that $G(\mathbf{r}, \mathbf{r}', \varepsilon)$ possesses poles at the values of ε equal to ε_λ – the energies of the bound states, and a cut (continuum of poles) at the part of the real ε-axis, corresponding to the continuous spectrum.

Let us consider now a many particle system. Below we are only dealing with systems consisting of Fermions. For systems of Bose particles we may construct a similar approach, but we shall not discuss it here due to lack of space; a proper presentation can be found in [2, 29]. We shall start with the case of noninteracting Fermions (ideal Fermi gas). We have seen above, that elementary excitations in this system are formed by pairs of particles (above the Fermi surface) and holes (below the Fermi surface).

Let us find the explicit form of the Green's function $G_{\lambda\lambda'}(\tau)$, i.e. the transition amplitude of a single particle from state λ to state λ', in a system of noninteracting Fermions. We have to take into account the Pauli principle and exclude all transitions into occupied states. This is achieved by the introduction into the definition of the Green's function of an extra factor $(1 - n_\lambda)$, where

$$n_\lambda = \begin{cases} 1 & \text{for } \varepsilon_\lambda \le \varepsilon_F \\ 0 & \text{for } \varepsilon_\lambda > \varepsilon_F \end{cases} \tag{11.15}$$

is the number of particles in state λ (Fermi distribution at $T = 0$). Thus, we obtain:

$$G^+_{\lambda\lambda'}(\tau) = (1 - n_\lambda)\delta_{\lambda\lambda'} \begin{cases} e^{-i\varepsilon_\lambda \tau} & \text{for } \tau > 0 \\ 0 & \text{for } \tau < 0. \end{cases} \tag{11.16}$$

Let us find similar expression for a hole. The number of "free" places for holes in state λ is proportional to n_λ, so that

$$G^-_{\lambda\lambda'}(\tau) = n_\lambda \delta_{\lambda\lambda'} \begin{cases} e^{i\varepsilon_\lambda \tau} & \text{for } \tau > 0 \\ 0 & \text{for } \tau < 0, \end{cases} \tag{11.17}$$

where we have taken into account that the energy of the hole, calculated with respect to the Fermi level, has the sign opposite to that of a particle.

It is convenient to introduce the Green's function $G_\lambda(\tau)$, defined both for $\tau > 0$ and $\tau < 0$:

$$G_\lambda(\tau) = \begin{cases} G^+_\lambda(\tau) & \text{for } \tau > 0 \\ -G^-_\lambda(-\tau) & \text{for } \tau < 0. \end{cases} \tag{11.18}$$

The Fourier transform of this function is easily calculated as:

$$\begin{aligned} G_\lambda(\varepsilon) &= -i(1 - n_\lambda) \int_0^\infty d\tau e^{-i\varepsilon_\lambda \tau + i\varepsilon\tau} + i n_\lambda \int_{-\infty}^0 d\tau e^{i\varepsilon_\lambda \tau + i\varepsilon\tau} \\ &= \frac{1 - n_\lambda}{\varepsilon - \varepsilon_\lambda + i\delta} + \frac{n_\lambda}{\varepsilon - \varepsilon_\lambda - i\delta}, \end{aligned} \tag{11.19}$$

where $\delta \to +0$ is necessary to guarantee the convergence of the integrals. This expressions is conveniently rewritten as:

$$\begin{aligned} G_\lambda(\varepsilon) &= \frac{1}{\varepsilon - \varepsilon_\lambda + i\delta \operatorname{sign} \varepsilon_\lambda} \\ &= \begin{cases} \frac{1}{\varepsilon - \varepsilon_\lambda + i\delta} & \text{for } \varepsilon_\lambda > \varepsilon_F \\ \frac{1}{\varepsilon - \varepsilon_\lambda - i\delta} & \text{for } \varepsilon_\lambda < \varepsilon_F, \end{cases} \end{aligned} \tag{11.20}$$

where we have introduced the sign function: $\operatorname{sign}(x) = 1$ for $x > 0$ and $\operatorname{sign}(x) = -1$ for $x < 0$. Note that the Fourier transform of the Green's function possesses a pole at ε equal to the energy of the particle (hole).

Consider now the system of interacting Fermions (Fermi liquid). A single particle Green's function in the system of interacting particles is defined by the following expression:

$$G^+(\mathbf{r}t; \mathbf{r}'t')_{t>t'} = \langle 0|\hat{\psi}(\mathbf{r}t)\hat{\psi}^+(\mathbf{r}'t')|0\rangle, \tag{11.21}$$

where $|0\rangle$ is an exact eigenfunction of the ground state ("vacuum"), corresponding to the filled Fermi sphere, $\hat{\psi}(\mathbf{r}t)$ is the Fermion creation operator in Heisenberg representation:

$$\hat{\psi}(\mathbf{r}t) = e^{iHt}\hat{\psi}(\mathbf{r})e^{-iHt}, \tag{11.22}$$

where H is the total Hamiltonian of the many particle system, which includes the interactions. The operator $\hat{\psi}(\mathbf{r})$ can be expressed via the annihilation operators a_λ of particles in eigenstates λ and ($\hat{\psi}^+$ is similarly expressed via creation operators a_λ^+):

$$\hat{\psi}(\mathbf{r}) = \sum_\lambda a_\lambda \varphi_\lambda(\mathbf{r}). \tag{11.23}$$

Expression (11.21) obviously represents the amplitude of a particle propagation[3] from point $(\mathbf{r}'t')$ to point $(\mathbf{r}t)$.

For hole propagation we can similarly write:

$$G^-(\mathbf{r}t;\mathbf{r}'t')_{t>t'} = \langle 0|\hat{\psi}^+(\mathbf{r}t)\hat{\psi}(\mathbf{r}'t')|0\rangle, \tag{11.24}$$

where we have taken into account that (in a Fermion system) the annihilation of a particle in a given point is equivalent to the creation of a hole.

Expressions (11.21) and (11.24) are defined for $t > t'$. It is convenient to define a single Green's function, which for $t > t'$ describes a particle, while for $t < t'$ it describes a hole (similar to (11.18)):

$$G(\mathbf{r}t;\mathbf{r}'t') = \begin{cases} G^+(\mathbf{r}t;\mathbf{r}'t') & \text{for } t > t' \\ -G^-(\mathbf{r}'t';\mathbf{r}t) & \text{for } t < t'. \end{cases} \tag{11.25}$$

Another way to write this definition is:

$$G(x,x') = \langle 0|T\hat{\psi}(x)\hat{\psi}^+(x')|0\rangle, \tag{11.26}$$

where we have denoted $x = (\mathbf{r}t)$ and introduced an operator of T-ordering, which places all operators to the right of T in order of diminishing times in their arguments, taking also into account the change of signs due to (possible) permutations of Fermion operators. The formal definition of T-ordering (originating in quantum field theory) is written as:

$$T\{F_1(t_1)F_2(t_2)\} = \begin{cases} F_1(t_1)F_2(t_2) & \text{for } t_1 > t_2 \\ -F_2(t_2)F_1(t_1) & \text{for } t_1 < t_2 \end{cases} \tag{11.27}$$

for Fermion operators, and

$$T\{B_1(t_1)B_2(t_2)\} = \begin{cases} B_1(t_1)B_2(t_2) & \text{for } t_1 > t_2 \\ B_2(t_2)B_1(t_1) & \text{for } t_1 < t_2 \end{cases} \tag{11.28}$$

for Boson operators.

[3] Green's functions are often called *propagators*.

The Green's function defined according to (11.26) is called Feynman or casual (T-ordered)[4].

Let us limit our consideration to an infinite (translationally invariant) system, when $G(\mathbf{r}t;\mathbf{r}'t') = G(\mathbf{r}-\mathbf{r}', t-t')$. Accordingly, it is convenient to introduce the Fourier representation over $t - t'$ and $\mathbf{r} - \mathbf{r}'$:

$$G(\mathbf{p}\tau) = \int d^3r G(\mathbf{r}\tau)e^{-i\mathbf{p}\mathbf{r}}, \qquad (11.29)$$

where

$$G(\mathbf{p}\tau) = \begin{cases} \langle 0|a_{\mathbf{p}}e^{-iH\tau}a^+_{\mathbf{p}}|0\rangle e^{iE_0\tau} & \tau > 0 \\ -\langle 0|a^+_{\mathbf{p}}e^{iH\tau}a_{\mathbf{p}}|0\rangle e^{-iE_0\tau} & \tau < 0 \end{cases} \qquad (11.30)$$

and E_0 is the ground state energy.

Quasiparticles in our system can be introduced, if the one particle Green's function can be written in the following form ($\tau > 0$):

$$G(\mathbf{p}\tau) \approx Ze^{-i(\varepsilon(\mathbf{p})-\gamma(\mathbf{p}))\tau} + \cdots \quad \text{and} \quad \gamma(\mathbf{p}) \ll \varepsilon(\mathbf{p}) \qquad (11.31)$$

i.e. it contains a contribution, resembling the Green's function of an ideal Fermi gas, which we derived above. Equation (11.31) means, that the state $|0>$ contains a wave packet with amplitude Z, representing a quasiparticle with energy $\varepsilon(\mathbf{p})$ and *damping* $\gamma(\mathbf{p})$. The necessary requirement is the weakness of this damping $\gamma(\mathbf{p}) \ll \varepsilon(\mathbf{p})$, i.e. the requirement for a quasiparticle to be "well-defined"[5]. Similarly, for $\tau < 0$ we can define the Green's function of a quasihole. Thus, in a system with well-defined quasiparticles, the Fourier transform of the Green's function (11.26) can be written as:

$$\begin{aligned} G(\mathbf{p}\varepsilon) &= Z\left\{\frac{1-n_{\mathbf{p}}}{\varepsilon-\varepsilon(\mathbf{p})+i\gamma(\mathbf{p})} + \frac{n_{\mathbf{p}}}{\varepsilon-\varepsilon(\mathbf{p})-i\gamma(\mathbf{p})}\right\} + G_{\text{reg}}(\mathbf{p}\varepsilon) \\ &= \frac{Z}{\varepsilon-\varepsilon(\mathbf{p})+i\gamma(\mathbf{p})\,\text{sign}(p-p_F)} + G_{\text{reg}}(\mathbf{p}\varepsilon). \end{aligned} \qquad (11.32)$$

We see that the pole of this expression defines the spectrum of quasiparticles and their damping. This is the most important property of Green's functions, allowing us to determine the spectrum of elementary excitations in a many-particle system. The value

[4] Let us stress, that this definition is different from the definition of the double time Green's function given in (9.22), and naturally appearing in linear response theory, even in the limit of zero temperature. The advantage of the use of Feynman Green's functions is the possibility to construct a diagram technique, much simplifying all calculations. There is no diagram technique to calculate double time Green's functions (9.22). There are certain exact relations and methods, allowing us to express the Green's functions of linear response theory via Feynman functions at $T = 0$ [29], as well as appropriate generalizations for the case of finite temperatures, to be considered below [2, 29]

[5] This condition is valid in Landau Fermi liquids, where close to the Fermi surface we have $\varepsilon(\mathbf{p}) \approx v_F(|\mathbf{p}| - p_F)$, and $\gamma(\mathbf{p}) \sim (|\mathbf{p}| - p_F)^2$.

of the nonsingular term $G_{\rm reg}$ in (11.32) is determined by multiparticle excitations and, in most cases, is not of great interest. At the same time, we have to note that in systems with strong interactions (correlations) there are cases, when we can not separate a singular pole-like contribution to the Green's function, related to single particle elementary excitations (quasiparticles). Then, all physics is in fact determined by $G_{\rm reg}$ and situation becomes more complicated.

What else do we need Green's functions for? Actually, with their help we can calculate the averages (over the ground state) of different physical characteristics of our system. Using the *one particle* Green's function, introduced above, we can calculate the ground state averages of operators, which are represented by sums over all particles (one particle operators):

$$\hat{A} = \sum_i \hat{A}_i(\xi_i, \mathbf{p}_i), \tag{11.33}$$

where ξ_i is the set of spatial and spin variables, while $\mathbf{p}_i$ are the momenta of all the particles of the system. Typical examples are:

$$n(\mathbf{r}) = \sum_i \delta(\mathbf{r} - \mathbf{r}_i) \tag{11.34}$$

– particle density at point $\mathbf{r}$,

$$\mathbf{j}(\mathbf{r}) = \frac{e}{m} \sum_i \mathbf{p}_i \delta(\mathbf{r} - \mathbf{r}_i) \tag{11.35}$$

– current density at point $\mathbf{r}$ etc.

Operator $\hat{A}$ in secondary quantization representation can be written as:

$$\hat{A} = \int d\xi \psi^+(\xi) A(\xi, \mathbf{p}) \psi(\xi). \tag{11.36}$$

Consider the Green's function (11.25), (11.26) at $t = t' - 0$:

$$G(\xi, \xi', \tau)|_{\tau \to -0} = -\langle 0|\psi^+(\xi')\psi(\xi)|0\rangle. \tag{11.37}$$

Then, the ground state average value of operator $\hat{A}$ is given by:

$$\langle A \rangle = \int d\xi A(\xi, \mathbf{p}) G(\xi, \xi', \tau = -0)|_{\xi=\xi'} = -\mathrm{Sp} A G|_{\tau=-0}. \tag{11.38}$$

We conclude, that $G|_{\tau=-0}$ just coincides (up to a sign) with the one particle density matrix (cf. (1.163)) at $T = 0$:

$$\rho(\xi', \xi) = \langle 0|\psi^+(\xi')\psi(\xi)|0\rangle = -G|_{\tau=-0}. \tag{11.39}$$

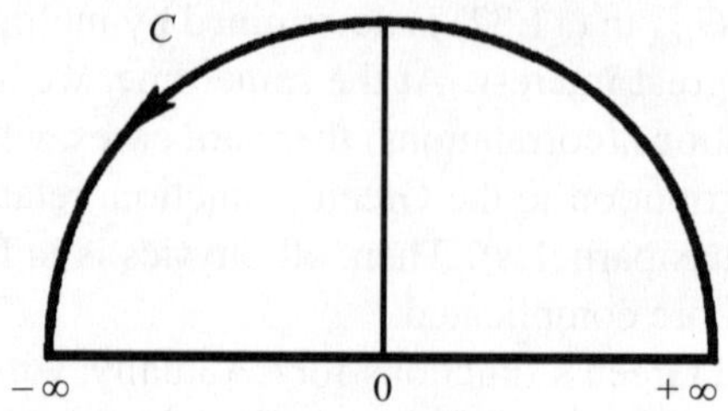

Figure 11.1. Integration contour used in calculations of the distribution function.

To find the averages of two-particle operators like:

$$\hat{B} = \sum_{ik} B_{ik}(\xi_i \mathbf{p}_i; \xi_k \mathbf{p}_k) \tag{11.40}$$

we need to calculate two-particle Green's function:

$$G_2(1,2;3,4) = \langle 0|T\psi(1)\psi(2)\psi^+(3)\psi^+(4)|0\rangle \tag{11.41}$$

etc.

From (11.37) we immediately obtain the particle momentum distribution function as:

$$n(\mathbf{p}) = -i \int_{-\infty}^{\infty} \frac{d\varepsilon}{2\pi} G(\mathbf{p}\varepsilon) e^{-i\varepsilon\tau}|_{\tau\to-0}. \tag{11.42}$$

Here, we can not simply take the limit of $\tau = 0$, as $G \sim \frac{1}{\varepsilon}$ and for $\varepsilon \to \infty$ the integral $\int d\varepsilon G(\mathbf{p}\varepsilon)$ diverges. For finite and negative τ we can transform the integral over the real axis of ε to an integral over the closed contour C, shown in Figure 11.1. After that we can put $\tau = 0$, so that:

$$n(\mathbf{p}) = -i \int_C \frac{d\varepsilon}{2\pi} G(\mathbf{p}\varepsilon). \tag{11.43}$$

Consider a Green's function like that of Equation (11.32) (quasiparticles!):

$$G(\mathbf{p}\varepsilon) = \frac{Z}{\varepsilon - \varepsilon(\mathbf{p}) + i\gamma(\mathbf{p})\,\mathrm{sign}(p - p_F)} + G_{\mathrm{reg}}(\mathbf{p}\varepsilon). \tag{11.44}$$

We see that the damping (imaginary part in the denominator of the first term) changes sign at $p = p_F$: it is positive for $p > p_F$ and negative for $p < p_F$. Thus, for $p < p_F$ we have a pole inside the contour C, so that the integral is equal to Z, while for $p > p_F$ the pole is in the lower half-plane and the integral over C is equal to zero. Neglecting the regular many-particle contribution G_{reg} we have:

$$n(p_F - 0) - n(p_F + 0) = Z. \tag{11.45}$$

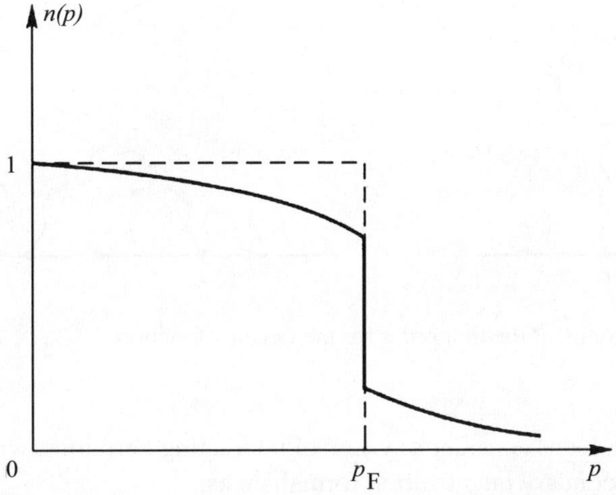

Figure 11.2. Qualitative form of the distribution function of particles in a Fermi liquid at $T = 0$.

As $0 \leq n(p) \leq 1$, it follows that $0 < Z < 1$. Now it is clear, that the qualitative form of the distribution function of Fermions at $T = 0$ (interacting Fermions, Fermi liquid!) has the form, shown in Figure 11.2. Thus, despite the presence of interactions (not necessarily weak!), which "smears" the momentum distribution of particles, there is still a "trace" of the Fermi distribution for an ideal gas. Even in a Fermi liquid there is a finite discontinuity in the distribution function at $p = p_F$. This important result was first derived by Migdal and gives a major microscopic justification of one of the most important assumptions of phenomenological Fermi liquid theory, introduced by Landau. Surely, our analysis is valid only for momenta p close enough to p_F, where the concept of quasiparticles "works" due to $\gamma \sim (p - p_F)^2$, making damping small in comparison to the real part of the spectrum $\varepsilon(\mathbf{p}) \approx v_F(|\mathbf{p}| - p_F)$.

11.2 Feynman diagrams for many-particle systems

The Feynman diagram technique is an elegant and compact formulation of the perturbation theory rules to calculate Green's functions. Unfortunately, we are unable to present here the detailed derivation of these rules and the reader should reger to [29, 2] for the complete presentation of Feynman's approach. An elementary, though detailed enough, presentation can be found in [30]. Many examples of the practical use of Feynman diagrams are given in [32]. Here we shall limit ourselves to the formulation of the elementary rules of the diagram technique (without derivation), which is sufficient to get some impression of the method and to not be "frightened" by the appearance of Feynman diagrams, which are rather ubiquitous in modern literature.

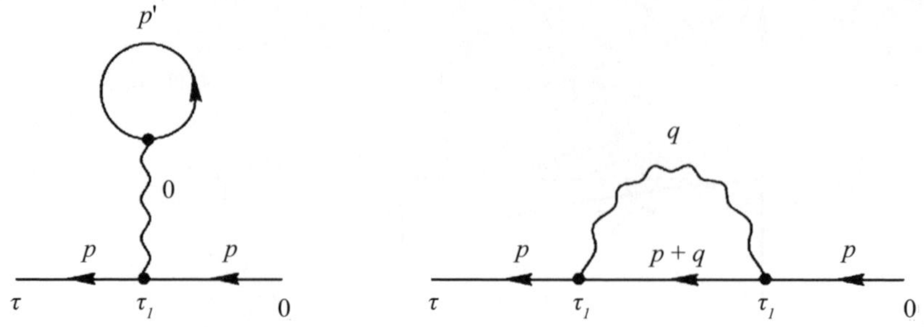

Figure 11.3. Diagrams of the first order for the Green's function.

To be concrete, let us consider a system of interacting Fermions, with the Hamiltonian, written in secondary quantization formalism as:

$$H = \sum_{\mathbf{p}} \varepsilon(\mathbf{p}) a_{\mathbf{p}}^{+} a_{\mathbf{p}} + \frac{1}{2} \sum_{\mathbf{pqk}} V_{\mathbf{k}} a_{\mathbf{p}+\mathbf{k}}^{+} a_{\mathbf{q}-\mathbf{k}}^{+} a_{\mathbf{q}} a_{\mathbf{p}}. \tag{11.46}$$

By definition, the Green's function $G(\mathbf{p}\tau)$ is dealing with the motion of a single particle. In the absence of interactions (free Green's function) we can represent this motion by a straight line, e.g. directed from right to left. Since the unperturbed ground state of the system is the filled Fermi sphere, there is a possibility of a hole motion, which we shall represent by a straight line, directed from left to right. Thus, a directed line represents the free Green's function $G(\mathbf{p}\tau)$, corresponding to a free particle with momentum $\mathbf{p}$.

The interaction corresponds to the scattering of one particle by another. In first/order perturbation theory over $V_{\mathbf{k}}$, we have two types of scattering processes, represented by the Feynman diagrams, shown in Figure 11.3. The process corresponding to the first diagram corresponds to a particle moving freely until it is directly scattered by the particles inside the Fermi sphere (surface) at time τ_1, while afterwards it continues the free motion from time τ_1 to time τ. The act of interaction (scattering) is represented by the wavy line and a closed circle describes the process, where a particle is scattered from the state with some momentum below the Fermi surface and returns to the same state. The process corresponding to second diagram represents the so called exchange scattering on the particles below the Fermi surface; its meaning is obvious from the diagram itself – the free motion after scattering is continued by a particle excited from below the Fermi surface, while the initial particle has gone to a state below.

In second order perturbation theory, the number of possible scattering processes increases. Examples of appropriate Feynman diagrams are shown in Figure 11.4. All diagrams here, except the last one, show different combinations of scattering of the first order, considered above. The last diagram describes something new – at time

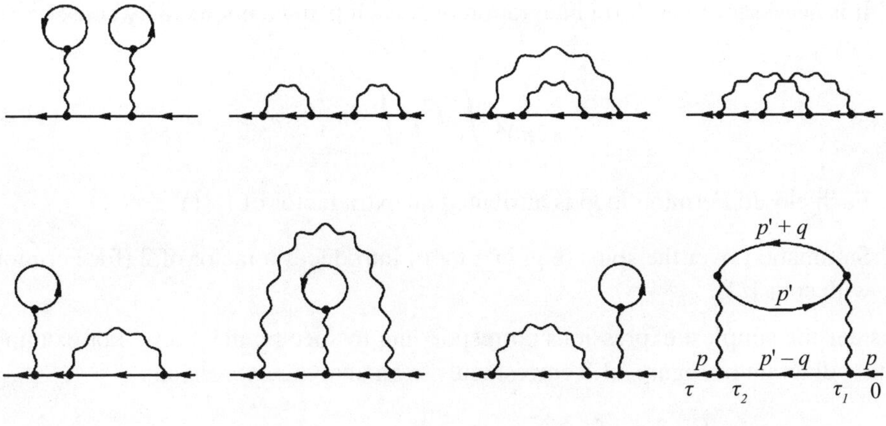

Figure 11.4. Examples of diagrams of the second order for the Green's function.

τ_1 the particle is scattered, creating a particle-hole pair, exciting it from states below the Fermi surface. At time τ_2 the particle is scattered again, as a particle-hole pair annihilates, returning to the initial state. Physically, this process corresponds to the polarization of particles below the Fermi surface.

Most conveniently, the rules of the diagram technique are formulated for calculations of the Fourier transform of Green's function $G(\mathbf{p}\varepsilon)$. In this case the arrows on lines, representing Green's functions, do not denote the direction of time, but correspond simply to incoming and outgoing "energies" ε and momenta $\mathbf{p}$, which are conserved in each vertex (interaction point). The rules to construct an analytic expression, corresponding to a given Feynman diagram, are formulated as follows:

1. To each straight line we attribute a momentum $\mathbf{p}$ and an "energy" ε, and write the corresponding analytic expression:

$$iG_0(\mathbf{p}\varepsilon) = \frac{i}{\varepsilon - \varepsilon(\mathbf{p}) + i\delta \operatorname{sign} \varepsilon(\mathbf{p})}. \tag{11.47}$$

2. To each interaction (wavy) line corresponds the factor $-iV_{\mathbf{q}}$ (in the case of instantaneous interaction) or $-iV(\mathbf{q}\omega)$ for retarded interaction.

3. In each vertex (a point where the wavy line is attached to the Green's function lines) energy and momentum are conserved, with energies and momenta attributed to the lines directed toward the vertex, taken with plus-sign, while energies and momenta attributed to outgoing lines are taken with minus-sign.

4. It is necessary to perform integration over each $\mathbf{p}$ and ε not fixed by conservation laws:

$$\frac{1}{(2\pi)^4}\int d^3\mathbf{p}\int d\varepsilon\cdots. \tag{11.48}$$

5. Each closed Fermion loop is attributed an extra factor of (-1).

6. Summation over the spins (e.g. in a loop) introduces a factor of 2 (for Fermions with spin 1/2).

Consider the simplest expressions corresponding to specific diagrams. For example, the first diagram of Figure 11.3 corresponds to an analytic expression:

$$i^2 G_0(\mathbf{p}\varepsilon)(-iV_0)\left\{\frac{2}{(2\pi)^3}\int d^3p'(-n(\mathbf{p}'))\right\} iG_0(\mathbf{p}\varepsilon) = G_0(\mathbf{p}\varepsilon)(-iV_0)NG_0(\mathbf{p}\varepsilon), \tag{11.49}$$

where in the first expression we have already taken into account (11.43) and N is the total number of particles. This gives the so called Hartree correction. The second diagram of Figure 11.3 gives:

$$i^2 G_0(\mathbf{p}\varepsilon)\frac{1}{(2\pi)^3}\int d^3q(-iV_\mathbf{q})(-n(\mathbf{p}+\mathbf{q}))G_0(\mathbf{p}\varepsilon), \tag{11.50}$$

which is the Fock correction. The last diagram of Figure 11.4 corresponds to:

$$G_0(\mathbf{p}\varepsilon)\frac{1}{(2\pi)^4}\int d^3q\int d\omega\, iG_0(\mathbf{p}-\mathbf{q}\varepsilon-\omega)(-iV_\mathbf{q})^2[-i\Pi_0(\mathbf{q}\omega)]G_0(\mathbf{p}\varepsilon), \tag{11.51}$$

where we have introduced the so-called polarization operator, corresponding to the loop in this graph:

$$\begin{aligned}-i\Pi_0(\mathbf{q}\omega) &= 2(-i)(-1)\int\frac{d^3p'}{(2\pi)^3}\int\frac{d\varepsilon'}{2\pi}(i)^2 G_0(\mathbf{p}'+\mathbf{q}\varepsilon+\omega)G_0(\mathbf{p}'\varepsilon')\\ &= i\int\frac{d^3p}{(2\pi)^3}\frac{n(\mathbf{p})-n(\mathbf{p}-\mathbf{q})}{\varepsilon(\mathbf{p}-\mathbf{q})-\varepsilon(\mathbf{p})+\omega+i\delta\,\mathrm{sign}\,\omega}.\end{aligned} \tag{11.52}$$

Note that this expression gives only the simplest contribution to the polarization operator, in the general case we have to deal with higher-order corrections, e.g. of the type shown in Figure 11.5.

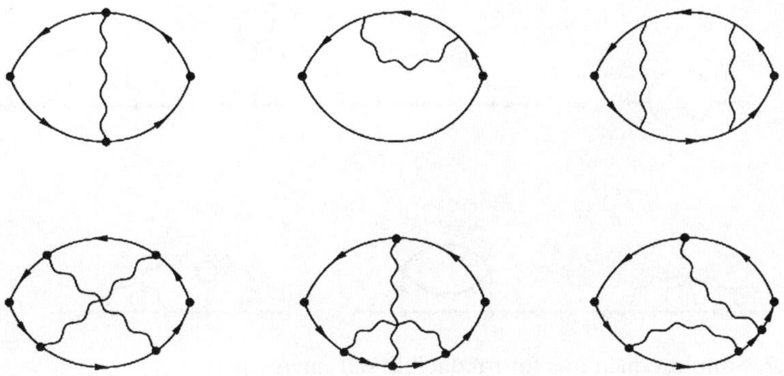

Figure 11.5. Higher corrections for polarization operator.

11.3 Dyson equation

A remarkable property of the Feynman diagram technique is the possibility to perform an intuitively clear *graphical summation* of an infinite series of diagrams. Let us denote an *exact* Green's function (taking into account all interaction corrections) by a "fat" (or "dressed") line, while the free-particle Green's function is denoted by a "thin" line as above. The total transition amplitude from point 2 to point 1 is, obviously, equal to the sum of all possible transition amplitudes, appearing in all orders of perturbation theory, i.e. to the sum of *all* diagrams of the type shown in Figure 11.6. Now we can classify these diagrams in the following way: first of all we separate the single graph, corresponding to free-particle motion. All the remaining diagrams have the following form: up to some point the particle moves freely, then it is scattered, which leads to creation and annihilation of several particles and holes (or it is scattered by particles below the Fermi surface), then again it performs free motion, then it is scattered again etc. Let us denote as Σ the sum of all diagrams *which cannot be cut over the single particle line*. Examples of such diagrams are shown in Figure 11.7.

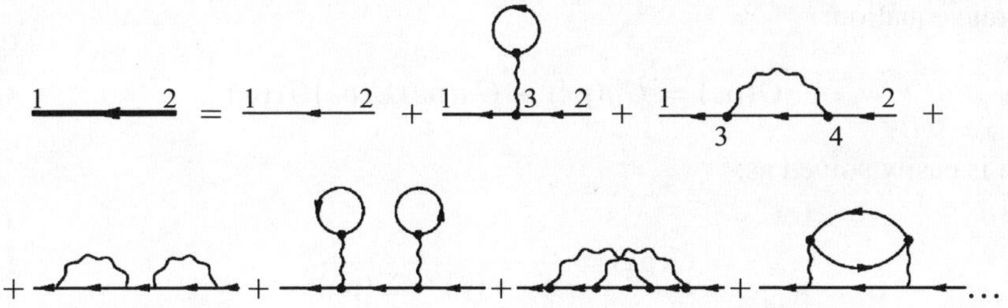

Figure 11.6. Diagrammatic series for the total (exact) Green's function.

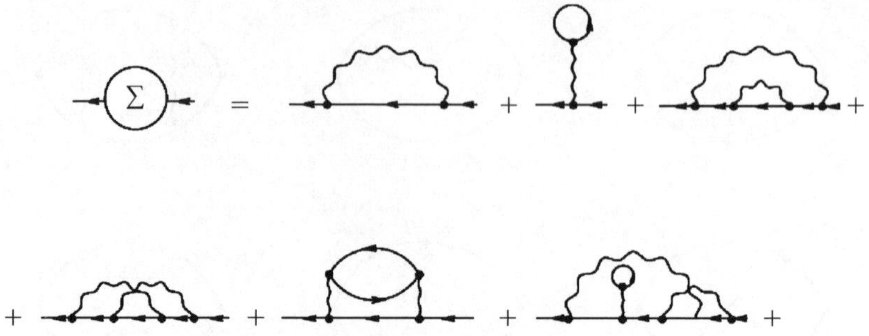

Figure 11.7. Simplest diagrams for irreducible self-energy part.

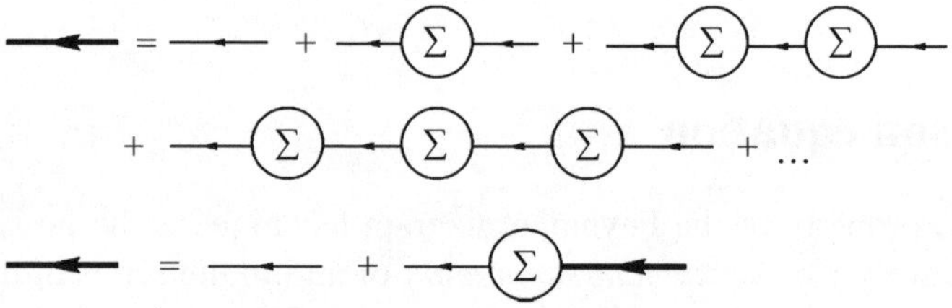

Figure 11.8. Diagrammatic derivation of the Dyson equation.

Σ is called the *irreducible self-energy part*, or simply self-energy. It is easily seen, that the full ("dressed") Green's function is determined by the so called Dyson equation, derived graphically in Figure 11.8. Analytically, it corresponds to the following integral equation:

$$G(1,2) = G_0(1,2) + \int d\tau_3 d\tau_4 G_0(1,3)\Sigma(3,4)G(4,2). \tag{11.53}$$

Iterating this equation, we obviously obtain the complete perturbation series for the Green's function. After Fourier transformation the Dyson equation becomes a simple algebraic equation:

$$G(\mathbf{p}\varepsilon) = G_0(\mathbf{p}\varepsilon) + G_0(\mathbf{p}\varepsilon)\Sigma(\mathbf{p}\varepsilon)G(\mathbf{p}\varepsilon), \tag{11.54}$$

which is easily solved as:

$$G(\mathbf{p}\varepsilon) = \frac{1}{\varepsilon - \varepsilon(\mathbf{p}) - \Sigma(\mathbf{p}\varepsilon)}, \tag{11.55}$$

where we have taken into account the explicit form of $G_0(\mathbf{p}\varepsilon)$. It is clear that the self-energy part $\Sigma(\mathbf{p}\varepsilon)$ describes, in a compact way, all the changes in particle motion due to its interactions with all other particles in the system. In the general case, the self-energy is a complex function consisting of real and imaginary parts (this is why in Equation (11.55) we have dropped the infinitesimally small imaginary contribution from the free particle Green's function $i\delta\,\mathrm{sign}(\varepsilon-\varepsilon_F)$). The energy of the quasiparticle can now be determined as a solution of the equation determining the pole of the total Green's function:

$$\varepsilon = \varepsilon(\mathbf{p}) + \Sigma(\mathbf{p}\varepsilon). \tag{11.56}$$

In the real case, the solution of this equation for ε may be quite complicated.

For the examples given above in Equations (11.49), (11.50) and (11.51), the appropriate contributions to the self-energy part are:

$$\Sigma_H = NV_0, \tag{11.57}$$

$$\Sigma_F = -\int \frac{d^3q}{(2\pi)^3} V_{\mathbf{q}} n(\mathbf{p}+\mathbf{q}), \tag{11.58}$$

$$\Sigma_{\mathrm{pol}} = \int \frac{d^3q}{(2\pi)^3} \int \frac{d\omega}{2\pi} V_{\mathbf{q}}^2 \Pi_0(\mathbf{q}\omega) G_0(\mathbf{p}-\mathbf{q}\varepsilon-\omega). \tag{11.59}$$

Let us return once more to the question of the possibility to introduce well-defined quasiparticles, i.e. to reduce the *exact* Green's function to the form given by Equation (11.32). In a Fermi system, it is convenient to count all energies from the chemical potential μ. For free particles we have $\varepsilon(\mathbf{p}) = \frac{p^2}{2m} - \mu$. In an isotropic system (Fermi liquid) $\Sigma(\mathbf{p}\varepsilon)$ depends only on the absolute value of $\mathbf{p}$. Let us define the value of the Fermi momentum p_F (radius of the Fermi sphere) for the system of *interacting* Fermions by the following equation:

$$\frac{p_F^2}{2m} + \Sigma(p_F, 0) = \mu. \tag{11.60}$$

This definition assumes, of course, that $\mathrm{Im}\,\Sigma(p,0) \to 0$ for $p \to p_F, \varepsilon \to 0$ (Fermi liquid behavior!). For the system of interacting Fermions we can prove in rather general form, that $\mathrm{Im}\,\Sigma(p\varepsilon) \sim \mathrm{Max}\,\{\varepsilon^2, (p-p_F)^2\}\,\mathrm{sign}\,\varepsilon$. Then, expanding $\Sigma(p\varepsilon)$ in a power series over $p-p_F$ and ε, we obtain the following expression for $G(p\varepsilon)$ close to the Fermi surface:

$$\begin{aligned} G^{-1} &= \varepsilon - \frac{p^2}{2m} + \mu - \Sigma(p\varepsilon) \\ &\approx \varepsilon - \frac{p^2}{2m} + \mu - \Sigma(p_F,0) - \left(\frac{\partial\Sigma}{\partial p}\right)_F (p-p_F) - \left(\frac{\partial\Sigma}{\partial\varepsilon}\right)_F \varepsilon + i\alpha|\varepsilon|\varepsilon \\ &= \left[1 - \left(\frac{\partial\Sigma}{\partial\varepsilon}\right)_F\right]\varepsilon - \left[\frac{p_F}{m} + \left(\frac{\partial\Sigma}{\partial p}\right)_F\right](p-p_F) + i\alpha'|\varepsilon|\varepsilon, \end{aligned} \tag{11.61}$$

where $\alpha' = \mathrm{const}$. From Equation (11.61) we can see, that Green's function can be written in the required form:

$$G(p\varepsilon) = \frac{Z}{\varepsilon - v_F(p-p_F) + i\alpha|\varepsilon|\varepsilon} + G_{\mathrm{reg}}, \tag{11.62}$$

where G_{reg} contains all contributions dropped in (11.61), and we defined:

$$Z^{-1} = 1 - \left(\frac{\partial \Sigma}{\partial \varepsilon}\right)_F = \left(\frac{\partial G^{-1}}{\partial \varepsilon}\right)_F \tag{11.63}$$

$$v_F = \frac{p_F}{m^\star} = \frac{\frac{p_F}{m} + \left(\frac{\partial \Sigma}{\partial p}\right)_F}{\left(\frac{\partial G^{-1}}{\partial \varepsilon}\right)_F} = -\frac{\left(\frac{\partial G^{-1}}{\partial p}\right)_F}{\left(\frac{\partial G^{-1}}{\partial \varepsilon}\right)_F}, \tag{11.64}$$

where $\alpha = Z\alpha'$. Thus, we obtain the Green's function of Fermion quasiparticles with an effective mass $m^\star$, which, like everything else, is determined by the behavior of $\Sigma(p\varepsilon)$ close to the Fermi level (surface). Note that in a simplified case, when $\Sigma(p\varepsilon)$ does not depend on p, so that $\left(\frac{\partial \Sigma}{\partial p}\right)_F = 0$, we have:

$$\frac{p_F}{m^\star} = \frac{p_F}{m} Z \quad \text{i.e.} \quad \frac{m^\star}{m} = Z^{-1} \tag{11.65}$$

so that Z simply renormalizes the mass of the quasiparticle. Due to the general property of $Z < 1$, the effective mass in a Fermi liquid is larger than the mass of free particles.

All these properties of $\Sigma(p\varepsilon)$ are rather easily confirmed, if we limit ourselves to the contributions of the simplest Feynman diagrams, both for the case of point-like or Coulomb interactions. A rigorous choice and summation of "dominating" (sub)series of diagrams can be made for the cases of high or (inversely) low density of Fermions, when there exist appropriately small parameters, allowing the use of perturbation theory [29, 2, 30]. All basic assumptions of Fermi liquid theory are thus explicitly confirmed by microscopic calculations. In the general case, when there is no small parameter and no "dominating" subseries of diagrams (a typical example are electrons in metals!), formally all diagrams have to be taken into account and we can only limit ourselves to a rather general analysis, of the kind briefly illustrated above, which constitutes the basis of the microscopic version of Landau Fermi liquid approach.

In recent years, the number of models of so-called strongly correlated systems demonstrated non-Fermi liquid behavior, breaking the main assumptions of Landau theory, such as the possibility to introduce the well-defined quasiparticles. This is a rather typical situation in low-dimensional systems, especially in the one-dimensional case. Two-dimensional systems apparently form a kind of borderline between non Fermi liquid and Fermi liquid behavior. The situation here is under active discussion at present, e.g. with respect to the properties of high-temperature superconductors in the normal state.

A similar diagram technique can be constructed for all other basic types of interactions in many-particle systems, such as electron–phonon interaction, scattering by impurities etc. Depending on the type of interaction, we can have different topologies of Feynman diagrams and diagram rules. For example, in the case of electron–phonon interaction wavy lines denote phonon (Bose) Green's functions, while in the case of random impurity, scattering diagrams do not have closed Fermion loops etc. Details of all these cases can be found e.g. in [29, 32].

11.4 Effective interaction and dielectric screening

As another example of the use of a diagrammatic approach, below we shall discuss diagram summation, leading to the concept of an effective (screened) interaction in Fermion system. Let us define the effective (renormalized or full) interaction by the diagrams shown in Figure 11.9. In Figure 11.10 we show the diagrams for the full polarization operator (containing higher-order corrections) and for the so called *vertex parts*, representing complicated "blocks" of diagrams, describing the processes of multiple scatterings of Fermions. Unfortunately, for vertex parts we can, in general, not find closed integral equations similar to the Dyson equation discussed above. It is only possible in some specific approximations and models. The screened effective interaction ("fat" wavy line in Figure 11.9) can be related to a frequency- and wave-vector-dependent dielectric function of the system $\epsilon(\mathbf{q}\omega)$. Using the diagrams shown in Figure 11.9, we get the screened interaction as:

$$\begin{aligned}
-i\mathcal{V}(\mathbf{q}\omega) &\equiv -\frac{iV_{\mathbf{q}}}{\epsilon(\mathbf{q}\omega)} \\
&= -iV_{\mathbf{q}} + (-iV_{\mathbf{q}})[-i\Pi(\mathbf{q}\omega)](-iV_{\mathbf{q}}) \\
&\quad + (-iV_{\mathbf{q}})[-i\Pi(\mathbf{q}\omega)](-iV_{\mathbf{q}})[-i\Pi(\mathbf{q}\omega)](-iV_{\mathbf{q}}) + \cdots \\
&= -iV_{\mathbf{q}} + (-iV_{\mathbf{q}})[-i\Pi(\mathbf{q}\omega)](-i\mathcal{V}(\mathbf{q}\omega)) \\
&= -iV_{\mathbf{q}} + (-iV_{\mathbf{q}})[-i\Pi(\mathbf{q}\omega)](-iV_{\mathbf{q}})\frac{1}{\epsilon(\mathbf{q}\omega)} \\
&= -iV_{\mathbf{q}}\left\{1 - V_{\mathbf{q}}\Pi(\mathbf{q}\omega)\frac{1}{\epsilon(\mathbf{q}\omega)}\right\}
\end{aligned}$$

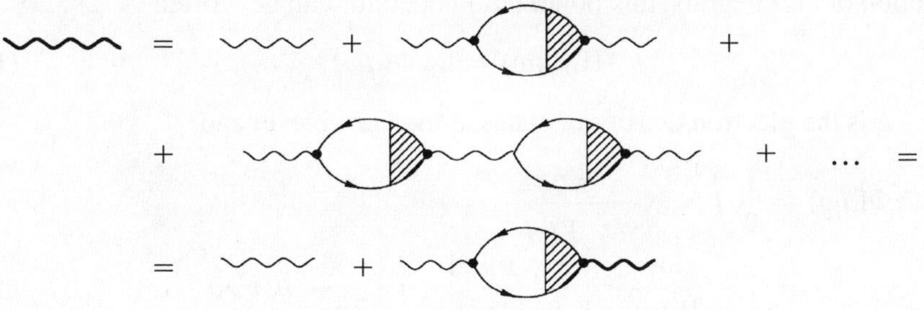

Figure 11.9. Feynman diagrams for effective interaction between particles.

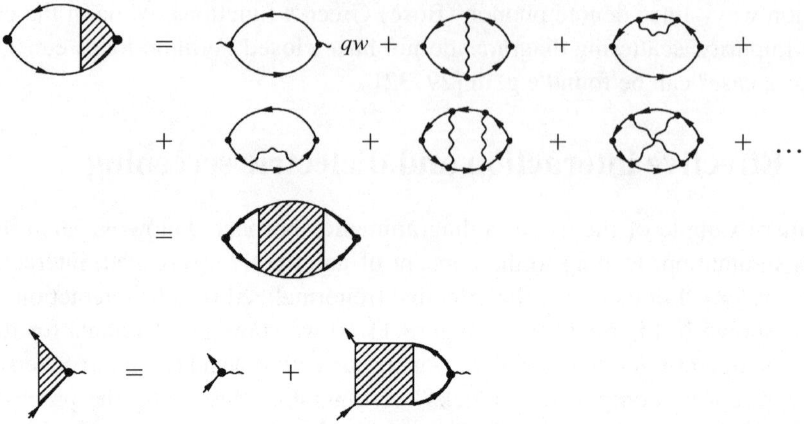

Figure 11.10. Full polarization operator and vertex parts.

so that:

$$\frac{1}{\epsilon(\mathbf{q}\omega)} = 1 - V_{\mathbf{q}}\Pi(\mathbf{q}\omega)\frac{1}{\epsilon(\mathbf{q}\omega)}. \tag{11.66}$$

From here, we can obtain the general expression for the dielectric function (permeability) of a many-particle system via the polarization operator:

$$\epsilon(\mathbf{q}\omega) = 1 + V_{\mathbf{q}}\Pi(\mathbf{q}\omega). \tag{11.67}$$

In the case of Coulomb interaction in the system of electrons we have $V_q = \frac{4\pi e^2}{q^2}$, so that:

$$\epsilon(\mathbf{q}\omega) = 1 + \frac{4\pi e^2}{q^2}\Pi(\mathbf{q}\omega). \tag{11.68}$$

Consider the simplest approximation for the polarization operator (11.52)[6]. After calculation of all integrals, this polarization operator can be written as [29, 30, 32]:

$$\Pi_0(q\omega) = \nu_F \Phi(q\omega), \tag{11.69}$$

where ν_F is the electron density of states at the Fermi level and

$$\Phi(q\omega) = \frac{1}{2}\int_{-1}^{1} dx \frac{v_F q x}{\omega - v_F q x} = 1 - \frac{\omega}{2 v_F q}\ln\left|\frac{\omega + v_F q}{\omega - v_F q}\right| + i\pi\frac{\omega}{2 v_F q}\theta(v_F q - \omega). \tag{11.70}$$

[6] This approximation is justified in the limit of a high enough density of electrons, when the Coulomb interaction can be considered weak. Appropriate estimates were given above during the discussion of the basic properties of Fermi gases.

In particular, $\Phi(q0) = 1$, which gives:

$$\Pi(q0) = \nu_F. \tag{11.71}$$

Then we obtain:

$$\epsilon(q0) = 1 + \frac{4\pi e^2}{q^2}\nu_F = 1 + \frac{\kappa^2}{q^2}, \tag{11.72}$$

where

$$\kappa^2 = 4\pi e^2 \nu_F. \tag{11.73}$$

Accordingly:

$$\mathcal{V}(q0) = \frac{4\pi e^2}{q^2 \epsilon(q0)} = \frac{4\pi e^2}{q^2 + \kappa^2}, \tag{11.74}$$

which describes the so called Debye screening of the Coulomb potential in a quantum plasma of electrons at temperature $T = 0$. Obviously, in coordinate space we have $\mathcal{V}(r) = \frac{e^2}{r}e^{-\kappa r}$, so that Equation (11.73), in fact, determines the screening radius κ^{-1}.

In the inverse limit of high frequencies of $\omega \gg v_F q$, we can show that $\Phi(q\omega) = -\frac{v_F^2 q^2}{3\omega^2}$, so that:

$$\epsilon(\omega) = 1 - \frac{4\pi e^2 v_F^2}{3\omega^2}\nu_F = 1 - \frac{4\pi n e^2}{m\omega^2} = 1 - \frac{\omega_p^2}{\omega^2}. \tag{11.75}$$

Here we used $\nu_F = \frac{3}{2}\frac{n}{\varepsilon_F}$, where n is the density (concentration) of electrons. We also introduced the square of the plasma frequency:

$$\omega_p^2 = \frac{4\pi n e^2}{m}. \tag{11.76}$$

Equation $\epsilon(q\omega) = 0$ determines the frequency of the plasma oscillations (plasmons) for the whole range of q. In particular, for small values of q, when plasmon damping is absent, we can find the dispersion (spectrum) of the plasmons as:

$$\omega^2 = \omega_p^2 + \frac{3}{5}v_F q^2. \tag{11.77}$$

In fact, the frequency of plasmons is very weakly dependent on their wavelength and this dispersion is just a small correction.

11.5 Green's functions at finite temperatures

The Feynman diagram technique discussed above was generalized by Matsubara to the case of finite temperatures [29]. Below we shall briefly discuss this generalization, limiting the discussion to Fermi systems only. The thermodynamic Green's function of a Fermi particle is defined according to Matsubara as:

$$\mathcal{G}(\mathbf{p}, \tau_2 - \tau_1) = -i\langle T a_{\mathbf{p}}(\tau_2) a_{\mathbf{p}}^{+}(\tau_1)\rangle, \tag{11.78}$$

where, by definition:

$$a_{\mathbf{p}}(\tau) = e^{(H-\mu N)\tau} a_{\mathbf{p}} e^{-(H-\mu N)\tau} \tag{11.79}$$

and $0 < \tau_1, \tau_2 < \beta = \frac{1}{T}$ are real variables, while the angular brackets denote averaging over the grand canonical Gibbs distribution, which is convenient to write here as:

$$\langle A\rangle = \frac{\mathrm{Sp}\,\rho A}{\mathrm{Sp}\,\rho} \quad \text{where} \quad \rho = e^{-\beta(H-\mu N)}. \tag{11.80}$$

Taking into account that $Z = \mathrm{Sp}\,\rho$, this is equivalent to the definition used above. The reason why the Green's function $\mathcal{G}$ can be represented by the same diagrammatic series as the Green's function G, previously defined for the case of $T = 0$, can be seen as follows: we have seen that diagrammatic expansion for G is a fundamental consequence of the time-dependent Schroedinger equation (11.1). The statistical operator ρ, written in the form of (11.80), satisfies the so-called Bloch equation:

$$\frac{\partial\rho}{\partial\beta} = -(H - \mu N)\rho, \tag{11.81}$$

which is easily verified by direct differentiation. Now we see the direct correspondence between the time-dependent Schroedinger equation (11.1):

$$\psi \leftrightarrow \rho \quad H \leftrightarrow H - \mu N \quad it \leftrightarrow \beta. \tag{11.82}$$

Thus, making the substitution

$$H \to H - \mu N \quad it \to \tau \tag{11.83}$$

in all expressions of the previous paragraphs, we can obtain the diagrammatic technique for $\mathcal{G}$, which is practically of the same form as in the case of $T = 0$. Substitution $H \to H - \mu N$ only shifts the energy scale of single particle energy by μ:

$$H_0 - \mu N = \sum_{\mathbf{p}} (\varepsilon(\mathbf{p}) - \mu) a_{\mathbf{p}}^{+} a_{\mathbf{p}}. \tag{11.84}$$

Though Matsubara's Green's functions $\mathcal{G}$ depend on "imaginary time" τ [7], we can always perform a transformation to real time in the final expression putting $\tau \to it$, or more precisely, making an analytic continuation to the real axis of time.

We noted above that the values of τ_1 and τ_2 in (11.78) vary over the interval from 0 to β. Thus, to make a transformation to the $(\mathbf{p}, \omega)$ representation, we have to introduce the periodically continuous function $\mathcal{G}$, obtained by periodic repetition of $\mathcal{G}$ on the interval from $-\infty$ to ∞. For this function we can write down an expansion into the Fourier *series*:

$$\mathcal{G}(\mathbf{p}\tau) = \frac{1}{\beta} \sum_{n=-\infty}^{\infty} e^{-i\omega_n \tau} \mathcal{G}(\mathbf{p}\omega_n), \tag{11.85}$$

where the summation is performed over the discrete (Matsubara) frequencies $\omega_n = \pi n T$. Accordingly

$$\mathcal{G}(\mathbf{p}\omega_n) = \frac{1}{2} \int_{-\beta}^{\beta} d\tau e^{i\omega_n \tau} \mathcal{G}(\mathbf{p}). \tag{11.86}$$

The "time" difference $\tau = \tau_2 - \tau_1$ varies in the interval $(-\beta, \beta)$, as the values of τ_1 and τ_2 vary in the interval $(0, \beta)$. The function $\mathcal{G}(\mathbf{p}\tau)$ periodically repeats itself in the intervals $(-\beta, \beta), (\beta, 3\beta), (3\beta, 5\beta), \ldots, (-3\beta, -\beta), \ldots$. For a system consisting of Fermions, the even values of n drop out of the series for $\mathcal{G}(\mathbf{p}\tau)$ due to the "quasi-periodic" boundary condition:

$$\mathcal{G}(\mathbf{p}, \tau) = -\mathcal{G}(\mathbf{p}, \tau + \beta) \quad \text{for } \tau < 0. \tag{11.87}$$

To see the validity of this relation, we can use the property $\text{Sp}AB = \text{Sp}BA$. Assuming $\tau' - \tau > 0$, we have:

$$\begin{aligned}
\mathcal{G}(\mathbf{p}, \tau - \tau') &= \frac{i}{Z} \text{Sp} e^{-\beta(H-\mu N)} a_{\mathbf{p}}^{+}(\tau') a_{\mathbf{p}}(\tau) \\
&= \frac{i}{Z} \text{Sp} a_{\mathbf{p}}(\tau) e^{-\beta(H-\mu N)} a_{\mathbf{p}}^{+}(\tau') e \\
&= \frac{i}{Z} \text{Sp} e^{-\beta(H-\mu N)} e^{\beta(H-\mu N)} a_{\mathbf{p}}(\tau) e^{-\beta(H-\mu N)} a_{\mathbf{p}}^{+}(\tau') \\
&= \frac{i}{Z} \text{Sp} e^{-\beta(H-\mu N)} a_{\mathbf{p}}(\tau + \beta) a_{\mathbf{p}}^{+}(\tau')
\end{aligned} \tag{11.88}$$

or

$$\mathcal{G}(\mathbf{p}, \tau - \tau') = -\mathcal{G}(\mathbf{p}, \tau - \tau' + \beta), \tag{11.89}$$

which for $\tau' = 0$ just coincides with (11.87). The minus sign appeared here due to the anti-commutation of the Fermi operators. Substituting (11.87) into (11.85) we can

[7] The variable τ is real, but Green's function $\mathcal{G}$ is obtained from G by the replacement $it \to \tau$, so that actually we are making a transformation to "imaginary time" $t = -i\tau$.

see that all terms with even n become zero. Thus, for Fermions we are always dealing with odd Matsubara frequencies:

$$\omega_n = \frac{(2n+1)\pi}{\beta} = (2n+1)\pi T. \tag{11.90}$$

In a similar way, for Bosons only even Matsubara frequencies remain:

$$\omega_n = \frac{2n\pi}{\beta} = 2n\pi T. \tag{11.91}$$

Remembering Equations (11.16), (11.17) and (11.18) for free particle Green's functions at $T = 0$, we can easily write down Matsubara's Green's function for free Fermions as:

$$\mathcal{G}_0(\mathbf{p}, \tau_2 - \tau_1) = -i\{\theta(\tau_2 - \tau_1)(1 - n(\mathbf{p})) - \theta(\tau_1 - \tau_2)n(\mathbf{p})\}e^{-(\varepsilon(\mathbf{p})-\mu)(\tau_2-\tau_1)}, \tag{11.92}$$

where $n(\mathbf{p}) = [e^{\beta(\varepsilon(\mathbf{p})-\mu)} + 1]^{-1}$ is the Fermi distribution for finite T. Thus, the step-like functions, entering the definition of G_0 at $T = 0$ are "smeared" by a finite T, so that the state with a given $\mathbf{p}$ can be filled either by a particle or a hole.

Substituting (11.92) into (11.86) we find:

$$\mathcal{G}_0(\mathbf{p}\omega_n) = \frac{i}{i\omega_n - \varepsilon(\mathbf{p}) + \mu} \qquad \omega_n = (2n+1)\pi T. \tag{11.93}$$

With the only change, related to the transition to discrete frequencies, which also "conserve" in the vertices, Matsubara's diagram technique for $T > 0$ is practically identical to the Feynman technique for $T = 0$. In particular, the full (exact) Green's function is determined by the Dyson equation:

$$\mathcal{G}(\mathbf{p}\omega_n) = \frac{i}{i\omega_n - \varepsilon(\mathbf{p}) + \mu - \Sigma(\mathbf{p}\omega_n)}, \qquad \omega_n = (2n+1)\pi T. \tag{11.94}$$

However, we must stress that Matsubara's Green's functions are not quantum propagators (transition amplitudes) at all!

Calculation of Matsubara's Green's functions allows us, in principle, to find arbitrary thermodynamic characteristics of the many-particle system at finite temperatures. In particular, it is possible to construct a diagrammatic expansion for the interaction correction to the thermodynamic potential Ω [29]. Appropriate diagrams of the lowest orders are shown in Figure 11.11. For concreteness we show here diagrams for the case of interacting Fermions. A perturbation series for $\Delta\Omega$ consists of loop diagrams, restricted to the case of connected diagrams. A certain difficulty here is related to the appearance in this series of an extra combinatorial factor of $\frac{1}{n}$ for every contribution of the n-th order. This makes a series for $\Delta\Omega$ rather inconvenient for summation. In particular, for $\Delta\Omega$ we can not derive any analogue of the Dyson

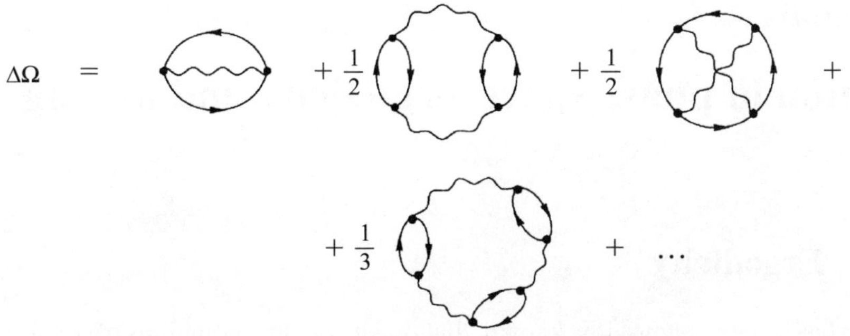

Figure 11.11. Diagrammatic expansion for the thermodynamic potential.

equation. As $\Omega = -VP(\mu, T)$, in fact here we are calculating the corrections to the pressure $\Delta P = P - P_0(\mu, T)$, where P_0 is the pressure in a system of free particles (ideal gas), so that we are actually dealing with quantum corrections to the equation of state.

Finally, we shall mention the diagram technique, proposed by Keldysh, which is applicable to finite temperatures and, more importantly, to the analysis of nonequilibrium processes in many-particle systems in real time, including the derivation of the kinetic equations. A detailed enough presentation of this technique can be found in [15].

Appendix A

Motion in phase space, ergodicity and mixing

A.1 Ergodicity

From classical mechanics it is known, that the differential equations of motion for any conservative mechanical system can be written in Hamilton form:

$$\dot{q}_k = \frac{\partial H}{\partial p_k} \quad \dot{p}_k = -\frac{\partial H}{\partial q_k}, \tag{A.1}$$

where q_k, p_k are the generalized coordinates and momenta ($k = 1, 2, \ldots, n = 3N$, i.e. in total we have $2n = 6N$ equations, where N is the number of particles in the system and n is the number of degrees of freedom),

$$H(p,q) = H(p_1, p_2, \ldots, p_n; q_1, q_2, \ldots, q_n) \tag{A.2}$$

is the Hamiltonian of the system, which is equal to the total energy, expressed as a function of the generalized coordinates and momenta. The Hamiltonian is related to the Lagrangian L by the well known relation:

$$H = \sum_{k=1}^{n} p_k \dot{q}_k - L. \tag{A.3}$$

The equations of motion (A.3) can be integrated and their solutions can be written in the following form[1]:

$$p_k = \varphi_k(q_l^0, p_l^0, t) \quad q_k = \psi_k(q_l^0, p_l^0, t), \tag{A.4}$$

where q_l^0, p_l^0 are the initial values of the coordinates and momenta. The functions φ_k, ψ_k represent (according to the Cauchy theorem) single-valued and continuous functions of the arguments q_l^0, p_l^0.

To obtain (conserving) integrals of motion we can use the following procedure. Divide all the other $2n - 1$ equations (A.1) by equation $\dot{p}_1 = -\frac{\partial H}{\partial q_1}$. Then we get:

$$\frac{dq_1}{dp_1} = -\frac{\frac{\partial H}{\partial p_1}}{\frac{\partial H}{\partial q_1}}, \ldots, \frac{dp_n}{dp_1} = -\frac{\frac{\partial H}{\partial q_n}}{\frac{\partial H}{\partial q_1}}. \tag{A.5}$$

---

[1] Below we mainly follow [10].

This system of equations does not contain time t (for H independent of t) and defines conserving quantities. In total it gives $2n - 1$ integrals of motion, obviously including energy, which we denote as:

$$\Phi_1(q,p) \equiv H(p,q) = \alpha_1 = E. \tag{A.6}$$

Then, the rest of the $2n - 2$ integrals of motion can be written as:

$$\begin{aligned} \Phi_2(q,p) &= \alpha_2, \ldots, \Phi_n(q,p) = \alpha_n \\ \Psi_2(q,p) &= \beta_2, \ldots, \Psi_n(q,p) = \beta_n, \end{aligned} \tag{A.7}$$

where $\alpha_1, \ldots, \alpha_n; \beta_2, \ldots, \beta_n$ are integration constants. One more integral of motion is obtained by solving the equation $\dot{p}_1 = -\partial H / \partial q_1$ and using Equations (A.6),(A.7). This can be written as:

$$\Psi_1(q,p) = t + \beta_1. \tag{A.8}$$

Adding an arbitrary constant to t does not change the equations of motion, as the time t enters only through differentials.

Consider the simplest example of a system with one degree of freedom – the harmonic oscillator. Then (setting the mass $m = 1$) the Hamiltonian is written as:

$$H = \frac{1}{2}(p^2 + \omega^2 q^2). \tag{A.9}$$

Hamilton's equations of motion now are:

$$\dot{q} = \frac{\partial H}{\partial p} = p \quad \dot{p} = -\frac{\partial H}{\partial q} = -\omega^2 q, \tag{A.10}$$

which give the following solutions (integrals):

$$q = q^0 \cos \omega t + \frac{p^0}{\omega} \sin \omega t, \quad p = -\omega q^0 \sin \omega t + p^0 \cos \omega t, \tag{A.11}$$

which can be rewritten as an energy integral:

$$2H = p^2 + \omega^2 q^2 = 2E \tag{A.12}$$

and the relation, determining the dependence of p and q on time:

$$\frac{1}{\omega} \arccos \frac{\omega q}{\sqrt{p^2 + \omega^2 q^2}} = t + \beta. \tag{A.13}$$

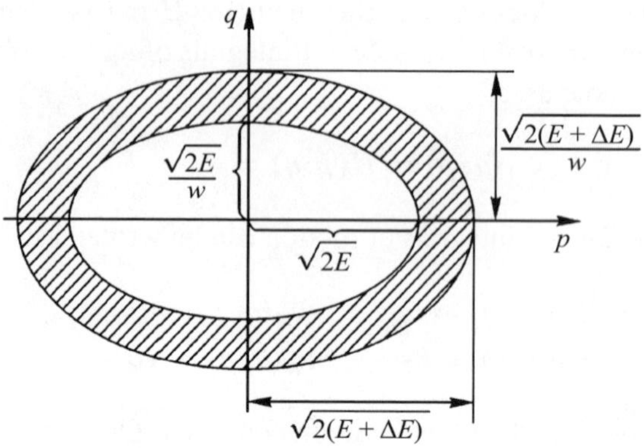

Figure A.1. Phase space of an harmonic oscillator. Shown are the iso-energetic "surfaces" – ellipses, corresponding to oscillators with energies differing by ΔE in energy. The micro-canonical distribution function is equal to a constant different from zero, in the area Ω between these ellipses.

An oscillator with one degree of freedom possesses these two integrals of motion. The mechanical state of the oscillator is represented by a point in the (p,q)-plane, which is the phase space for this simple system. The motion of the system is represented by the movement of the phase point over the "ergodic surface" (a line on (p,q)-plane), determined by the value of the energy E. These lines of constant energies, as is obvious from Equation (A.12), form ellipses like those shown in Figure A.1. The second integral (A.13) determines the velocity of the movement of the phase point over these ellipses. The integrals of motion for the oscillator (A.11) can be rewritten, using Equations (A.12),(A.13), as:

$$q = \frac{\sqrt{2E}}{\omega} \sin \omega(t+\beta) \quad p = \sqrt{2E} \cos(t+\beta). \tag{A.14}$$

For this simple system the *time average* can be calculated in an elementary way. Due to the periodicity of motion (A.14), the time average of an arbitrary function of the dynamic variables $F(q,p)$ on an infinite time interval, can be reduced to the average over the period of the motion $T = \frac{2\pi}{\omega}$:

$$\widetilde{F} = \frac{\omega}{2\pi} \int_0^{\frac{2\pi}{\omega}} dt F \left\{ \frac{\sqrt{2E}}{\omega} \sin \omega(t+\beta), \sqrt{2E} \cos \omega(t+\beta) \right\}. \tag{A.15}$$

This average depends on E, with E being fixed here. Without changing the value of (A.15), we can calculate its average over an infinitesimally small interval of energies:

$$\begin{aligned}\widetilde{F} &= \lim_{\Delta E\to 0}\frac{1}{\Delta E}\int_{E}^{E+\Delta E} dE\widetilde{F}\\ &= \lim_{\Delta E\to 0}\frac{\omega}{2\pi\Delta E}\int_{E}^{E+\Delta E} dE\int_{0}^{\frac{2\pi}{\omega}} dtF\left\{\frac{\sqrt{2E}}{\omega}\sin\omega(t+\beta), \sqrt{2E}\cos\omega(t+\beta)\right\}.\end{aligned} \tag{A.16}$$

Let us transform from variables E and t to q and p. Using (A.14), we can calculate the Jacobian of this transformation as:

$$\frac{\partial(q,p)}{\partial(t,E)} = \begin{vmatrix} \sqrt{2E}\cos\omega(t+\beta) & \frac{1}{\omega\sqrt{2E}}\sin\omega(t+\beta) \\ -\omega\sqrt{2E}\sin\omega(t+\beta) & \frac{1}{\sqrt{2E}}\cos\omega(t+\beta)\end{vmatrix} = 1. \tag{A.17}$$

Then we obtain:

$$\widetilde{F} = \lim_{\Delta E\to 0}\frac{\omega}{2\pi\Delta E}\int dq\int dpF(q,p), \tag{A.18}$$

where the integration is performed over the infinitesimally narrow area between the ellipses of constant energies E and $E+\Delta E$ with $\Delta E\to 0$.

On the other hand we can define the *microcanonical* distribution for an oscillator, with probability density $\rho(p,q)$ equal to a constant (independent of specific values of p and q) within the area Ω between the ellipse $p^2+\omega^2q^2 = 2E$ and ellipse $p^2+\omega^2q^2 = 2(E+\Delta E)$, and equal to zero outside this area (cf. Figure A.1):

$$\rho(p,q) = \begin{cases}\frac{\omega}{2\pi\Delta E} & \text{for } p,q\subset\Omega\\ 0 & \text{for } p,q\not\subset\Omega,\end{cases} \tag{A.19}$$

where to guarantee normalization of $\rho(p,q)$ to unity, we have taken into account that the actual area of Ω is:

$$\Delta(\pi ab) = \Delta\left(\frac{2\pi E}{\omega}\right) = \frac{2\pi\Delta E}{\omega}, \tag{A.20}$$

where a and b denote the semi-axes of the ellipse, corresponding to energy E. Then, the microcanonical average (over phase space) of $F(q,p)$ is equal to:

$$\langle F\rangle = \int dpdq\rho(p,q)F(q,p) = \lim_{\Delta E\to 0}\frac{\omega}{2\pi\Delta E}\iint_{p^2+\omega^2q^2=2(E+\Delta E)}^{p^2+\omega^2q^2=2E} dpdqF(q,p). \tag{A.21}$$

Comparing (A.18) and (A.21) we can see, that in this simplest case of a system with only one degree of freedom, the time average simply coincides with the microcanonical average.

In the general case the, as we have seen above, the integrals of the Hamilton equations can be written as:

$$\begin{aligned} p_k &= \varphi_k(t+\beta_1,\beta_2,\dots,\beta_n,\alpha_1,\alpha_2,\dots\alpha_n) \\ q_k &= \psi_k(t+\beta_1,\beta_2,\dots,\beta_n,\alpha_1,\alpha_2,\dots\alpha_n) \end{aligned} \tag{A.22}$$

or in shortened form:

$$X = \Phi(t+\beta_1,\beta_2,\dots,\beta_n,\alpha_1,\alpha_2,\dots\alpha_n). \tag{A.23}$$

The time average of an arbitrary dynamic variable $F(X)$ is determined by:

$$\begin{aligned} \widetilde{F} &= \lim_{T\to\infty}\frac{1}{T}\int_0^\infty dtF(X) \\ &= \lim_{T\to\infty}\frac{1}{T}\int_0^\infty dtF\{\Phi(t+\beta_1,\beta_2,\dots,\beta_n,\alpha_1,\alpha_2,\dots\alpha_n)\}. \end{aligned} \tag{A.24}$$

This average, in general, depends on *all* $2n-1$ integration constants (integrals of motion) $\beta_2,\dots,\beta_n,\alpha_1,\alpha_2,\dots\alpha_n$, except β_1, on which it does not depend. At the same time, we have shown before that statistical mechanical averages of any dynamic variables in *equilibrium* depend only on one integral of motion – that of energy[2]. Thus, the many-particle systems under consideration should satisfy the special property that the time averages of any *single valued* dynamical variable is dependent only on the energy $\alpha_1 = E$:

$$\widetilde{F}(X) = f_F(E). \tag{A.25}$$

Such systems are called *ergodic*. For ergodic systems the time average of any single-valued dynamical variable is equal to its average over the microcanonical ensemble.

The proof of this statement is rather simple. Consider the microcanonical average:

$$\langle F\rangle = \int dXF(X)w_E(X), \tag{A.26}$$

where

$$w_E(X) = \frac{\delta\{H(X)-E\}}{\Omega(E)}. \tag{A.27}$$

As the value of $\langle F\rangle$ does not depend on time, its time average is equal to itself, so that:

$$\langle F\rangle = \widetilde{\langle F\rangle} = \lim_{T\to\infty}\frac{1}{T}\int_0^T dt\int dXF(X)w_E(X). \tag{A.28}$$

[2] For fixed external parameters like volume, pressure, external fields etc.

Variables X determine the state of the system at time t, let us make a transformation to variables X_0, determining the state of the system at $t = 0$. These variables are related through the solutions of the Hamilton equations, which can be written as:

$$X = \Phi(t, X_0). \tag{A.29}$$

Then

$$F(X) = F\{\Phi(t, X_0)\}. \tag{A.30}$$

Obviously $H(X) = H(X_0)$, so that

$$w_E(X) = \frac{\delta\{H(X) - E\}}{\Omega(E)} = \frac{\delta\{H(X_0) - E\}}{\Omega(E)} = w_E(X_0), \tag{A.31}$$

and according to the Liouville theorem $dX = dX_0$. Thus, after changing the variables we have:

$$\langle F \rangle = \lim_{T\to\infty} \frac{1}{T} \int_0^T dt \int dX_0 w_E(X_0) F\{\Phi(t, X_0)\}. \tag{A.32}$$

Let us change the order of integration over t and X_0, then:

$$\langle F \rangle = \int dX_0 w_E(X_0) \lim_{T\to\infty} \frac{1}{T} \int_0^T dt F\{\Phi(t, X_0)\} = \int dX_0 w_E(X_0) \widetilde{F}. \tag{A.33}$$

Due to the assumed ergodicity the time average $\widetilde{F}$ depends only on the energy $H(X_0)$, so that:

$$\widetilde{F} = f_F[H(X_0)]. \tag{A.34}$$

Thus

$$\langle F \rangle = \int dX_0 w_E(X_0) f_F[H(X_0)]. \tag{A.35}$$

But $w_E(X_0)$ is different from zero only for $H = E$, so that $f_F(H)$ can be taken out of the integral, putting $H = E$. Then we get:

$$\langle F \rangle = f_F(E) \int dX_0 w_E(X_0) = f_F(E) = \widetilde{F}, \tag{A.36}$$

where we have taken into account that the integral is equal to unity, due to the renormalization condition. This ends the proof of the equivalence of time and microcanonical averaging for ergodic systems.

It may seem that ergodic mechanical systems just do not exist at all, as the general time average (A.24) definitely depends on other integrals of motion $\alpha_2, \alpha_3, \ldots, \beta_n$, besides energy. Consider one of them, e.g. $\Phi_2(X) = \alpha_2$. The time average of $\Phi_2(X)$ is obviously equal to α_2 and depends not on the energy integral $E = \alpha_1$, but on α_2. However, for ergodic systems the left parts of all integrals of motion $\Phi_k = \alpha_k, \Psi_k = \beta_k$ $(k = 2, \ldots, n)$, besides energy, momentum and angular momentum are *multivalued* functions of the coordinates and momenta (and can not be transformed to single-valued functions). This is always so for systems with inseparable variables. Systems with separable variables are, in this sense, trivial – they are exactly solvable and are also called *integrable*, their motion is regular (non-random) and we do not need statistics at all to describe their properties[3]. The restriction to single-valued functions $F(p,q)$ is quite natural from a physical point of view; for systems at rest we can drop the momentum and angular momentum integrals of motion. Statistical mechanics is dealing with complicated *non-integrable* systems (performing nontrivial motion). In recent decades a number of explicit examples of such systems, sometimes consisting of rather few particles, were demonstrated to show all the properties of ergodic motion [12].

A.2 Poincare recurrence theorem

Let us continue our discussion of the system motion in phase space using more abstract language. Consider the phase point (p,q). Let us define the operator of time translation $\hat{T}(t)$ as:

$$(q(t), p(t)) = \hat{T}(t)(q(0), p(0)). \tag{A.37}$$

This operator gives a complete description of the phase point motion and is implicitly defined by the Hamilton equations. We shall not try to construct such operators explicitly for specific systems, but it is clear that, in principle, they always exist. The Liouville theorem corresponds to the conservation of an arbitrary phase volume Γ under the action of the operator $\hat{T}$:

$$\Gamma(t) = \hat{T}(t)\Gamma(0) = \Gamma(0). \tag{A.38}$$

Using the Liouville theorem, it is rather easy to prove the so called Poincare recurrence theorem [12]. Consider a conservative (H is time independent) mechanical system, performing motion in a finite region of its phase space. Let us take some region (set of points) of the phase space A and chose an initial point $z_0 = (q_0, p_0)$ in it. We shall now show that, after a certain (finite) time, the system will necessarily return to the region A (Poincare theorem), except probably a set of initial points of measure zero. The proof can be done through *reductio ad absurdum*. Let us denote as B the

[3] More details on this can be found in [11], where it is shown that in the general case of systems with inseparable variables, the set of single-valued integrals of motion is limited to those, which are directly related to general properties of time and translational invariance, as well as to the isotropy of space, i.e. to energy, momentum and angular momentum conservation laws.

subset of points in A, which never return to A. Suppose that after some large time t_1 the set B moves to B_1:

$$\hat{T}(t_1)B = B_1. \tag{A.39}$$

According to the definition of B the intersection of B_1 and A is an empty set:

$$B_1 \cap A = \emptyset. \tag{A.40}$$

After time interval $t_2 = 2t_1$ we have:

$$\hat{T}(2t_1)B = \hat{T}(t_1)B_1 \equiv B_2. \tag{A.41}$$

Then also

$$B_2 \cap B_1 = \emptyset. \tag{A.42}$$

If this is not so, there exist points, which have not left B_1. However, due to time reversibility of the Hamilton equations that would mean, that these points could not have entered B_1. This contradicts their past: at $t = 0$, according to our assumption, they belonged to A. Continued application of $\hat{T}(nt_1)$-operator to B leads to an infinite sequence $B_1, B_2, \ldots$ of non-intersecting images of B. According to Liouville theorem:

$$\Gamma(B) = \Gamma(B_1) = \Gamma(B_2) = \cdots, \tag{A.43}$$

so that during the motion, the points from B cover the phase volume $\Gamma = \infty$. However, due to the finite nature of the motion of our system, this volume is to be finite. This is possible only in case of $\Gamma(B) = 0$, which proves the Poincare recurrence theorem.

From Poincare's theorem it follows, that the system will return to the initial region A infinitely many times. It may seem that this result contradicts the irreversible evolution of many-particle systems, observed in experiments, and the possibility of its description along the lines of statistical mechanics. Actually, this is not so. To understand this situation, we have to consider the average recurrence time or the duration of the *Poincare cycle*. Let us make a rough estimate of this time for the simplest many-particle system – an ideal gas [33]. Consider N molecules of the gas moving in volume V. We may understand the recurrence in the sense of a repetition of the state of each molecule with some finite accuracy Δv for its velocity and some Δx for its coordinate. This accuracy corresponds to an element of the phase volume $\Delta\Gamma = [m\Delta v\Delta x]^{3N}$, while the total set of possible states of the gas, with fixed energy

$E = \sum_i \frac{mv_i^2}{2} = \frac{3}{2}NT$, corresponds the phase volume[4]:

$$\Gamma \approx C_{3N}(m^2 \sum_i v_i^2)^{3N/2} V^N \approx C_{3N} (3NTm)^{3N/2} V^N. \qquad (A.44)$$

It is clear, that before returning (with the given accuracy) to the initial position, the phase point, representing our system, is to pass through $\sim \frac{\Gamma}{\Delta\Gamma}$ states. Let τ be some characteristic time for the gas, e.g. mean free time of the molecule. Then, the recurrence time can be roughly estimated as:

$$\tau_R \sim \tau \frac{\Gamma}{\Delta\Gamma} \sim C_{3N} \left(\frac{V}{\Delta x^3}\right)^N \left(\frac{3NT}{m\Delta v^2}\right)^{3N/2} \tau \sim \left(\frac{V}{\Delta x^3}\right)^N \left(\frac{T}{m\Delta v^2}\right)^{3N/2} \tau. \qquad (A.45)$$

Let us take $\Delta x \sim 0.1(V/N)^{1/3}$, i.e. of the order of 10% of the interparticle distance in our gas, and $\Delta v \sim 0.1(T/m)^{1/2}$, i.e. of the order of 10% of an average velocity (so that the conditions for its "return" are rather crude). Then, we obtain:

$$\tau_R \sim \tau (10N)^N (10^2)^{3N/2} \sim \tau N^N. \qquad (A.46)$$

For 1 cm^3 of a gas in normal conditions we have $N \sim 10^{18}$, so that

$$\frac{\tau_R}{\tau} \sim (10^{18})^{10^{18}} \sim 10^{2\ 10^{19}} \qquad (A.47)$$

and the ratio of the recurrence time τ_R to the mean free time $\tau \sim 10^{-6}$ sec, or to one second, one year, or even to a characteristic "lifetime" of our Universe ($\sim 10^{10}$ years $\sim 10^{17}$ sec), with a logarithmic accuracy the same (!) and of the order of $10^{2\ 10^{19}}$. Thus, the typical time of the Poincare cycle, even for such a simple system, is immensely large, and the probability of such a return is immensely small. This leads to an obvious conclusion, that the *most probable* behavior of a many-particle system is, in fact, the irreversible behavior, observed in reality.

A.3 Instability of trajectories and mixing

Consider the motion of a drop of "phase liquid" in the phase space. The character of this motion may be very complicated and, as time grows, the borders of the drop may become irregular, with the drop becoming "amoeba"-like (cf. Figure A.2), filling different regions of the phase space. The volume of the drop is conserved (Liouville theorem). Such motion is called *mixing*. The phase points, which were close to each

---

[4] Here $C_{3N} \approx \left(\frac{2\pi e}{3N}\right)^{3N/2}$ is related to a constant in the expression for the volume of an n-dimensional sphere $V_n = CR^n$, the exact value of this constant being $C_n = \frac{2\pi^{n/2}}{n\Gamma(n/2)}$. For $n \gg 1$, using the asymptotic expression for Γ-function $\Gamma(n/2) \approx (2\pi)^{1/2}(n/2)^{(n-1)/2}e^{-n/2}$, we get $C_n \approx \left(\frac{2\pi e}{n}\right)^{n/2}$.

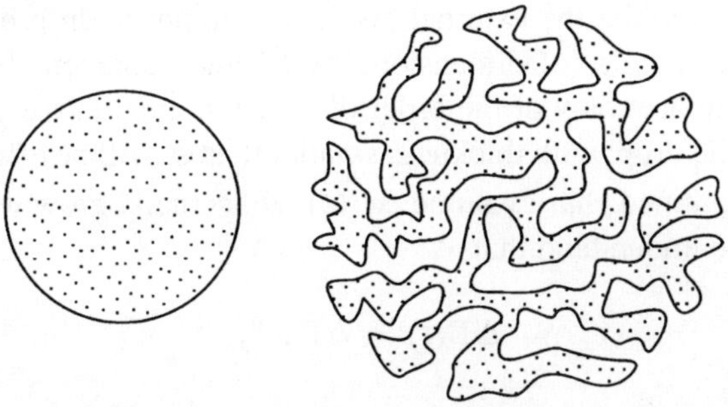

Figure A.2. Qualitative evolution of a phase drop during mixing.

other initially, may become very far from each other during this time evolution, and move in practically independent ways. The property of mixing is natural to expect for systems, characterized by *unstable* motion, when phase trajectories, initially close to each other, become exponentially far away from each other with the growth of time, i.e. small perturbations of the initial conditions lead to arbitrarily large deviations of the phase trajectory from unperturbed motion. If the phase space is finite (and we are interested just in this case – the system moves over the hypersurface of constant energy!), the phase trajectories can not deviate more than the characteristic size of this space and begin to intermix in a very complicated way. Denoting by $D(t)$ the *distance* between two points in the phase space, belonging to two different trajectories at time t, we can formally define the local instability of motion in the following way [12] – there exists a direction in phase space for which:

$$D(t) = D_0 e^{h_0 t}, \tag{A.48}$$

where the increment of instability (Lyapunov exponent, $h_0 > 0$) is, in general, a function of a point in phase space and has the statistical meaning [12] of an inverse time of "decoupling" of correlations between trajectories during mixing. It is obvious, that this picture can be directly related to an idea of the description of entropy growth, using the coarse-grained distribution function, which we used previously. The question arises – whether we can define entropy in such a way, that will allow its use for dynamical systems, using only the properties of phase trajectories (not distribution functions)? This problem was solved by Kolmogorov, who introduced the notion of dynamic of K-entropy. Consider again the evolution of some initial element of the phase volume $\Delta\Gamma_0$. According to the Liouville theorem:

$$\Delta\Gamma(t) = \Delta\Gamma_0 \tag{A.49}$$

but the structure of the phase drop changes with time (cf. Figure A.2). There may appear "bubbles", empty regions etc. As t grows, the "bubble" structure becomes more and more fine, while the external border of the phase drop becomes wider and wider. Let us take some ε (of dimensionality Γ) and "coarsen" the structure of the phase drop up to an accuracy of the order of ε. Then, qualitatively it is clear that all thin structures of the drop, with thickness smaller than ε, will be effectively "dressed", so that the coarse-grained phase volume $\widetilde{\Delta\Gamma(t)}$ will actually grow with time. Knowing (A.48), it is easy to understand that

$$\widetilde{\Delta\Gamma(t)} = \Delta\Gamma_0 e^{ht}, \tag{A.50}$$

where h is some quantity, related to the increment of instability of phase trajectories h_0. Then we may define entropy as:

$$S = \ln \widetilde{\Delta\Gamma(t)} = \ln(\Delta\Gamma_0 e^{ht}) = ht + \ln \Delta\Gamma_0. \tag{A.51}$$

We are interested in defining physical characteristics, including entropy S, with highest possible accuracy. If coarse graining is defined by ε, then it is obvious, that there is no sense in taking $\Delta\Gamma_0$ less than ε. Thus, we can put $\Delta\Gamma_0 = \varepsilon$ and go to the limit of $\varepsilon \to 0$. Consider:

$$\lim_{\varepsilon\to 0}\lim_{t\to\infty}\frac{1}{t}\ln \widetilde{\Delta\Gamma(t)} = \lim_{\varepsilon\to 0}\lim_{t\to\infty}\frac{1}{t}(ht + \ln\varepsilon) = h. \tag{A.52}$$

This expression is the definition of K-entropy h. Let us stress the importance of the order of taking the limits here. The basic properties of K-entropy are:

1. K-entropy h determines the velocity of the entropy S change due to the purely dynamic process of the mixing of phase trajectories in phase space.

2. K-entropy h, the increment of local instability h_0 and the inverse time of decoupling of time correlations are of the same order of magnitude.

These properties explain the physical meaning of Kolmogorov's entropy.

How does the physical entropy S reach its maximum? For $\varepsilon \to 0$, i.e. defining the entropy $S(t) = ht$ $(t \to \infty)$ with arbitrarily large accuracy, the entropy S does not reach a maximum, but the situation changes if we fix the finite accuracy of the coarse graining ε_0. Then, from (A.50) it is easy to find the characteristic time t_0, during which the region $\Delta\Gamma_0 = \varepsilon_0$ is expanded up to the value $\widetilde{\Delta\Gamma} = 1$:

$$t_0 = \frac{1}{h}\ln\frac{1}{\varepsilon_0}. \tag{A.53}$$

During this time the phase drop of the size ε_0 homogeneously fills the whole phase volume and the further growth of entropy stops.

Appendix B

Statistical mechanics and information theory

B.1 Relation between Gibbs distributions and the principle of maximal information entropy

Information entropy

The notion of entropy in statistical mechanics is closely related to the similar notion in information theory [34]. There exists a wide range of literature, where this relation is discussed in detail [35, 36]. Below, we shall deal with some of the problems, illustrating the basic principles, and connecting these fundamental concepts.

In a narrow sense, information theory represents the statistical theory of communications, i.e. transmission of signals, texts etc. [34]. The main concept in this theory is that of *information* entropy, which acts as a measure of the information, contained in a given communication, text, set of signals etc., which are considered as a more or less random sequence of symbols or events. More precisely, information entropy gives the measure of indeterminacy of information, corresponding to a given statistical distribution of such events. Let p_k be some discrete probability distribution of events, enumerated by index k. Information entropy is defined as[34][1]:

$$H = -\sum_{k=1}^{n} p_k \ln p_k; \quad \sum_{k}^{n} p_k = 1. \tag{B.1}$$

In fact, the value of H equals zero if some of $p_k = 1$, while the remaining $p_k = 0$, i.e. when the result can be predicted with certainty and there is no indeterminacy in the information at all. H acquires its maximum value, when all p_k are equal, i.e. for $p_k = 1/n$. It is obvious, that this limiting case corresponds to maximal indeterminacy – we do not know anything about specific events, all are equally probable (i.e. letters of the text appear absolutely randomly, in physics this corresponds to an absolutely random realization of different states of the system etc.). The maximum of information entropy corresponds to the maximum of our *ignorance* about events and in this case our information on these is minimal.

[1] For us it is irrelevant here, that in information theory this definition normally uses the logarithm with base 2, i.e. $\log_2$, which is related to measuring information in *bits*, instead of ln

The entropy H is additive for independent events, realized with probabilities u_i and v_i, when $p_{ik} = u_i v_k$, so that

$$H = -\sum_{ik} p_{ik} \ln p_{ik} = -\sum_i u_i \ln u_i - \sum_k v_k \ln v_k; \quad \sum_i u_i = 1; \quad \sum_k v_k = 1. \tag{B.2}$$

For the continuous distribution of events x, characterized by a probability density $f(x)$, the information entropy is given by:

$$H = -\int dx f(x) \ln f(x); \quad \int dx f(x) = 1. \tag{B.3}$$

For independent events, again we have additivity. If $f(x, y) = f_1(x) f_2(y)$, we obtain:

$$H = -\int dx \int dy f(x, y) \ln f(x, y) = -\int dx f_1(x) \ln f_1(x) - \int dy f_2(y) \ln f_2(y). \tag{B.4}$$

The Gibbs entropy defined by the distribution function $\rho(p, q)$ in phase space is essentially also the information entropy:

$$S = -\int d\Gamma \rho \ln \rho; \int d\Gamma \rho = 1 \tag{B.5}$$

and can be considered as a measure of our ignorance (absence of information) of the details of the microscopic states of the macroscopic system.

For ensembles with a variable number of particles Equation (B.5) is generalized as:

$$S = -\sum_{N\geq 0} \int d\Gamma_N \rho_N \ln \rho_N; \quad \sum_{N\geq 0} \int d\Gamma_N \rho_N = 1. \tag{B.6}$$

Below we consider extremal properties of Gibbs ensembles, which were established long before the formulation of information theory. The proofs will be given using the Gibbs inequality (1.187):

$$\int d\Gamma \rho' \ln \left(\frac{\rho'}{\rho} \right) \geq 0, \tag{B.7}$$

where ρ and ρ' are two normalized distributions, defined in the same phase space. Equality here holds only in the case of $\rho = \rho'$.

Extremal properties of microcanonical distribution

Let us prove that the microcanonical distribution corresponds to the maximal information entropy among all distributions with the same number of particles in the same energy layer. Let ρ be the distribution function of the microcanonical ensemble, while ρ' is an arbitrary distribution function, defined in the same phase space and in the same energy layer, with both satisfying the normalization condition:

$$\int d\Gamma\rho' = \int d\Gamma\rho = 1. \tag{B.8}$$

Substituting ρ and ρ' into inequality (B.7), we obtain:

$$-\int d\Gamma\rho' \ln\rho' \leq -\int d\Gamma\rho' \ln\rho = -\ln\rho \int d\Gamma\rho' = -\int d\Gamma\rho\ln\rho \tag{B.9}$$

and the proof is complete. In Equation (B.9) we used the constancy of the microcanonical distribution ρ in its energy layer and the normalization conditions for ρ and ρ'.

Extremal properties of the canonical distribution

Let us show that the Gibbs canonical distribution corresponds to maximal information entropy at fixed average energy of the system:

$$\langle H\rangle = \int d\Gamma H\rho \tag{B.10}$$

with the normalization condition:

$$\int d\Gamma\rho = 1. \tag{B.11}$$

Consider the canonical distribution:

$$\rho = Z^{-1}\exp(-\beta H); \quad Z = \int d\Gamma\exp(-\beta H), \tag{B.12}$$

where $\beta = 1/T$. Consider ρ' – another normalized distribution, corresponding to the same average energy as the canonical distribution ρ:

$$\int d\Gamma\rho',\ H = \int d\Gamma\rho H, \tag{B.13}$$

while in all other respects ρ' is arbitrary. Substituting (B.12) to (B.7), we get:

$$-\int d\Gamma\rho' \ln\rho' \leq -\int d\Gamma\rho' \ln\rho = \ln Z + \beta\int d\Gamma\rho' H = \ln Z + \beta\int d\Gamma\rho H$$

$$\text{i.e.} \quad -\int d\Gamma\rho' \ln\rho' \leq -\int d\Gamma\rho\ln\rho \tag{B.14}$$

which completes the proof.

Extremal properties of the grand canonical distribution

Let us give an example of the proof for the quantum case. The entropy of a quantum ensemble is defined as:

$$S = -\mathrm{Sp}\,\rho \ln \rho, \tag{B.15}$$

where ρ is the density matrix. In diagonal representation (cf. (1.175)):

$$S = -\sum_k w_k \ln w_k, \tag{B.16}$$

which has the explicit form of (B.1) – the information entropy for a discrete sequence of events (in our case quantum states).

Extremal properties of quantum ensembles can be derived using the inequality:

$$\mathrm{Sp}\,\rho' \ln \rho' \geq \mathrm{Sp}\,\rho' \ln \rho, \tag{B.17}$$

where ρ and ρ' are arbitrary normalized statistical operators. Equality again holds only for the case of $\rho = \rho'$. This general inequality follows from $\ln x \geq 1 - 1/x$, which is valid for $x > 0$ (equality holds for $x = 1$). Substituting $x = \rho'\rho^{-1}$ and averaging over ρ', we have:

$$\mathrm{Sp}\,\rho' \ln(\rho'\rho^{-1}) \geq \mathrm{Sp}\,\rho'(1 - \rho\rho'^{-1}) = 0 \tag{B.18}$$

as both density matrices are normalized to unity and we can make permutations of operators under Sp.

Let us demonstrate that the grand canonical quantum ensemble corresponds to the maximum of information entropy at fixed average energy:

$$\langle H \rangle = \mathrm{Sp}\,\rho H \tag{B.19}$$

and average number of particles:

$$\langle N \rangle = \mathrm{Sp}\,\rho N \tag{B.20}$$

with the normalization:

$$\mathrm{Sp}\,\rho = 1. \tag{B.21}$$

Let us write the grand canonical ensemble as:

$$\rho = \exp\left(\frac{\Omega - H + \mu N}{T}\right); \quad e^{-\frac{\Omega}{T}} = \mathrm{Sp}\exp\left(-\frac{H - \mu N}{T}\right). \tag{B.22}$$

Then, from inequality (B.17) we obtain (assuming that ρ' is an arbitrary density matrix with the same averages (B.19), (B.20),(B.21)):

$$-\mathrm{Sp}\,\rho'\ln\rho' \le -\mathrm{Sp}\,\rho'\ln\rho = -\mathrm{Sp}\left[\rho'\left(\frac{\Omega}{T}-\frac{H}{T}+\frac{\mu N}{T}\right)\right] = -\mathrm{Sp}\,\rho\ln\rho, \quad \text{(B.23)}$$

which proves our statement. Here we used (B.19),(B.20),(B.21), which are valid for ρ and ρ', i.e.

$$\mathrm{Sp}\,\rho' H = \mathrm{Sp}\,\rho H, \quad \mathrm{Sp}\,\rho' N = \mathrm{Sp}\rho N. \quad \text{(B.24)}$$

These extremal properties of Gibbs ensembles can be used as their definitions. This gives another approach to the justification of *equilibrium* statistical mechanics[2]. From our discussion it becomes clear that the physical entropy describes the lack of information on the real microscopic structure of a multi-particle system. This lack of information leads to the possibility of different microscopic states, which we can not discern from each other, which corresponds to real randomness in hidden degrees of freedom of the system. It is maximal, when the system is in equilibrium, and we know almost nothing about the details of its microscopic organization, and its state is completely determined by a few thermodynamic parameters. Attempts to clarify the microscopic details of the internal organization of the system will inevitably perturb the equilibrium state and lead to lower values of the entropy.

B.2 Purging Maxwell's "demon"

An interesting relation between statistical thermodynamics and information theory can be studied by analyzing the problem of Maxwell's demon [35]. We have just noted that attempts to get information on the details of the microscopic organization of the system by interfering with microscopic processes within the system can move it out of the equilibrium state. Probably the first example of such interference was proposed by Maxwell, introducing the paradox of a "demon", which "works" against the second law of thermodynamics. The simplest variant of such a demon can work as follows. Consider a vessel with a gas in equilibrium state, with a wall inside, separating the vessel into parts A and B, and a hole in the wall with a door. We can imagine, that our demon is sitting near this door and can let fast molecules passing through the hole, say from A to B, while from B to A it allows the passage of slow molecules. Then, after some time interval, since the start of these activities, in part B we shall collect more fast molecules, than in part A. The thermodynamic equilibrium will be broken, the temperature of the gas in part B will become higher than in part A. This situation explicitly contradicts the second law, it is not difficult now to make

[2] In fact, we have just shown, that different versions of the Gibbs distribution correspond to the maximum of thermodynamic entropy with specific additional conditions. This naturally defines the corresponding equilibrium states.

Figure B.1. Maxwell's demon.

heat pass from a colder part of the vessel to a warmer part. Obviously, we may replace the demon by some automatic device, which will violate the second law in this way. More so, it seems much more probable that a kind of "intellectual being" will deal with this process even more effectively. At the same time, we are sure that the second law is of universal nature and all processes in Nature should obey it. In fact, this is a correct conclusion and we shall see shortly, that no demon will be able to overcome this law via decreasing entropy in a closed system, which includes himself (itself). The paradox of Maxwell's demon was first resolved by Szillard, who used clear and simple arguments, as will be discussed below [35].

The essence of Szillard's argument is, that the demon has to *observe* separate molecules, to separate "fast" molecules from "slow". This observation can be made using some physical methods, e.g. he can shine on molecules using electric light, so that he can see them and start to act. Thus, the closed system to be analyzed may consist of:

- a gas at finite temperature $T = T_0$, contained in a vessel with a wall and a door,
- a demon, operating the door,
- an electric light with a charged cell, giving energy to an electric bulb.

The cell energy heats the wire in the bulb up to some high enough temperature $T_1 > T_0$. This allows us to obtain the light with quantized energy $\hbar\omega_1 > T_0$, which is necessary for these quanta to be recognized on the background of the "black body" radiation, which in turn is always present within the vessel with a gas with temperature T_0. During the experiment, the cell gives energy E to the bulb, the bulb wire radiates this energy and looses entropy. This change in entropy is estimated as:

$$S_f = -\frac{E}{T_1} \tag{B.25}$$

and it is introduced to the gas as a negative entropy. With no interference from the demon's side, the energy E is absorbed by gas at temperature T_0, and we observe the total growth of entropy:

$$S = \frac{E}{T_0} + S_f = \frac{E}{T_0} - \frac{E}{T_1} > 0. \tag{B.26}$$

Consider now the demon at work. It (or he) can find a molecule only in the case where it will scatter at least one quantum of energy $\hbar\omega_1$ from the molecule to its (his) "eye" (or to photomultiplier). This inevitably leads to the growth of demon's entropy:

$$\Delta S_d = \frac{\hbar\omega_1}{T_0}. \tag{B.27}$$

The obtained information can be used to decrease the entropy of the system. The initial entropy of the system is given by:

$$S_0 = \ln \Omega_0, \tag{B.28}$$

where Ω_0 is the statistical weight of the (closed) system. After getting the information, the system is defined in more detail, Ω_0 is decreased by some value p_0:

$$\Omega_1 = \Omega_0 - p. \tag{B.29}$$

This leads to a decrease in entropy:

$$\Delta S_i = S_1 - S_0 = \ln(\Omega_0 - p) - \ln \Omega_0 \approx -\frac{p}{\Omega_0} \tag{B.30}$$

as in most practical cases we have $p \ll \Omega_0$. The total balance of entropy is expressed by:

$$\Delta S_d + \Delta S_i = \frac{\hbar\omega_1}{T_0} - \frac{p}{\Omega_0} > 0 \tag{B.31}$$

as $\hbar\omega_1/T_0 > 1$, but $p/\Omega_0 \ll 1$. Thus, as a result, the entropy of the closed system increases, in accordance with the second law.

Let us consider this situation in more detail. Suppose that after some time, the demon has created the temperature difference ΔT between parts A and B of the vessel:

$$T_B > T_A; \quad T_B - T_A = \Delta T$$

$$T_B = T_0 + \frac{1}{2}\Delta T; \quad T_A = T_0 - \frac{1}{2}\Delta T. \tag{B.32}$$

After that, the demon chooses a fast molecule in the region A with kinetic energy $\frac{3}{2}T(1 + \varepsilon_1)$ and sends it to the region B. Then he chooses a slow molecule in B with kinetic energy $\frac{3}{2}T(1 - \varepsilon_2)$ and allows it to pass to the region A. To observe both molecules, the demon needs at least two light quanta, which leads to a decrease of his entropy:

$$\Delta S_d = 2\frac{\hbar\omega_1}{T_0} > 2. \tag{B.33}$$

The exchange of molecules leads to a transfer of energy from A to B:

$$\Delta Q = \frac{3}{2}T(\varepsilon_1 + \varepsilon_2), \tag{B.34}$$

which, taking into account (B.32), corresponds to a decrease of total entropy:

$$\Delta S_i = \Delta Q\left(\frac{1}{T_B} - \frac{1}{T_A}\right) \approx -\Delta Q\frac{\Delta T}{T^2} = -\frac{3}{2}(\varepsilon_1 + \varepsilon_2)\frac{\Delta T}{T}. \tag{B.35}$$

The values of ε_1 and ε_2 are, most probably, small and $\Delta T \ll T$, then:

$$\Delta S_i = -\frac{3}{2}\eta; \quad \eta \ll 1, \quad \text{so that}$$

$$\Delta S_d + \Delta S_i = \left(2\frac{\hbar\omega_1}{T_0} - \frac{3}{2}\eta\right) > 0 \tag{B.36}$$

in agreement with the second law.

In principle, we can analyze another situation, that of the demon at low temperature, when its temperature $T_2 \ll T_0$. In this case it can absorb quanta $\hbar\omega$, radiated by molecules of the gas at temperature T_0. Then, instead of conditions $T_1 > T_0$ and $\hbar\omega_1 > T_0$ used above, we have $\hbar\omega > T_2$ and $T_2 < T_0$, and we can repeat our arguments. We always need some difference of temperatures, or the demon will not be able to operate. But in any case it will not be able to overcome the second law.

These results lead to an important conclusion: physical measurements of rather general nature can lead to an increase in entropy. There is some low limit, below which most measurements become impossible. A rough estimate for this limit corresponds to a decrease in entropy of $\sim 1 (\sim k_B)$. A more accurate estimate gives the value of this limit as $k_B \ln 2 \approx 0.7 k_B$, per one bit of acquired information [35].

However, this is not the end of the story of Maxwell's demon. Though all arguments, given above, are undoubtedly valid for typical physical measurements, more recent studies demonstrated the specific ways to determine the positions of the molecules, not leading to an appropriate increase in entropy [37]. It was also discovered that some operations with information data, e.g. writing data from one device to the other, can under certain conditions be performed without thermodynamic limitations. However, there is still a deep reason why the appropriate demon will not be able to break the second law. The thing is that it first has to "forget" the results of the previous measurement, i.e. destroy information (and thus "pay" in thermodynamic sense). Any memory state (e.g. of a computer) is represented by appropriate physical states (electric current, voltages, magnetizations etc.). The corresponding cleaning of memory, as was first noted by Landauer, is a *thermodynamically irreversible* operation, leading to a general increase in entropy of the closed system[3].

Below we briefly explain the Landauer principle of information erasure, analyzing the so called Szillard engine model[4]. The Szillard engine consists of a one-dimensional cylinder, whose volume is V_0, containing a one-molecule gas and a partition that works as a movable piston. The operator, e.g. a demon, of the engine inserts the partition into the cylinder, measures the position of the molecule, and connects to the partition a string with a weight at its end. These actions by the demon are ideally performed without energy consumption [37]. The demon's memory is also modeled as a one-molecule gas in a box with a partition in the middle. Binary information, 0 and 1, is represented by the position of the molecule in the box, on the left and on the right, respectively.

The following is the protocol to extract work from the engine through information processing performed by the demon (see Figure B.2), where we denote "SzE" for the Szillard engine and "DM" for the demon's memory at each step of the protocol. Initially, the molecule in the cylinder moves freely over the volume V_0.

Step 1 (SzE) The partition is inserted at the center of the cylinder.

Step 2 (SzE, DM) The demon measures the location of the molecule, either the left ("L") or the right ("R") side of the partition. The demon records the measurement outcome in his memory. When it is L (R), his memory is recorded as "0" ("1").

Step 3 (SzE) Depending on the measurement outcome, the demon arranges the device differently. That is, when the molecule was found on the left (right) hand side, i.e., the record is 0 (1), he attaches the string to the partition from the left (right). In either case, by putting the cylinder in contact with the heat bath of temperature T, the molecule pushes the partition, thus exerting work on the weight, until the partition reaches the end of the cylinder. The amount of work extracted by the engine is

$$W = k_B T \ln 2, \tag{B.37}$$

as can be seen by applying the combined gas law in one dimension.

Step 4 (SzE) The demon removes the partition from the engine, letting the molecule return to its initial state.

Step 5 (DM) The demon removes the partition from his memory to erase information.

Step 6 (DM) In order to reset the memory to its initial state, the demon compresses the volume of the gas by half.

---

[3] If the demon possesses a very large memory, it can surely simply remember the results of all measurements, so that there are no irreversible actions. However, this situation does not correspond to the thermodynamic cycle. Demon just increases the entropy of its memory to decrease the entropy of surrounding medium.

[4] This example is taken from A. Hosoya, K. Maruyama, Y, Shikano. Phys, Rev. E **84**, 061117 (2011).

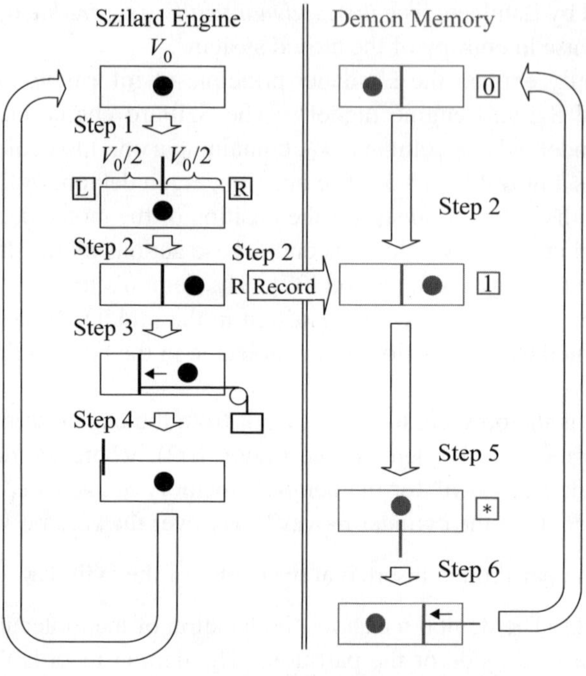

Figure B.2. A protocol of Szillard engine (left side) and demon's memory (right side). This figure shows an example in which the molecule was found on the right-hand side of the cylinder. In the demon's memory, the state after removing the partition is denoted by "$*$".

In order to complete the cycle for both the Szillard engine and the memory, the demon has to reset the memory, which follows the erasure of one bit of information. More precisely, the physical process of information erasure and memory resetting described in Steps 5 and 6, goes as follows. The box is in contact with the thermal bath at the same temperature T as that of the engine. The record in memory can be erased simply by removing the partition, since the location of the molecule becomes completely uncertain. To bring the memory back to its initial state, e.g., 0, one has to compress the gas by a factor two, by sliding a piston from the right end to the middle. The necessary work for this compression is $k_B T \ln 2$, which exactly cancels out the work gain by the engine (B.37).

Let us look at the same process in terms of thermodynamic entropy. By Steps 1 and 2, the volume of the gas in the engine is halved, regardless of the outcome of the measurement. As the entropy change of an ideal gas under the isothermal process is given by $\Delta S = S(V') - S(V) = k_B \ln(V'/V)$, the entropy of the engine is lowered by $k_B \ln 2$. The isothermal expansion in Step 3 increases the entropy of the gas by $k_B \ln 2$, while that of the heat bath is decreased by the same amount. As far as the Szillard engine and its heat bath are concerned, the net result is an entropy decrease of $k_B \ln 2$. This is exactly canceled out by the entropy increase due to information erasure and the reset performed in Steps 5 and 6.

These last two steps are of crucial importance when closing a cycle of memory. Information erasure in Step 5 is an irreversible process and increases the thermodynamic entropy by $k_B \ln 2$. The isothermal compression to reset the memory in Step 6 requires work and dissipates an entropy of $k_B \ln 2$ to its heat bath. This is the essence of the Landauer-Bennett mechanism that finally resolves the Maxwell's demon paradox.

Appendix C

Nonequilibrium statistical operators

C.1 Quasi-equilibrium statistical operators

There have been many attempts to construct a general formulation of nonequilibrium statistical mechanics along the lines of the general Gibbs approach to equilibrium statistical mechanics. Below, we briefly discuss one of the most popular formulations, developed essentially by Zubarev and coworkers [3, 38].

In classical nonequilibrium statistical mechanics we have to analyze solutions of the Liouville equation (1.50) for the general statistical distribution function ρ:

$$\frac{\partial \rho}{\partial t} = \{H, \rho\}, \tag{C.1}$$

where $\{H, \rho\}$ denote the Poisson brackets (1.49) for H and ρ.

The quantum Liouville equation (1.128) for the general density matrix ρ (statistical operator) in operator form is written as:

$$i\hbar \frac{\partial \rho}{\partial t} = [H, \rho]. \tag{C.2}$$

Below we consider only the quantum case, since the classical equations can be formulated in a similar way.

The formal solution of the Liouville equation (C.2) can be written as:

$$\rho(t) = U(t, t_0)\rho(t_0)U^{+}(t, t_0), \tag{C.3}$$

where $\rho(t_0)$ is an arbitrary statistical operator at the initial time t_0, while $U(t, t_0)$ is the operator of time evolution, determined by the equation:

$$\frac{\partial U(t, t_0)}{\partial t} = \frac{1}{i\hbar} H U(t, t_0) \tag{C.4}$$

with initial condition $U(t_0, t_0) = 1$. However, this solution can be useful only in case of an appropriate choice for the statistical operator $\rho(t_0)$ and initial moment t_0. A typical example is linear response theory, where we choose $t_0 \to -\infty$ and $\rho(-\infty)$ is assumed to be an equilibrium Gibbs ensemble. Thus, the main problem of nonequilibrium statistical mechanics is not reduced to finding the formal solutions of the Liouville equation, but to the proper choice of initial conditions.

Note that, depending on the specific problem, the number of parameters, necessary to describe the nonequilibrium state of a system, depends on the characteristic time-scale of interest to us. For larger time-scales we actually need a smaller number of such parameters. For example, at the hydrodynamic stage of a nonequilibrium process it is sufficient to deal only with the average values of energy, momentum and particle densities. This idea of a reduced description of nonequilibrium processes at large enough intervals of time is basic for almost all theories of nonequilibrium processes (cf. our discussion of the derivation of the kinetic equations in Chapter 10). It was clearly formulated first by Bogolyubov.

We are interested in solutions of the Liouville equation for not very short time intervals, when the description of nonequilibrium state can be achieved with some set of operators P_m, where the index m may be both discrete or continuous. We shall look for those solutions of the Liouville equation, which depend on these operators and its conjugated parameters $F_m(t)$, which will be explained a bit later. Depending on the choice of the operators P_m, such an approach is possible for both the kinetic or hydrodynamic stage of a nonequilibrium process. For the hydrodynamic stage we can choose P_m as operators of energy, momentum and particle densities $H(\mathbf{r})$, $\mathbf{p}(\mathbf{r})$ and $n(\mathbf{r})$. For the kinetic stage P_m may be chosen as the appropriate one-particle density matrices.

To formulate a proper initial condition for the Liouville equation, we now introduce the notion of the *quasi-equilibrium* statistical operator. It can be defined similarly to that we have used in our discussion of the equilibrium statistical operators in Appendix B. Let us assume that our nonequilibrium state is characterized by the set of the averages of operators P_m. The quasi-equilibrium statistical operator can be defined as corresponding to the extremum of information entropy:

$$S = -\mathrm{Sp}\, \rho \ln \rho \tag{C.5}$$

under additional conditions of fixing the average values of P_m:

$$\mathrm{Sp}\rho P_m = \langle P_m \rangle^t \tag{C.6}$$

and the normalization condition:

$$\mathrm{Sp}\rho = 1. \tag{C.7}$$

To solve this problem we can look for the extremum of the following functional:

$$L(\rho) = -\mathrm{Sp}\, \rho \ln \rho - \sum_m F_m(t) \mathrm{Sp}\, \rho P_m - (\Phi(t) - 1)\mathrm{Sp}\, \rho, \tag{C.8}$$

where $F_m(t)$ and $\Phi(t) - 1$ are the appropriate Lagrange multipliers. Demanding

$$\delta L(\rho) = -\mathrm{Sp}\, \{[\ln \rho + \Phi(t) + \sum_m F_m(t) P_m]\delta\rho\} = 0 \tag{C.9}$$

for arbitrary variations $\delta\rho$, we get the quasi-equilibrium statistical operator as:

$$\rho_l = \exp\{-\Phi(t) - \sum_m F_m(t) P_m\} \equiv \exp\{-S(P_m, t)\}, \tag{C.10}$$

where

$$\Phi(t) = \ln \mathrm{Sp} \exp\{-\sum_m F_m(t) P_m\} \tag{C.11}$$

and

$$S(P_m, t) = \Phi(t) + \sum_m F_m(t) P_m \tag{C.12}$$

is the entropy operator for a quasi-equilibrium ensemble.

Conjugate parameters F_m are determined by demanding that the physical averages from the total density matrix coincide with the averages, calculated with the quasi-equilibrium statistical operator:

$$\mathrm{Sp}\, \rho P_m = \mathrm{Sp} \rho_l P_m \tag{C.13}$$

or

$$\langle P_m \rangle^t = \langle P_m \rangle_l^t. \tag{C.14}$$

The entropy of the quasi-equilibrium ensemble is:

$$\begin{aligned} S = -\langle \ln \rho_l \rangle_l &= \langle S(P_m, t) \rangle_l^t = \Phi(t) + \sum_m F_m(t) \langle P_m \rangle_l^t \\ &= \Phi(t) + \sum_m \langle P_m \rangle^t. \end{aligned} \tag{C.15}$$

Thus, by construction, the quasi-equilibrium statistical operator (C.10) corresponds to the extremum (in fact maximum!) of information entropy, at fixed values of the averages $\langle P_m \rangle$ and with normalization, in the same way as equilibrium Gibbs ensembles correspond to the maximum of information entropy, at fixed average values of the appropriate integrals of motion[1] (cf. Appendix B). In the particular case of the hydrodynamic regime we can take:

$$\begin{aligned} F_0(\mathbf{r}, t) &= \beta(\mathbf{r}, t), & P_0 &= H(\mathbf{r}) \\ F_1(\mathbf{r}, t) &= -\beta(\mathbf{r}, t)\mathbf{v}(\mathbf{r}, t), & P_1 &= \mathbf{p}(\mathbf{r}) \\ F_2(\mathbf{r}, t) &= -\beta(\mathbf{r}, t)\left[\mu(\mathbf{r}, t) - \frac{m}{2}\mathbf{v}^2(\mathbf{r}, t)\right], & P_2(\mathbf{r}) &= n(\mathbf{r}), \end{aligned} \tag{C.18}$$

[1] In the equilibrium state (C.10) naturally reduces to either the canonical distribution

$$\rho = \exp\{-\Phi - \beta H\}, \quad \Phi = \ln Z = \ln \mathrm{Sp} e^{-\beta H} \tag{C.16}$$

with $\Phi = -F/T$, or to the grand canonical distribution

$$\rho = \exp\{-\Phi - \beta(H - \mu N)\}, \ \Phi = \ln Z = \ln \mathrm{Sp} e^{-\beta(H - \mu N)}, \tag{C.17}$$

where $\Phi = -\Omega/T$.

where $\beta^{-1}(\mathbf{r},t)$, $\mu(\mathbf{r},t)$ and $\mathbf{v}(\mathbf{r},t)$ are the (local!) temperature, chemical potential and velocity.

The quasi-equilibrium statistical operator (C.10) guarantees the validity of the thermodynamic relations between Φ, F_m and S[2]:

$$\frac{\delta\Phi}{\delta F_m(t)} = -\langle P_m\rangle_l^t, \qquad \frac{\delta S}{\delta\langle P_m\rangle_l^t} = F_m(t) \tag{C.19}$$

so that $F_m(t)$ and $\langle P_m\rangle_l^t$ are each other's conjugate.

However, the quasi-equilibrium statistical operator, defined as in Equation (C.10), *does not* satisfy the Liouville equation and does not describe nonequilibrium processes. At the same time, as we shall see below, it can be used as a proper initial condition to the Liouville equation, to find the general form of the nonequilibrium statistical operator.

C.2 Nonequilibrium statistical operators and quasi-averages

Both the classical (C.1) and quantum Liouville equations (C.2) are symmetric with respect to time inversion (in the classical case this corresponds to $t \to -t$, reversal of the momenta (velocities) of all particles and of the direction of the magnetic field). However, the solution of the Liouville equation is unstable to small perturbations, breaking this symmetry.

Let us introduce into the Liouville equation an infinitesimal "source", which satisfies the following requirements:

1. the source is breaking time reversal invariance of the Liouville equation and goes to zero for $\varepsilon \to 0$ (after the thermodynamic limit);

2. the source selects *retarded* solutions of the Liouville equation. This requirement determines the sign of $\varepsilon > 0, \varepsilon \to +0$. Advanced solutions, corresponding to the opposite sign, will lead to a decrease of the entropy with time;

3. the source becomes zero for ρ equal to the quasi-equilibrium statistical operator ρ_l (C.10). For the equilibrium state the source is just absent.

We may consider two ways to introduce the source into the Liouville equation. The first one is to introduce this infinitesimal source directly into the r.h.s. of Liouville equation:

$$\frac{\partial\rho_\varepsilon}{\partial t} + \frac{1}{i\hbar}[\rho_\varepsilon, H] = -\varepsilon(\rho_\varepsilon - \rho_l), \tag{C.20}$$

[2] If index m is discrete, the functional derivatives in (C.19) are replaced by the usual partial derivatives.

where $\rho \to +0$, after taking the thermodynamic limit (during calculations of statistical averages). This infinitesimal source breaks the time reversal invariance of Liouville equation, as the l.h.s. changes sign under this reversal, while the r.h.s. does not change.

Let us rewrite Equation (C.20) in the following form:

$$\frac{d}{dt}(e^{\varepsilon t}\rho_\varepsilon(t,t)) = \varepsilon e^{\varepsilon t}\rho_l(t,t), \tag{C.21}$$

where

$$\begin{aligned} \rho_\varepsilon(t,t) &= U^+(t,0)\rho_\varepsilon(t,0)U(t,0) \\ \rho_l(t,t) &= U^+(t,0)\rho_l(t,0)U(t,0) \\ U(t,0) &= \exp\left\{-i\frac{Ht}{\hbar}\right\} \end{aligned} \tag{C.22}$$

(H is assumed to be time independent) and we introduced the notations:

$$\rho_\varepsilon = \rho_\varepsilon(t,0), \quad \rho_l = \rho(t,0). \tag{C.23}$$

Integrating Equation (C.21) from $-\infty$ to t and assuming that $\lim_{t\to-\infty}\rho(t,t) = 0$, we get:

$$\rho_\varepsilon(t,t) = \varepsilon\int_{-\infty}^{t} e^{\varepsilon(t_1-t)}\rho_l(t_1,t_1)dt_1 = \varepsilon\int_{-\infty}^{t} e^{\varepsilon t'}\rho_l(t+t',t+t')dt'. \tag{C.24}$$

Finally, the solution of Liouville equation (C.20) gives the *nonequilibrium statistical operator* in the following form:

$$\rho_\varepsilon = \rho_\varepsilon(t,0) = \widetilde{\rho_l(t,0)} = \varepsilon\int_{-\infty}^{t} e^{\varepsilon t'}\rho_l(t+t',t')dt'. \tag{C.25}$$

Integrating by parts, we can rewrite Equation (C.25) as:

$$\rho_\varepsilon = \rho_l + \int_{-\infty}^{0} dt' e^{\varepsilon t'}\int_0^1 d\tau e^{-\tau S(t+t',t')}\dot{S}(t+t',t')e^{(\tau-1)}S(t+t',t'), \tag{C.26}$$

where

$$\begin{aligned} \dot{S}(t,0) &= \frac{\partial S(t,0)}{\partial t} + \frac{1}{i\hbar}[S(t,0),H] \\ \dot{S}(t,t') &= U^+(t,0)\dot{S}(t,0)U(t',0) \end{aligned} \tag{C.27}$$

defines the operator of entropy production, which can be proved to be positive definite [38].

The parameters $F_m(t)$, entering the expression for the entropy operator are chosen so that the average values of P_m, calculated with the nonequilibrium statistical operator (C.25), coincide with the averages over the quasi-equilibrium statistical operator (C.10):

$$\langle P_m \rangle^t = \langle P_m \rangle_l^t, \tag{C.28}$$

where

$$\langle \cdots \rangle^t = \lim_{\varepsilon \to +0} \mathrm{Sp}(\rho_\varepsilon \cdots). \tag{C.29}$$

Then $\langle P_m \rangle^t$ and $F_m(t)$ become conjugate parameters, so that:

$$\frac{\delta \Phi}{\delta F_m(t)} = -\langle P_m \rangle_l^t = -\langle P_m \rangle^t. \tag{C.30}$$

The nonequilibrium statistical operator (C.25) can be used to calculate the average value of an arbitrary operator A as:

$$\langle A \rangle = \lim_{\varepsilon \to +0} \mathrm{Sp}\, \rho_\varepsilon A \equiv \prec A \succ, \tag{C.31}$$

which is a typical *quasi-average*, as introduced by Bogolyubov (cf. discussion in Chapter 8). Applying (C.31) to the operators $\dot{P}_m$ and taking into account (C.28) we obtain the transport equations:

$$\frac{\partial}{\partial t} \langle P_m \rangle_l^t = \langle \dot{P}_m \rangle^t = \lim_{\varepsilon \to +0} \mathrm{Sp}\, \rho_\varepsilon \dot{P}_m = \prec \dot{P}_m \succ. \tag{C.32}$$

The second way to introduce infinitesimal sources uses the fact that the logarithm of a statistical operator satisfying the Liouville equation, also satisfies the Liouville equation:

$$\frac{\partial \ln \rho}{\partial t} + \frac{1}{i\hbar}[\ln \rho, H] = 0. \tag{C.33}$$

We may introduce an infinitesimal source directly into Equation (C.33) as:

$$\frac{\partial \ln \rho_\varepsilon}{\partial t} + \frac{1}{i\hbar}[\ln \rho_\varepsilon, H] = -\varepsilon(\ln \rho_\varepsilon - \ln \rho_l), \tag{C.34}$$

where $\varepsilon \to +0$ is again taken after the thermodynamic limit. Once again we see, that this extra source breaks the time reversal symmetry of Equation (C.33).

Let us rewrite Equation (C.34) as:

$$\frac{d}{dt}(e^{\varepsilon t} \ln \rho_\varepsilon(t,t)) = \varepsilon e^{\varepsilon t} \ln \rho_l(t,t). \tag{C.35}$$

Integrating Equation (C.35) from $-\infty$ to t, we obtain:

$$\ln \rho_\varepsilon(t,t) = \varepsilon \int_{-\infty}^{t} e^{\varepsilon(t_1-t)} \ln \rho_l(t_1,t_1)dt_1 = \varepsilon \int_{-\infty}^{0} e^{\varepsilon t'} \ln \rho_l(t+t',t+t')dt' \tag{C.36}$$

so that this version of the nonequilibrium statistical operator is written as:

$$\rho_\varepsilon = \rho_\varepsilon(t,0) = \widetilde{\exp\{\ln \rho_l(t,0)\}} = \exp\left\{-\varepsilon \int_{-\infty}^{0} dt' e^{\varepsilon t'} \ln \rho_l(t+t',t')\right\}, \tag{C.37}$$

where again $\varepsilon \to +0$ after taking the thermodynamic limit. After partial integration, we can rewrite (C.37) as:

$$\rho_\varepsilon = \exp\widetilde{\{-S(t,0)\}} = \exp\left\{-S(t,0) + + \int_{-\infty}^{0} dt' e^{\varepsilon t'} \dot{S}(t+t',t')\right\}. \tag{C.38}$$

The parameters $F_m(t)$, entering the expressions for $S(t,0)$ and entropy production $\dot{S}(t,0)$, are defined, as above, by Equations (C.28).

It can be shown that the nonequilibrium statistical operator (C.38) corresponds to the extremum of information entropy (C.5) under the additional conditions of fixing $\langle P_m(t')\rangle^t = \mathrm{Sp}\, \rho P_m(t')$ for any previous moment in time $-\infty \le t' \le 0$ and the usual normalization condition.

Nonequilibrium statistical operators (C.25), (C.38) were used by different authors to derive equations for hydrodynamics, relaxation equations and kinetic equations [38]. It can be shown that in the lowest orders of interactions, or in case of small thermodynamic perturbations, both (C.25) and (C.38) lead to the same transport equations (C.32). However, the question of the equivalence or nonequivalence of these forms of nonequilibrium statistical operators is still open. A detailed discussion of nonequilibrium statistical operators and applications to various physical problems can be found in [38].

Bibliography

[1] L. D. Landau, E. M. Lifshitz. Statistical Physics. Part I. Pergamon Press, Oxford 1980

[2] L. D. Landau, E. M. Lifshitz. Statistical Physics. Part II. Pergamon Press, Oxford, 1980

[3] D. N. Zubarev. Nonequilibrium Statistical Thermodynamics. Consultants Bureau, NY, 1974

[4] J. W. Gibbs. Elementary Principles in Statistical Mechanics. Charles Scribner's Sons, NY, 1902

[5] L. D. Landau, E. M. Lifshitz. Quantum Mechanics. Pergamon Press, Oxford, 1977

[6] L. D. Landau, E. M. Lifshitz. The Classical Theory of Fields. Pergamon Press, Oxford, 1973

[7] M. Gelfer, V. L. Luboshitz, M. I. Podgoretskii. Gibbs Paradox and Identity of Particles in Quantum Mechanics. Nauka, Moscow, 1975 (in Russian)

[8] A. I. Khinchin. Mathematical Foundations of Statistical Mechanics. Dover, NY, 1960

[9] Ya.G. Sinai. Introduction to Ergodic Theory. Princeton University Press, Princeton, 1977

[10] M. A. Leontovich. Introduction to Thermodynamics, Statistical Physics. Nauka, Moscow, 1983 (in Russian).

[11] L. D. Landau, E. M. Lifshitz. Mechanics. Pergamon Press, Oxford, 1976

[12] G. M. Zaslavsky. Chaos in Dynamic Systems. Harwood Academic Publishers, Amstredam, 1985

[13] N. N. Bogolyubov. Problems of Dynamical Theory in Statistical Physics. Interscience, NY, 1962.

[14] K. P. Gurov. Fundamentals of Kinetic Theory. The Bogolyubov Method, Nauka, Moscow, 1966 (in Russian)

[15] E. M. Lifshitz, L. P. Pitaevskii. Physical Kinetics. Pergamon Press, Oxford, 1981

[16] A. I. Akhiezer, S. V. Peletminskii. Methods of statistical physics. Pergamon Press, Oxford, 1981

[17] F. M. Kuni. Statistical physics and thermodynamics. Nauka, Moscow, 1981 (in Russian)

[18] N. N. Bogolyubov. Lectures on Quantum Statistics. Gordon and Breach, NY, 1967

[19] I. M. Lifshitz, M. Azbel, M. I. Kaganov, Electronic Theory of Metals. Plenum, NY, 1974.

[20] C. J. Pethick, H. Smith. Bose–Einstein Condensation in Dilute Gases. Cambridge University Press, Cambridge, 2002

[21] D. Pines, P. Nozieres. The Theory of Quantum Liquids. W. A. Benjamin, NY, 1966

[22] A. A. Abrikosov. Fundamentals of the Theory of Metals. North-Holland, Amsterdam, 1988

[23] N. N. Bogolyubov. Lectures on Quantum Statistics.Vol. 2, Quasi-Averages. Gordon and Breach, NY, 1971

[24] P. G. de Gennes. Superconductivity of Metals and Alloys. W. A. Benjamin, NY, 1966

[25] A. J. Leggett. Quantum Liquids. Oxford University Press, Oxford, 2006

[26] Shang-keng Ma. Modern Theory of Critical Phenomena. W. A. Benjamin, NY, 1976

[27] D. N. Zubarev. Double-time Green Functions in Statistical Physics, Sov. Phys - Uspekhi **3**, 320 (1960).

[28] P. S. Zyrianov, M. I. Klinger. Quantum Theory of Electron Transport Phenomena in Crystalline Semiconductors. Nauka, Moscow, 1976 (in Russian)

[29] A. A. Abrikosov, L. P. Gor'kov, I. E. Dzyaloshinskii. Quantum Field Theoretical Methods in Statistical Physics. Pergamon Press, Oxford, 1965

[30] R. D. Mattuck. A Guide to Feynman Diagrams in the Many-Body Problem. McGraw-Hill, NY, 1974

[31] A. B. Migdal. Qualitative Methods in Quantum Theory. Westview Press, NY, 2000

[32] M. V. Sadovskii. Diagrammatics. World Scientific, Singapore, 2006

[33] Y. B. Rumer, M.Sh. Rivkin. Thermodynamics, statistical physics and kinetics, Nauka, Moscow. 1977 (in Russian)

[34] C. E. Shannon. A mathematical theory of communication. Bell System Technical Journal **27** 379, 623 (1948)

[35] L. Brillouin. Science and Information Theory. Academic Press, NY, 1956

[36] B. B. Kadomtsev. Dynamics and information. Physics-Uspekhi **37**, 425 (1994)

[37] C. H. Bennett. Demons, Engines and the Second Law. Scientific American **257**, 108 (1987)

[38] I. I. Lyapilin, V. P. Kalashnikov. Nonequilibrium Statistical Operator. UB RAS, Ekaterinburg. 2008 (in Russian)

Index

H-theorem 214, 216
K-entropy 258
N-particle distribution function 19
ε-expansion 193
s-particle density matrices 31
s-particle distribution function in second quantized form 32
$u - v$-transformation 123

Abrikosov's vortices 157
additive integral of motion 14, 28
anomalous averages 122, 137, 179
average value 5

bath 43
BCS model 136
BCS ratio 145
Bloch equation 244
Bogolyubov's $u - v$-transformation 137
Bogolyubov's chain 21, 219, 223
Bogolyubov's sound 124
Boltzmann distribution 60
Boltzmann equation 208
Boltzmann's collision integral 213
Boltzmann's entropy 64
Boltzmann's kinetic equation 213
Bose condensate 121
Bose condensation 96, 117
Bose distribution 76, 77
Brillouin function 171

canonical distribution 43, 46
chain of equations for partial distribution functions 209
chain of equations of motion for Green's functions 200
chemical potential 53, 54, 76, 77, 83
chemical potential of a Boltzmann gas 62
classical ideal gas 60
"coarse grained" distribution function 36, 37
coherence length 143
collision integral 209
condensate wave function of Cooper pairs 151
Cooper instability 133, 135
Cooper pair 133, 148
correlation function of the order parameter 186
correlation length critical exponent 184
correlation length of fluctuations 184, 186
Coulomb pseudo-potential 149
criterion for validity of the Boltzmann statistics 73
critical exponent of susceptibility 181
critical exponents 175, 176, 181, 187, 193
critical region 186
critical temperature of the superconducting transition 142
Curie constant 173

damping 230, 232
de Haas–van Alphen effect 94
Debye frequency 106
Debye screening 243
Debye temperature 106
Debye theory 105
decoupling" 219
degenerate Bose gas 94
degenerate gas of electrons 82
degenerate interacting Bose gas 121
density matrix 21, 25
density of electronic states at the Fermi surface 87
diamagnetic susceptibility 91
dielectric function 242
dielectric screening 241

dimensionless phase space element 5
dispersion relations 204
distribution function 4
distribution function of particles in a Fermi liquid 233
double-time Green's functions 198
Dulong–Petit's law 105, 108
Dyson equation 237, 238, 246

effective mass of a quasiparticle 128
Ehrenfest's model 41
electron – phonon interaction 134
electron liquid in metals 130
electron–electron interaction 222
electron-phonon interaction 218
elementary quantum of magnetic flux 157
energy gap 139, 141
energy spectrum of superconductors 135
ensemble 4
ensemble average 6
entropy 32
entropy growth 35
entropy of a nonequilibrium Fermi gas 78
entropy of the nonequilibrium Bose gas 78
equation of state for ideal quantum gases 80
equation of state of Boltzmann gas 67
equipartition theorem 70
ergodic hypothesis 17
ergodic surface 4
ergodic systems 252

factorization of the distribution function 7
Fermi distribution 75, 76
Fermi energy 82
Fermi liquid 125, 126
Fermi liquid constants 129
Fermi momentum 82, 125
"Fermi step" function 83
Fermi surface 82, 127
Feynman diagram 225, 233–235
fluctuation–dissipation theorem 205
fluctuations 160
fluctuations of basic physical properties 164
fluctuations of the order parameter 180
flux quantization 156
free energy 51
free energy of Boltzmann gas 66

gap Δ in BCS theory 146
gap equation of BCS theory 140
Gaussian distribution 162
general properties of Fermi and Bose gases 79
Gibbs distribution 43, 51
Gibbs ensemble 4
Gibbs entropy 33
Ginzburg criterion 185
Ginzburg–Landau equation 154
Ginzburg–Landau theory 151
grand canonical distribution 53
grand partition function 54
graphical summation 237
Green's function 198, 200, 225, 232
Green's functions at finite temperatures 244

Heisenberg model 173

ideal gas with constant specific heat 69
information entropy 259
interacting electrons 222
irreducible self-energy part 238
Ising model 173

Kadanoff construction 188
kinetic equation 209
kinetic equation for electrons 224
Kolmogorov's entropy 258
Kramers–Kronig relations 207
Kubo formula for conductivity 202
Kubo formulas 199
Kubo identity 197

Landau diamagnetism 91
Landau expansion 180
Landau function 129
Landau levels 90
Landau's criterion of superfluidity 114
Landau–Silin theory 131
Landauer principle 267
limits of applicability of GL theory 158

linear response to mechanical perturbation 195
Liouville equation 12, 13, 196
Liouville theorem 9
London's equation 155
long range order 184
lower critical dimensionality 185
Luttinger theorem 125
Lyapunov exponent 257

magnetic equation of state 175
magnetism of an electron gas 88, 92
Matsubara frequencies 245, 246
Matsubara's Green's functions 245
Maxwell distribution 48
Maxwell's demon 263, 264, 266
mean field theory 171
mean square fluctuation 8
mean-field theory of magnetism 170
microcanonical distribution 13, 44
microcanonical distribution of quantum statistics 28
mixed ensemble 24
mixing 256
"mixing" of phase points 37
molecular 171

Nernst's theorem 58
non-equilibrium correction to the density matrix 197
nonequilibrium entropy 38
nonequilibrium Fermi and Bose gases 77
nonequilibrium ideal gas 63
nonequilibrium statistical operator 274, 276
normal Fermi liquid 125
normal oscillators 102
normalization condition 4, 5

off-diagonal long-range order 117
one-atom ideal gas 72
one-particle density matrix 217
order parameter 137, 151, 186
Ornstein–Zernike correlator 183

partial density matrices 29
partial distribution functions 17, 19
partition function 46–48

Pauli paramagnetic susceptibility 89
Pauli specific heat 87
penetration depth 155
phase average 6
phase diagram of a type II superconductor in a magnetic field 158
phase point 2
phase space 2
phase trajectory 3
phonon density of states 106
phonons 103, 111
phonons in a Bose liquid 117
physical kinetics 208
Planck distribution 97
Planck function 98
Planck's law 98
plasma frequency 243
plasmons 243
Poincare cycle 255
Poincare recurrence theorem 254
Poisson bracket 12
Poisson summation formula 92
Poisson–Gauss integral 49
polarization operator 236
projection operator 23
pure ensemble 22

quantum ideal gases 75
quantum kinetic equations 216, 222
quantum kinetic equations for electrons and phonons 222
quantum Liouville equation 26, 27
quantum Poisson brackets 27
quantum version of the H-theorem 224
quasi-averages 177, 179, 275
quasi-equilibrium statistical operator 271
quasiparticles 103, 109, 124, 139

radiation pressure 100
Rayleigh–Jeans law 98
recurrence time 256
renormalization group 192
retarded solutions of the Liouville equation 273
rotons 111

Sakura–Tetrode formula 73
scale invariance 187
scaling 186, 187

scaling relations 192
self-energy 238
solid state at high temperature 104
solid state at low temperature 101
specific heat of a degenerate electron gas 86
spectrum of elementary excitations in liquid He^4 112
spectrum of quasiparticles 230
spontaneous symmetry breaking 122
statistical independence 7
statistical meaning of entropy 34
statistical mechanics of free spins in an external magnetic field 170
statistical operator 25
statistical weight 16, 28
statistics of photons 97
Stefan–Boltzmann constant 99
superfluidity 113
susceptibility 186
symmetry breaking 152
Szillard engine 267, 268
Szillard's argument 264

temperature (energy) of degeneracy 74
temperature dependence of the gap in BCS theory 146
temperature dependence of the specific heat in Debye model 108
temperature of Bose condensation 95
temperature of superconducting transition 142, 143
the quasiparticles in a Fermi liquid 127
theory of Fermi liquids 125
thermodynamic limit 38
thermodynamic relations 55
thermostat 43
time average 6
type I superconductors 156
type II superconductors 156

ultra-relativistic gas 85
unstable motion 257
upper critical dimension 193

Wien's displacement law 99
Wigner's distribution 32